Percy H. Dougherty
Editor

The Geography of Wine

Regions, Terroir and Techniques

Springer

Editor
Percy H. Dougherty
Lehigh County Government Center
17 South Seventh Street
18101-2401 Allentown
Pennsylvania
USA
PercyDougherty@lehighcounty.org

ISBN 978-94-007-0463-3 e-ISBN 978-94-007-0464-0
DOI 10.1007/978-94-007-0464-0
Springer Dordrecht Heidelberg London New York

Library of Congress Control Number: 2011941750

Cover illustration: Illustrations from www.photos.com.

Printed on acid-free paper

Springer is part of Springer Science+Business Media (www.springer.com)

Preface

Throughout history, wine has been an essential part of culture; the Greeks and Romans even had gods of wine, Dionysus and Bacchus. Today, wine is the beverage of choice for many Americans, and it is challenging beer as the leading adult beverage. Because of its popularity, there are numerous producers and varieties of wine, resulting in both a blessing and a curse. The latter occurs when some consumers face confusion and anxiety when purchasing wine. People are interested in trying new wines from many different regions of the world, but with thousands of varieties and the endless geographical nomenclature so popular in European wines, it is an intimidating process. Making the process of choosing a wine easier will lead to better-informed decisions on the purchase of wine. Knowing the geography of wine gives the customer an advantage.

In order to become a more informed consumer, it is important that one knows as much as possible about the wine, including the physical and socioeconomic conditions that influence the grapes in the vineyards and the production of the resulting wine. It is the purpose of this book to highlight how geography influences the unique taste and aroma of wines from different parts of the world, and how the field of geography is making a difference by scientifically studying the process of grape growing and wine production.

Knowledge of a wine enhances its enjoyment because wine provides the palate and olfactory senses with a myriad of tastes and smells. To the uninitiated, this is confusing and interferes with the hedonistic pleasure of analyzing a wine's characteristics. The more we know about a wine, the more we are able to appreciate it. This applies to every aspect of its production – even the region of production, as well as the physical characteristics of the weather, climate, soil, drainage, and solar prospect; or, what the French call *terroir*, or its geographical imprint.

At the 1998 Association of American Geographers (AAG) Annual Meeting in Philadelphia, the Wine Specialty Group of the AAG was formally recognized. Several paper sessions, workshops, and a panel session initiated a fruitful exchange of information that has continued yearly. Based on conversations at the convention, and with the encouragement of Robert Doe of Springer Publishing, this book is the result of many requests to publish a book to illustrate the variety of research geographers are doing on the geography of wine. Although an overview of the work done in the field of geography, it should also be of interest to winemakers, viticulturists, and individuals interested in drinking wine. This volume is a valuable contribution to the study of viticulture and wine because it is the first book to report on the research of geographers with the intent of showing what geographers study, as well as the techniques they use. There is no better way to comprehend a region's geography than by studying its wine, and there is no better way to understand a wine than to understand its geography.

The following chapters provide an overview of the ways in which geographers study grape growing and wine production. Travel through the field of geography and join geographers as they study how the environment influences spatial variations in wine, ranging from the impact of soil and rock types to the impact that climate change will exact upon the world's vineyards. Delve into the social and economic implications of race relations in the South African wine producing districts to the economics of California wine producing districts and the spatial implications. See how geographers are using advanced spatial technology of remote sensing and Geographical Information Systems to improve the understanding of grape growing and wine production.

Wine is undoubtedly the agricultural product that best mirrors the environmental, social, and economic conditions under which the grapes grow. Join us as we look at the impact of geography on the variety and style of wine produced in a region, and learn how geographical factors influence the wine's aroma and taste. Drinking wine is "tasting the geography" of the area from which a wine comes.

Acknowledgments

Any edited volume is the hard work of many people operating as a team. Just as in any other endeavor, the final product is only as strong as the collective input of the team. The team for this project is a collection of professors and government workers who demonstrate a professional demeanor, a strong work ethic, and a love of the geography of wine. I thank the authors for making short deadlines and for their many revisions that enhanced the quality of this volume. The authors are the foremost authorities in their subjects, and they have brought a truly inter-disciplinary breadth to the collection of chapters.

The Wine Specialty Group of the Association of American Geographers deserves particular recognition. It is from their annual meetings that the idea for this book originated, and most of the chapters in this volume are modifications of papers presented at the annual meetings. The Wine Specialty Group has awakened a new interest in the geographical analysis of wine and viticulture without which this publication would not have been possible.

Many members of the AAG Wine Specialty Group played a prominent role by reading chapters and making many constructive criticisms that helped the authors and the editor. Although some of the suggestions were difficult to implement and resulted in significant revisions, the authors and the editor benefitted from the many years of experience and subject matter expertise of the reviewers. A special debt of gratitude goes to the following for reviewing materials in the book and offering advice: Sheryl Lazludder Beach, John Bodenman, Teresa Bulman, Bill Crowley, Robert Czerniak, Nick Dunning, Tom Feeney, Daniel Gade, Franklin Gossette, Fritz Gritzner, Grieg Guthey, Nancy Hultquist, Robert Hutton, Ken Martis, William Monfredo, James Newman, Zok Pavlovic, Gary Peters, Phyllis Petersen, Greg Pope, Tony Shaw, Marilyn Silberfein, Frances Sternberg, Karen Omara Voytas, and Robert Ziegenfus.

Robert Doe, the Publisher from Springer, who guided this work through the publication effort, deserves special recognition for his patience in waiting for the herd of authors and an over-extended editor to submit materials. Many erstwhile deadlines came and went, but he persevered with the project. In addition, a debt of gratitude goes to Nina Bennink, the Administrative Assistant to Robert Doe, who helped guide the preliminary review of the manuscript until her untimely death. Hermine Vloemans capably filled in and did the heavy lifting that saw the project to its completion.

Another valuable person in the work leading to production is Valerie Hildebeitel. Trained as an English major and employed in the newspaper business in a former professional life, Val made several suggestions on how to improve the readability of the text and caught many errors that got by the editor.

I thank everyone who participated in the production of this book and the many geographers whose pioneering research forms the backdrop for these chapters. All problems with the format and content of the book are the sole responsibility of the editor.

Finally, and most importantly, I wish to acknowledge the loving support and the invaluable help of my wife, Anne Zinn Dougherty. She is my partner in many journeys to wine producing regions around the world, and she has shared many bottles of wine with me in my quest of learning about the impact of geography on wine. She is a sounding board for many ideas and a valuable resource.

Allentown, PA, USA Percy H. Dougherty, Ph.D.

Contents

Contributors

Wim Bastiaanssen WaterWatch, Wageningen, The Netherlands, w.bastiaanssen@waterwatch.nl

David Bek Department of Geography, University of Durham, Durham, UK, dave.bek@durham.ac.uk

Tony Binns Department of Geography, University of Otago, Dunedin, New Zealand, jab@geography.otago.ac.nz

Scott Burns Department of Geology, Portland State University, Portland, OR 97205-0751, USA, burnss@pdx.edu

Barbara A. Carmichael Department of Geography and Environmental Studies, NeXt Research Centre, Wilfrid Laurier University, Waterloo, ON, Canada N2L 3C5, bcarmichael@wlu.ca

Percy H. Dougherty Professor Emeritus, Department of Geography, Kutztown University, Kutztown, PA 19530, USA, percydougherty@lehighcounty.org

Deborah L. Elliott-Fisk Graduate Group in Geography, University of California Davis, One Shields Ave., Davis, CA 95616, USA, debbieef@wildblue.net

David R. Green Department of Geography and Environment, University of Aberdeen, St. Mary's, Elphinstone Road, Aberdeen AB24 3UF, Scotland, UK, d.r.green@abdn.ac.uk

Greig Tor Guthey Department of Liberal Studies, California State University San Marcos, San Marcos, CA, USA, gguthey@csusm.edu

John Hornbuckle CSIRO Land and Water, Griffith, NSW, Australia, John.Hornbuckle@csiro.au

Lee F. Johnson Division of Science and Environmental Policy, California State University, Monterey Bay, Seaside, CA, USA

NASA Ames Research Center, Moffett Field, CA, USA, Lee.F.Johnson@nasa.gov

Gregory V. Jones Departments of Geography and Environmental Studies, Southern Oregon University, Ashland, OR, USA, GJones@sou.edu

David Kasserman Department of Anthropology and Geography, Rowan University, Glassboro, NJ 08028, USA, Kasserman@rowan.edu

Denyse Lemaire Department of Anthropology and Geography, Rowan University, Glassboro, NJ 08028, USA, Lemaire@rowan.edu

Cheryl McEwan Department of Geography, University of Durham, Durham, UK, Cheryl.McEwan@durham.ac.uk

Bernard Momer Community Culture and Global Studies, IK Barber School of Arts and Sciences, University of British Columbia Okanagan, Kelowna, BC V1V 1V7, Canada, bernard.momer@ubc.ca

Rama Nemani NASA Ames Research Center, Moffett Field, CA, USA, rama.nemani@nasa.gov

Lars Pierce Division of Science and Environmental, Policy California State University, Monterey Bay, Seaside, CA, USA, lpierce@csumb.edu

Ryan Reid Departments of Geography and Environmental Studies, Southern Oregon University, Ashland, OR, USA, RReid@sou.edu

Robert Sechrist Department of Geography and Planning, Indiana University of Pennsylvania, Indiana, PA 15705, USA, rpsecrst@iup.edu

Donna M. Senese Community Culture and Global Studies, IK Barber School of Arts and Sciences, University of British Columbia Okanagan, Kelowna, BC V1V 1V7, Canada, donna.senese@ubc.ca

Bryan Thoreson SEBAL North America, Davis, CA, USA, bryant@de-water.com

Bruno Tisseyre UMR ITAP Montpellier SupAgro/Cemagref, Montpellier, France, tisseyre@supagro.inra.fr

Tim Unwin Department of Geography, Royal Holloway, University of London, Egham, Surrey TW20 0EX, UK, tim.unwin@rhul.ac.uk

Aleksander Vilks Departments of Geography and Environmental Studies, Southern Oregon University, Ashland, OR, USA, AVilks@sou.edu

Wayne Wilson Kelowna Museums, 470 Queensway, Kelowna, BC V1Y 6S7, Canada, wwilson@kelownamuseums.caDepartment of Geology, Portland State University, Portland, OR 97207-0751, USA

Part I

Overview

Introduction to the Geographical Study of Viticulture and Wine Production

1

Percy H. Dougherty, Ph.D.

Witness the water which falls from the skies upon our vineyards; and then passes into the roots of the vine, to be converted into wine; a constant proof that God loves us, and that he is pleased to see us happy

—Benjamin Franklin, 1779 (Bruce 1917).

Abstract

The study of viticulture and the geography of wine have intrigued humans from ancient civilizations to the modern world. Ancient Egyptians, Greeks, Romans, and other early cultures had a keen appreciation of the influence of geography upon the characteristics of wine produced within an area and an understanding of differences in wine coming from various regions. Throughout history, grape growing and wine production have been significant economic activities and have had a profound impact on culture and the resulting landscapes. The distribution of grapes and wine production are concentrated in certain areas where the "*terroir*" or geography is favorable. Because of the uneven geographical distribution, the geographer is uniquely qualified to study the spatial distribution as it relates to physical, cultural, and economic factors. In addition, the geographer is well equipped to provide the maps and the remote sensing imagery for the analysis of grape growing and wine production and, in turn, create the geographical information systems (GIS) that enable us to undertake applied research. This chapter investigates how the field of geography is ideally qualified to engage in these studies, and it presents examples of research carried out by geographers.

Introduction

Have you ever wondered why a 750-ml bottle of Yellow Tail Merlot from Australia costs $7.99, while a bottle of 2005 Chateau Petrus from Bordeaux, France costs $2,500 (2011 prices at the Pennsylvania State Store), even though the vintners use the same grape? Why does a Chablis from France, made from the Chardonnay grape, have a steely, minerally, austere taste while a California Chardonnay often has a fruit forward nose with hints of vanilla and spice? Or even more importantly, why do American wines have the grape variety prominently displayed on the label, while European wines rarely do, preferring to cite the region of production, e.g., Chablis, Bordeaux, Burgundy, Champagne, Brunello di Montalcino, and Rioja. The answer is clear, GEOGRAPHY!

A tour of your local wine shop is a vicarious field trip to the world of wine, and there is no need for a passport or an airline ticket to take a serendipitous trip through the vineyards of the world. Read the labels on the wine bottles and visualize the many alluring regions where grapes are grown and wine production occurs.

P.H. Dougherty, Ph.D. (✉)
Professor Emeritus, Department of Geography,
Kutztown University, Kutztown, PA 19530, USA
e-mail: percydougherty@lehighcounty.org

P.H. Dougherty (ed.), *The Geography of Wine: Regions, Terroir and Techniques*,
DOI 10.1007/978-94-007-0464-0_1,

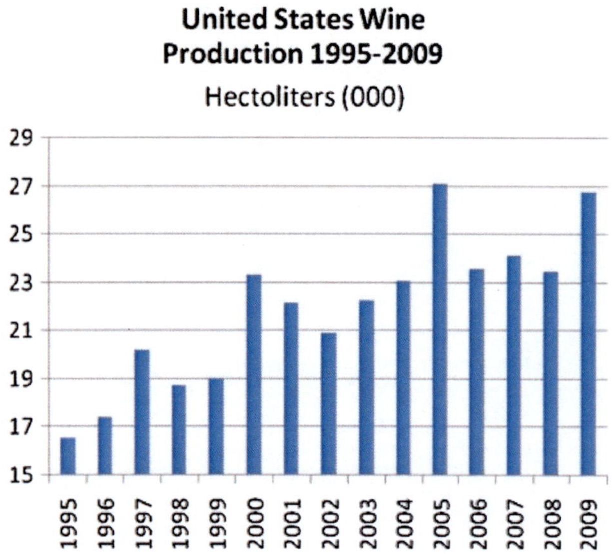

Fig. 1.1 Trend in United States wine production, 1995–2009 in thousands of gallons (Data from The Wine Institute 2010)

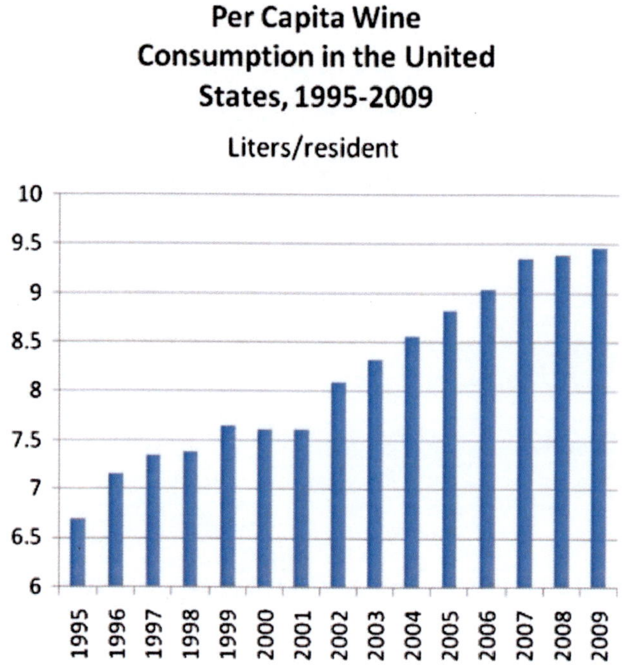

Fig. 1.2 Per capita wine consumption in the United States, 1995–2009 (Data from The Wine Institute 2010)

In addition to the standard wines from America, whole sections of the store provide wines from France, Italy, Spain, Portugal, Germany, Argentina, Chile, Australia, New Zealand, and South Africa, representing thousands of producers. If it were not for geographical differences in grapes and wine, a couple of behemoth producers in large factories would make the product, e.g., such as Budweiser does for beer, and Coca Cola and Pepsi do for soda.

The study of the geography of wine is becoming more important as Americans become more interested in wine. In the United Sates, all 50 states have at least 2 wineries, and the total number of wineries in 2010 stands at 6,785 (Fisher 2011). Figure 1.1 shows there is a steady increase in wine production over the past decade in the United States, resulting from the increasing infatuation with wine. The increase is attributable to both the increase in US exports of wine to other countries as shown in Fig. 1.2 and the increasing consumption of wine by Americans as shown in Fig. 1.3. Although beer and spirits consumption is increasing in many countries, Vins de Provence (2011) reports the following for the United States: "…in 2010, retail wine sales were well ahead of retail beer and spirit sales, both on volume and on value."

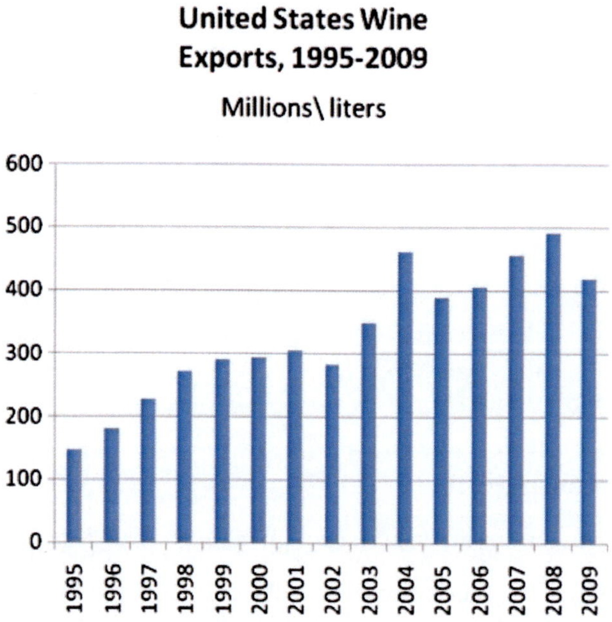

Fig. 1.3 Trend in United States wine exports, 1995–2009 (Data from The Wine Institute 2010)

Not all areas grow the same grape, and even if they do, there are distinct differences in the aroma and taste of wine from different geographical regions. It is astounding how the grape *Vitis vinifera,* the primary wine-producing grape in the world, has subdivided into so many subspecies. Of the more than 100 species of grapes in the world, *Vitis vinifera* is responsible for 99% of the world's wine today (McGovern 2007:1). Jancis Robinson in *The Oxford Companion to Wine* states that there are approximately 10,000 known varieties of vinifera, with Galicia in northwest Spain alone boasting of 1,000 indigenous species (Robinson 1994:1038). In

Vine, Grapes and Wines, a book on the most popular wine grapes, Robinson lists over 1,000 of the most commercially viable varieties in detail (Robinson 1986). As an additional indication of the diversity of grape types, Gold (2007:1) estimates that there are more than 3,000 registered Italian grape varieties.

Adding to the complexity is the fact that *Vitis vinifera* is subject to changes referred to as clones, that is, a growth defined as a group of closely related but not quite identical vine varieties sharing the same genetic code. An excellent example is Pinot Noir. Hawkins (2007) estimates that worldwide, there is a minimum of 200 Pinot Noir clones – including approximately 47 versions planted in the vineyards of Burgundy and nearby districts in France – resulting in a great variation of wine quality and aging potential. The fascinating study of the different varieties of grapes and their characteristics is the subject of ampelography. *Vitis vinifera* is a remarkably adaptable plant with many subvarieties, each with a preferred environment, resulting in a fertile area of study for geographers.

It is no wonder that the typical customer entering his or her favorite shop and purchasing a bottle of wine has an intimidating experience, to say the least. There are hundreds, if not thousands, of choices to make, and many people fall back on the advice of a store clerk. Since each person's tastes vary and his or her likes and dislikes differ, this may not be the best situation. A wine that is objectionable to the store worker or even to a wine expert writing for *Wine Spectator* or *Wine Enthusiast* may titillate your taste buds and set off an explosion of sensory pleasure. The answer to the oft-repeated question of "what is the best wine?" is the wine that *YOU* like. Often, this can only be found by extensive trial and error and tasting many varieties and styles of wine. A good way to accomplish this is through wine tastings at the local community college, or through local chapters of the American Wine Society, Tasters Guild, and similar organizations.

Knowing about the geography of wine can also help customers make decisions on selecting wine. Even within a single grape variety, e.g., Chardonnay, there are differences in its characteristics from one region to another region around the world. That crisp, steely, mineral laden Chablis made from Chardonnay will differ drastically from a fruity, heavily oaked California Chardonnay in both price and your personal preference. The same grape grown in a cool coastal zone will taste different from the same variety grown in a more continental location. Even in the same region, there may be differences in the grape variety depending on the soil or other conditions, e.g., the Cabernet Sauvignon growing on the gravelly soils on the "left bank" of the Gironde River in Bordeaux versus the Merlot growing on the silty and clay soils of the "right bank."

This chapter sets the stage for the rest of the book by looking at why geographers are interested in viticulture and wine and what academic basis generates this interest. To do so, it is necessary to travel back to the time of ancient cultures to note the involvement of geographers in researching viticulture and wine in order to understand what is happening today. Following the historical background, the chapter will answer the question of why geography as a field should have an interest in these topics. The remainder of the chapter focuses on an overview of publications in wine geography. It is a wide-ranging look at the types of geographical research – research that incorporates the physical, cultural, and economic plus modern geographic techniques such as satellite imagery and geographical information systems.

The remaining chapters illustrate examples of the great variety of work done by geographers today by looking at the philosophical background, the impact of the physical environment, the utilization of the regional approach, the cultural and economic factors, and the use of geographical techniques to study wine.

Geography is an important part of viticulture and oenology, and the more one understands geography, the more one understands wine. The geography of wine is a window into places, cultures, and times. Join us as we embark on an exciting journey into the geography of wine. As the chapters of this book unfold, lean back, drink a glass of your favorite wine, and taste the geography as we vicariously travel through the wine regions of the world – as seen through the eyes of the geographer.

Historical Background

Wine has been a research topic in geography since the ancient Greek and Roman geographers, but its origins date back much further in time. Geographers surmise that the home of the wine grape, *Vitis vinifera*, is the Transcaucasia and Georgia regions of Southwest Asia, and they describe its diffusion and history in works by Carl Sauer (1952), Harm de Blij (1983b), Jonathan Sauer (1993), and Tim Unwin (1996). Recent work by archeologists narrows the origin to areas in Georgia and

Fig. 1.4 Grape and wine scene from the tomb of Nachkt, 1550–1295 BC (Source: Egyptian Tourist Bureau 2011. Original from Davies 1917)

Iran (McGovern 2007). Remnants of pottery with wine residue found in Georgia date to 8,000 years ago (Keys 2003). The recent discovery of a 6,000-year-old winery in a cave in Armenia with all the necessary equipment, including a grape press, fermentation vats, and storage jars, confirms the area had the technology to produce wine (Barnard et al. 2011). Tartaric acid residue on 8,000-year-old pottery fragments in Greece indicates the high likelihood of wine production (Keys 2003).

The Egyptians were also important in the early production of wine. Although Vitis vinifera was not indigenous to the deserts of Egypt, grape growing and wine production were important to ancient Egyptians as early as the First Dynasty (3100–2890 BC). Evidence of an early appellation system is on amphorae in the tomb of King Tutankhamen. The amphorae contain inscriptions of labels identifying the name of the wine, its year of production and harvest, the source of the wine, and even who grew the grapevines (Estreicher 2006:18). Therefore, early in history, people realized that place or geography played an important role in differentiating wines from each other.

Figure 1.4 shows a well-preserved painting in the tomb of Nakht (Egyptian Tourist Bureau 2011), illustrating grapes growing on a pergola, people stomping grapes, the straining of the juice, and the storage of wine in amphorae. Nakht was most likely a government official from the middle class, who lived at the end of the Eighteenth Dynasty (about 1550–1295 BC) during the reign of Amenhotep II or early Thutmose IV (Davies 1917:50). The painting shows how advanced grape growing and wine making was in ancient Egypt.

The geographical study of grapes and wine has its roots in the classical Greek and Roman writings – particularly in the legendary works of Homer and Virgil. In Homer's Odyssey, one reads about the food and culture of the ancient world as Odysseus travels from adventure to adventure on his trip home to Ithaca from his victory in Troy (Homer, Translated by Ian Johnson 2006). Virgil dedicates his second book of Georgics to Bacchus, god of the winepress, with highly poetic instructions on the cultivation of the vine, particularly the vines of Italy. Virgil also includes the earliest extensive list of preferred wines in antiquity, many of them personified according to their distinctive qualities (Johnston 1999:207).

The great number of remaining wine amphorae and shards shows the importance of wine in the lives of people in the ancient world and the realization by Greeks and Romans that wine varies from place to place in both type and quality. In an era of polluted water supplies, and in many cases unsafe to drink, wine was the beverage of choice. The amphorae for shipping wine from each wine-producing district have their own unique amphora design and often have inscriptions about the wine's provenance (Clinkerbread 1982). The importance of wine to ancient Greeks and Romans carried over to their spiritual lives with the creation of the Greek God Dionysus and the Roman God Bacchus representing the god of wine.

Stanislawski (1975:441) postulates that wine was a key factor in the expansion of the Mycenaean, Phoenician, and Greek cultures. The Mycenaeans and Phoenicians used wine as a bartering tool to trade with the barbarians as far away as Spain. Stanislawski states that the major expansion of the Greek empire was by wealthy Greeks planting grain and vineyards for profit (Stanislawski 1975:400). More from this prolific geography of wine writer follows in later sections.

An interest in food and wine is evident in the oldest Greek literature. A pottery fragment with a quote by Alcman, lyric poet of Sparta, lists five fine wines of the southern Peloponnese. The Greek historian Thucydides also indicates the importance of viticulture in ancient Greek culture when he writes, "the peoples of the Mediterranean began to emerge from barbarism when they learned to cultivate the olive and the vine" (Johnston 1999). The Greek fourth century BC writer Theophrastus discussed Greek viticulture and grape growing with an emphasis on vineyard soils and matching them with specific grapevines (Hort 1916:133).

The greatest account of Greek and Roman grape varieties is that of Pliny the Elder who wrote a *Natural History* of the ancient world, describing grape varieties and styles of wine in detail. In Book XIV, he references 91 varieties of grapes, 50 kinds of "generous" wines, 38 foreign wines, and 18 sweet wines (Pliny, translated by Bostok and Riley 1892). He describes the growing conditions and care of grape vines and discusses the "culture of the vine." This is the definitive proof of Roman appreciation of differences in wines produced from disparate locales and their preference of some wines over others.

Other religions place a great importance on wine with the Hebrew Bible having over 200 mentions of wine and the Babylonian Talmud over 3,500 (Jordan 2002). Wine also played a prominent role in ceremonies in the Christian faith, such as Holy Communion. Indeed, the Catholic Church was responsible for the diffusion of the grape and the cultural inclusion of wine in the new world, especially where the Spanish and Portuguese colonized the land (Dickenson and Unwin 1992). California is an excellent example because Jesuit priests planted grapes whenever they built new missions; thus, the Church was responsible for the diffusion of the grape and wine industry (de Blij 1983b:59). Even the staunchly religious Pilgrims at Plymouth Rock were wine and beer drinkers, and it is likely the wine they consumed on the first Thanksgiving came from the local vines laden with a bountiful supply of grapes. Enthusiastically, they set about to produce wines in competition with French brands when in 1635 John Winthrop imported some European cuttings. However, it was not long before the colonists found that while the indigenous grape grew well in the pebble-strewn soil of New England, imported varieties languished (Albertson 1950:480).

Why Geography?

Even with a great demand for wine in the United States and worldwide, the common wine grape *Vitis vinifera* does not grow everywhere. Some locations are better suited for its cultivation. De Blij (1983b:12) states that the core area for grape growing is the zone between the annual isotherms of 10°C (50°F) and 20°C (68°F), both in the northern and southern hemispheres. Historically, the locations with Mediterranean climates, Koppen Csa and Csb, have been the best suited for grape cultivation and wine making, referred to as viticulture and oenology respectively. Today, viticulturalists are pushing the boundaries of grape growing poleward by growing more cold resistant varieties and by irrigation, extending the frontier into the arid steppe and desert climates. Table 1.1 shows the major wine-producing countries and indicates trends in wine production over the most recent 3 years for which data are available, whereas Table 1.2 lists the major wine-exporting countries of the world and their output over the same 3-year period. Both tables reinforce the regional concentration of grape growing and wine production and supply ample statistics and variables for geographical analysis.

As wine consumption grows, it is not surprising to see the scientific study of grapes and wines increase. The field of geography is a natural setting for research relating to location, the impact of the environment on grape and wine production, diffusion studies, mapping, remote sensing analysis of grape production, and many other topics that interest geographers. Over the years, the numerous paper presentations at the Association of American Geographers (AAG) Annual Meeting and the many scientific papers appearing in geographic journals attest to the importance of the topic to geographers. It is only natural that people with common interests join to form a special interest group. The Wine Specialty Group received its charter at the 1998 AAG Annual Meeting in Philadelphia, and continues to flourish.

The field of geography is well qualified to study grape and wine production. Most wines of the world have a name associated with the geographical region in which the grapes grow, not the grape variety. Each geographical location has a history of producing a

Table 1.1 Major wine-producing countries in 2006–2008 (Source: Food and Agriculture Organization of the United Nations 2006–2008)

Wine production by country 2006 (tonnes)			Wine production by country 2007 (tonnes)			Wine production by country 2008 (tonnes)		
Rank	Country	Production	Rank	Country	Production	Rank	Country	Production
1	France	5,349,333	1	Italy	5,050,000	1	France	4,711,600
2	Italy	4,711,665	2	France	4,711,600	2	Italy	4,609,554
3	Spain	3,643,666	3	Spain	3,645,000	3	Spain	3,400,000
4	USA	2,232,000	4	USA	2,300,000	4	USA	2,300,000
5	Argentina	1,539,600	5	Argentina	1,550,000	5	Argentina	1,520,000
6	Australia	1,410,483	6	China	1,450,000	6	China	1,500,000
7	China	1,400,000	7	S. Africa	1,050,000	7	Australia	1,244,780
8	S. Africa	1,012,980	8	Australia	961,972	8	Germany	1,026,100
9	Chile	977,087	9	Germany	891,600	9	S, Africa	1,026,000
10	Germany	891,600	10	Chile	827,746	10	Chile	850,000

Table 1.2 Wine-exporting countries, 2007 (Source: Food and Agriculture Organization of the United Nations 2008)

Wine-exporting countries 2007 (tonnes)		
Rank	Country	Tonnes
1	Italy	1,967,388
2	Spain	1,561,505
3	France	1,506,196
4	Chile	1,163,664
5	Australia	781,597
6	S. Africa	500,887
7	USA	431,689
8	Argentina	370,268
9	Germany	369,085
10	Portugal	343,112

unique style of wine made from a particular grape variety. There is no need to place the grape names of Pinot Noir, Pinot Meunier, and Chardonnay on a bottle of Champagne because most wine geeks know that is what is in the bottle. Much of the charm of drinking a bottle of wine is lost when the grape name "Cabernet Sauvignon" replaces a geographical location from which the wine comes and which has a unique imprint on the way the wine smells and tastes.

When non-geographers talk about the geography of an area, they are generally referring to landforms, rock types, weather, climate, hydrology, vegetation, and other characteristics of the physical landscape. All of these factors have a substantial impact on the type of grapes that can be grown and the style of wine, as we shall see later, but geographic analysis is much more.

Geography is the science of spatial organization and deals with the distribution of objects and phenomena across the face of the earth, as well as the interaction of the physical and cultural elements. Harm de Blij, a multifaceted geographer who also has a specialty in the geography of wine, states that the field of geography consists of four traditions:

> 1. Geography deals with the natural as well as the human world.
> 2. Geography assesses the complicated relationships between human societies and natural environments.
> 3. Geographers do research in and try to understand foreign cultures and distant regions.
> 4. Geographers practice the "location tradition," or why do activities and phenomena occur where they do?
> (de Blij 2005:7)

The cement that holds all of these traditions together is space or the spatial element for which de Blij states:

> To pull it all together, we need a word that telegraphs our main geographic preoccupation, and that word derives from space, not celestial space, but Earthly space. We geographers look at the world *spatially*. I sometimes try this concept on questioners: historians look at the world temporally or chronically, economists and political scientists come at it structurally, but we geographers look at it spatially. (de Blij 2005:7)

It is evident from the previous discussion that the geography of wine is commensurate with the mission of the field of geography. Geographers are interested not only in the location of vineyards and wine-producing facilities but also in studying the relationship between the physical environment and the cultural environment in order to see why certain areas produce different types of wine. Viticulture and wine production by their very nature are spatial and therefore geographical.

The science of geography has many subdisciplines. Look at the course offerings at a university to get a better understanding. Major subdivisions of geography contain regional, topical, and techniques courses. Regional geography studies the world and smaller regions including continents, countries, states, cities, or even places such as wine-producing regions or appellations. Other courses are topical in nature and can specialize in physical or environmental geography, cultural geography, or economic geography. In addition, there are techniques courses including cartography, remote sensing, and geographical information systems.

Examples of Geographic Research

This chapter is not a definitive survey of the subject of the geography of wine, but it gives examples of research by geographers on the geography of wine. Most of the citations and examples are from North American and English language publications. For research in foreign-language publications, especially in France, e.g., see the references in Gade (2004) and Pitte (2008) for many French-language sources. Since this chapter emphasizes what geographers are doing, the following material does not include the many excellent works of a geographical nature by non-geographers. Works cited below are by professional geographers, people trained as geographers, works in geographical journals that are geographical in nature but not by geographers, and works by multiple authors in which at least one is a geographer.

Geographers have published research on grapes and wine for many years, and there are several works highlighting the "state of the art." Dickenson (1990) presents a status report on the research done by geographers in the first issue of the *Journal of Wine Research*. In describing his objective, Dickenson states:

> This paper provides a bibliographic introduction to such work in English. It explores two major themes-the relationship between environment and viticulture, and between areas of production and wine quality, and outlines topics for further geographical research on vine-related themes as diverse as history, trade, disease and crime. (Dickenson 1990:5)

Dickenson presents arguments for and against the concept of *terroir* and states that many of the views in the literature are reminiscent of the philosophy of environmental determinism, a concept that was popular in geography in the past, but is frowned upon today. He notes that many geographers have become involved in wine as a side interest to their principal disciplinary expertise and that much geographical research is in a "popular" rather than a scientific style (Dickenson 1990:5). Additional comments on changes in the geography of wine are made by Dickenson (1992a) in the *Interdisciplinary Science Review*. In another state-of-the-art paper, Theresa Bulman presents an overview of the geography of wine in her presidential address to the Pacific Coast Geographers (Bulman 2003). She subdivides the field into a number of categories and gives a thorough review of the literature. Other early articles on the place of the geography of wine in the discipline of geography include Holtgrieve (1972) on the geography of the vine, de Blij (1983a, 1986c, 1992d) discussing the rationale behind geographers studying the geography of viticulture, Dickenson and Salt (1982) and Dickenson (1989, 1992a) further exploring how geography can contribute to the subject. In an application to geographic education, MacDonald and Lemaire (1995) present the possibility of using American Viticultural Areas in studying regional geography. Newman (2000) gives an overview of wine in The *Cambridge History of Food* and Dougherty (2010) summarizes the geography of wine in the *Encyclopedia of Geography*.

The study of viticulture and wine has advanced to the stage that a number of geography graduate students are writing their theses and dissertations on wine-related subjects. At each AAG meeting, eager graduate students are presenting their research for professional scrutiny. Recent doctoral dissertations on the geography of wine include Nichol's (2002) research on Mexican migrants and the impact on several areas including Napa Valley, Guthey's (2004) politics of the agro-industry of California's North Coast, Barker's (2004) comparison of the New Zealand and French wine industry, Velluzzi's (2007) study of the Washington wine industry in Walla Walla, Puleo's (2008) spread of parasites in the Vallis Tellina of Italy, and Goldberg's (2010) historical study of the German wine industry.

While geographers have been less than prolific in producing books, their contributions to journals and reports are exceptional. Geography journals have an excellent selection of geography of wine articles. A search of the *Annals, Professional Geographer, Economic Geography, Geographical Review, Journal of Geography, Geographical Journal, Geography, Geographical Magazine, Journal of Latin American Geography, Focus,* and regional geographic journals turns up many fascinating grape and wine articles. Although not a geographical journal per se, the *Journal of Wine Research* has a wealth of articles on geography and *terroir*, and in addition, the editor, Tim Unwin, is a geographer. Because of the number of geography of wine articles by geographers in non-geographical journals, the author apologizes to those geographers whose works do not appear in the following pages, but, rest assured, enough examples appear to provide anyone an appreciation of the types of research done by geographers.

For those interested in further reading, many books and journal articles on the geography of wine abound. In the following sections, a search of the literature of each of the following subdisciplines in geography illustrates the type of work geographers are publishing:

General Reference Books
Regional Geography
Physical Geography
Cultural Geography
Economic Geography
Terroir
Geographic Techniques.

General Reference Books

The classic books in the field include those by Harm de Blij on *Wine: A geographic appreciation* (1983b) and *Wine and the vine: An historical geography of viticulture and the wine trade* by Tim Unwin (1996). The de Blij book often serves as a text in introductory courses on the subject. It includes historical, physical, cultural, and economic overviews of viticulture and the production of wine, and explores the regions of the world in terms of major grape types and wines produced. In his review of the geographical literature, Dickenson (1990) states the following about the contributions of de Blij:

> (de Blij)...through his three books and a series of thematic articles has provided a firm foundation for the study of viticultural geography. He has provided not only a model approach to the field, but made available to other students of the subject insights as to the nature of the geographic perspective, and indication as to areas needing further research by both geographers and others. The two versions of his 'basic text' (the one an amplification of the other) provide a clear outline for the geographical study of wine (Dickenson 1990).

On the other hand, Tim Unwin takes a historical approach in terms of the origin and diffusion of viticulture and wine and highlights the historical, cultural, and economic aspects from ancient to the modern times. Unwin has also made a significant contribution to the geography of wine by being the editor of the *Journal of Wine Research*, one of the leading academic wine journals in the world.

A book by Brian Sommers (2008) entitled *The geography of wine: How landscapes, cultures, terroir, and the weather make a good drop* aims at the popular market and has met with success. He introduces a number of geographic concepts and presents both the physical and cultural background by case studies. Another book by John Boyer (2009) *Drink this now* is a general introduction to wine with case studies on the major wine-producing regions. Both of these books are interesting surveys of world wine regions but do not delve into the detail needed for a text in a university course without supplemental resources.

With the exception of the previous books, there has been a dearth of general wine books written by geographers. This is disappointing considering the number of geographers and their interest in spatial patterns worldwide. On the other hand, there are many excellent books on wine regions of the world by non-geographers including those by Clarke (1997), Kolpan et al. (2010), McCarthy and Ewing-Mulligan (2006), and Parker (2008). Most of these books are descriptions of the wine regions with little analysis of why the regions exist and how geographical aspects are important. There is a definite need for books with geographical analysis, but geographers are not filling that niche in the rapidly expanding viticulture and wine market.

Regional Geography

Geographers have better representation in regional books on viticulture and wine-producing regions. In his *Wine regions of the southern hemisphere*, Harm de Blij presents an analysis of the wine regions of South America, South Africa, Australia, and New Zealand (de Blij 1985d). These were overlooked wine-producing regions, which have become significant producers in recent years.

On the country level, the classic *Landscapes of Bacchus*, a book about grape and wine production in Portugal by Stanislawski (1970), is a classic in geographic scholarship. Stanislawski traces the development of the wine landscape from prehistoric times to the present by showing the relationship between cultivation techniques, physical characteristics, socioeconomic variables, and marketing to Portugal's wine-growing regions. The main contribution of the book is the detail given to the importance of the cultural geography of each region and the lucid view of how viticulture and wine have created a distinctive landscape.

The Wine regions of America: Geographical reflections and appraisals, by John J. Baxevanis (1992), *The wines of Bordeaux and western France* (Baxevanis 1987a), and *The wines of Champagne, Burgundy, eastern and southern France* (Baxevanis 1987b) are prime examples of in-depth geographical research and scholarly writing by a geographer. It is unfortunate that these books were prior to today's interest in wine, limited in distribution, and did not receive the attention they deserve in and outside the field of geography.

Another prominent regional book by a geographer is Gary L. Peter's work on *American winescapes: The cultural landscapes of America's wine country* (1997). Along with the usual origin of grapes and diffusion commentary, Peters develops the idea of winescapes in America, further developing a concept initiated by Stanislawski (1970) in his book on Portugal.

One of the ways in which geographers study regions is the comparative approach, where the geographer studies two or more regions in terms of how they are alike and how they differ. An outstanding example is the book by Jean-Robert Pitte, French geographer and president of the University of Paris-Sorbonne, in which he compares and contrasts two famous wine-producing regions in his *Bordeaux/Burgundy: A vintage rivalry* in terms of their physical, cultural, and economic geography (Pitte 2008). Although written by a French geographer, the book is now available in English. Following the French geographic tradition, Pitte emphasizes historical and cultural differences between the two regions, but with a balance for the physical factors that also define the regions.

Many regional articles appear in geographical journals with an early example of the comparative approach being an article by Blanchard and Blanchard (1929) on the grape industry in Spain and Portugal. Another example of the comparative approach is an article by Warren Moran (1993) on the comparison of wine production in France and California. It details how physical factors play a role in influencing political structure and criteria to delimit wine regions, resulting in France having a highly controlled methodology versus *laissez-faire* policies in California. Comparisons with France have been popular with Barker (2004) performing a comparative study on how laws have affected the changes in the geography of wine in France and New Zealand.

In addition to the Barker research on France and New Zealand, several other researchers have picked New Zealand as one of the countries in a comparative study. Gwynne (2006a) contrasts how the wine industry in New Zealand and Chile has developed. In another article, Gwynne (2006b) evaluates factors influencing the growth of the wine industry in New Zealand versus that in Chile. Barton et al. (2007) also use New Zealand and Chile in an article comparing their competition and cooperation.

De Blij (1983b) uses a comparison of Chile and Argentina to explore the wine production of the two countries. In another Chile and Argentina comparison, Rosas (2008) shows how government policies and the influence of outside companies have resulted in globalization of the Chilean wine industry, as opposed to a homegrown expansion in Argentina. Giuliani et al. (2010) shows the popularity of Chile as an area of comparative study with a discussion of the researchers who are working on Chile, South Africa, and Italy. The three countries are also the site for a study by Cusmano et al. (2010) comparing their status in world markets. Other examples of the comparative approach are studies of Los Haso, Zacatecas and Napa, California (Nichols 2002), Barolo and Barbaresco (Smith 2005), and Castile and Leon, Spain (Sánchez et al. 2010).

Many articles by geographers pertain to a single region and often show how a region has an advantage or disadvantage compared to other areas. Many of the early local regional studies by geographers did not pertain to sexy regions that captivate our interest today. Morrison (1936, 1950) wrote about the Ohio wine industry, which historically was a very important region until its decimation by phylloxera and prohibition. Dahlberg (1961) gives us a locational study of the Concord grape in the Chautauqua-Erie region. Newman (1986) illustrates how the physical and cultural characteristics define the Finger Lakes region, and in another study, discusses the decline and redevelopment of the Finger Lakes wine industry (Newman 1992). Other lesser-known vineyard regions discussed by geographers are the Ozarks (Hewes 1953), the northern Virginia wine regions (Gruber 1981), the semiarid area of Texas (Templer 1986), the Southeastern United States (de Blij 1987d), Pennsylvania (Dickenson 1992b), Umpqua Valley (Jones 2003b,; Jones et al. 2004), Rogue River Valley (Jones 2001, Jones et al. 2006), Oregon (Jones and Hellman 2003a, b; Jones 2003a), and Walla Walla (Velluzzi 2007).

Meigs (1941) was a pioneer in studies about California done by geographers with his economic study of the importance of orchards and vineyards. Other California studies include Peters (1984) investigation of trends in California viticulture, Peters (1987) analysis of cultivars that are best suited for California, and Peters and Gossette (1990) further refining this by looking specifically at Pinot Noir, Barbera, and Zinfandel. Baxevanis (1990) looks at the potential of the Sierra Foothills, Bulman (1991) investigates the development of grape monoculture in Napa, and Guthey (2004) uses *terroir* and politics in his study of the California North Coast Wine Districts.

Many geographical studies utilize France as their location, e.g., the work by Agnew (1945) on the vineyards of Bas-Languedoc, Weigend (1954) on viticulture in south-west France, Weigend (1955) on the changing port functions of Bordeaux, Gade (1978) on windbreaks in the lower Rhone Valley, Jones (1989a, b) on agricultural despecialization in Languedoc, Crowley (1983) presenting a study on the declining hectarage of grape growing in France as well as a change in the varietals grown in response to French government and EU pressure, Jones and Davis (2000) on Bordeaux, Jones and Storchmann (2001) on Bordeaux, Crowley (2001) on Languedoc versus AOC wines, and Gade (2004) with his history of wine production and the importance of *terroir* in Cassis, France. Other regional studies include soil erosion in Ardeche (Augustinus and Nieuwenhuyse 1986), soil reflectance (Vaudour et al. 2008; Vaudour 2008), and land use in southern France (Van Eetvelde and Antrop 2004).

Geographers interested in Latin America have written a number of articles, including Morris (1969) on migration and diffusion in Mendoza, Hansis (1977) on land tenure in Mendoza, Ackerman (1978) on central place theory in Argentina, de Blij on Brazil (1985c), Crowley (2000) on the historical character and changing geography of the Chilean wine industry, Gwynne (2008a, b) on value chains in Chile, Corby (2010) on exploring the impact of neoliberal economic reforms on the cooperative sector of the wine industry of Argentina and its relation to globalization, and Overton and Murray (2011) on rural space transformations in Chile.

New Zealand and Australia have been the location of much work by geographers on wine districts, e.g., an early study on wineries in Barossa by Smith (1970), a regional study of Coonawarra (Banks and Sharpe 2006); a marketing study of Australian and New Zealand wines (Banks et al. 2007); regulation and development in New Zealand (Barker et al. 2001); the reaction of boom to gloom in Australia (Kleeman 2006); regional dynamics of the New Zealand wine industry, with an emphasis on the Marlborough District (Hayward and Lewis 2008); Hunter River Valley (Holmes and Hartig 2007); land use change in the Hunter Valley (Manandhar et al. 2009); and vineyard land prices in Hawkes Bay and the impact of urbanization (Overton 2010).

Other regional studies include Dobby (1936) on the economic geography of the Port region, northern Portugal (Ackerman 1937); Herault (Stevenson 1980); China (de Blij 1981); Japan (de Blij 1982); England and Wales (Ilbery 1983, 1985; Dickenson 1985, 1991), the United Kingdom (Unwin 1994; Spellman and Field 2002; Turner 2010); Switzerland (Nachtergaele et al. 1998); Niagara, Canada (Shaw 1999, 2001, 2005); Ontario (Carmichael 2005); Barolo and Barbaresco (Smith 2005); Okanagan (Kingsbury and Hayter 2006); South Africa (McEwan and Bek 2006; Bek et al. 2007; McEwan and Bek 2009a, b; Vaudour et al. 2010); Priorat, Spain (López et al. 2006); Slovenia (Azman and Drago 2009); and Germany (Goldberg 2010).

Articles by geographers appear in the popular literature, e.g., the *Geographical Magazine* (Stevenson 1976, and a series of 19 articles found in the reference section of Dickenson (1990); Fisher 1975, 1978, 1981, 1984), and a series of articles by Harm de Blij in *Focus* (de Blij 1985a, b, 1986a, b, d, e, 1987a, b, c, e, 1988a, b, c, 1989a, b, c, 1990a, b, c, 1991a, b, c, d, 1992a, b, c). Geographers are involved in national and local wine newsletters in many cases. Of particular note is a series of articles by geographer Robert Hutton in the *Journal of the American Wine Society* including articles on Tokaji wines (1993), Hungarian wines (1997), and the wines of Romania (2004).

Another geographer who writes extensively in the general literature is Bruce Sanderson, senior editor and tasting director of *Wine Spectator* magazine which has a circulation of over three million monthly (Wine Spectator June 2011). Sanderson is an urban geography graduate from the University of Waterloo, with a master's degree from Queen's University. His frequent articles in *Wine Spectator* have a healthy dose of geography, complete with maps, e.g., his recent articles on Riesling (Sanderson 2010a), *terroir* at Domaine de la Romanée-Conti (Sanderson 2010b), and the Piemonte region of Italy (Sanderson 2011).

Physical Geography Research

For non-geographers, physical geography is what they often think about when someone mentions the word geography – landforms, soils, rivers, weather, climate, and vegetation. Many early American geographers were geologists, including the first president of the AAG, William Morris Davis. Physical geographers differ from geologists in their interest in the impact of earth materials and landforms on humans, as well as the interaction with the atmosphere, hydrosphere, and biosphere. Even today, physical geographers have much in common with their peers in geology, soil science, hydrology, biology, meteorology, and climatology. It is, therefore, reasonable to expect physical geographers to have a great interest in the impact of earth materials, landforms, soils, drainage, weather, and climate, and the impact of these elements on grapes and the resulting wines.

To the contrary, there has been a lack of research in physical geography related to viticulture and wine for a number of reasons. Although physical geography is the birthplace of the discipline of geography in the United States and most university geography departments still have courses in physical geography, physical geography has seen a major decline in the number of departments offering it as a specialty. The growth of cultural, human, and urban geography in the last century and the recent explosion of techniques courses, especially geographical information systems, have resulted in the field of geography shifting in that direction. Bierly and Gatrell (2004:342) in a study of graduate geography departments from 1991 to 2001 show that the number of geotechniques professors has grown substantially, the number of geomorphologists has remained flat, soil scientists are shrinking in numbers, while biogeographers and climatologists have shown a small increase. During this period, geography was experiencing great growth as a field, so as a percentage of the whole, physical geography has actually suffered a relative decrease.

Many current physical geography professors are older and have their degrees from a time when wine was not as popular as it is today, and many do not have an interest in wine research, whereas recent hires have a greater interest in wine and are doing more wine research. The dichotomy results from physical geography professors being in the older group and the younger hires being the specialists in geotechniques, economic, and socioenvironmental topics. This will continue to increase as older traditional faculty members retire and new professors who teach more job-oriented courses take the positions. Contributing to the disparity in publications between these two groups is the fact that most new hires are under pressure to "publish or perish," while the more established professors on tenure are less prolific in publishing.

The largest group of professionals in physical geography remains those in geomorphology (Bierly and Gatrell 2004:342). As a group, geomorphologists are more process oriented, whereas the geography of wine has been historically descriptive, regional, or based on the interaction of many variables and the study of *terroir*.

The future is not bright either, because at the K-12 level, earth science classes have absorbed most of physical geography and much of the rest of geography resides in social science classes. Few K-12 schools offer geography as an independent subject; hence, students do not see it as a potential university major or career path. Consolidation is also occurring at the university level where some colleges and universities are combining geography with earth sciences, anthropology, and/or environmental sciences because of the financial situation. An additional threat to physical geography has been the division of curricula at some universities into physical science and social science. This may result in reassignment of physical geographers to physical science departments along with their courses.

Much physical geography is taught in smaller state colleges and universities where teaching is the primary emphasis, not research, so the number of professors in physical geography may not be as important as where they teach. Even where there is a critical threshold of physical geographers in a department, it is often true that an interest in wine is a secondary interest. Dickenson (1990) states this is a reason why geographers, in general, are not more productive scholars in wine-related research. Many geographers pursue wine as a hobby first rather than as graduate coursework. Even in many large universities, one or two professors who have large teaching loads and little time for research other than in their primary interest do the teaching of physical geography. In addition, few graduate departments of geography have a concentration in geomorphology and soil science, and even fewer universities offer a credit course in the geography of wine.

Fig. 1.5 Typical Chateauneuf-du-Pape soil covered with galettes

Although not the primary emphasis of many of the research articles on the geography of wine, physical geography is still an important portion of many articles on regions, *terroir*, and cultural topics. Living up to its reputation as a synthesizing subject, physical geography often provides the stage for much regional and topical research. Even in weather and climate studies, landforms provide an important influence on atmospheric events.

The following section presents a sample of contributions in physical geography on the topics of landforms and soil, weather and climate, and climatic change. It is difficult to isolate one topic in physical geography that is more important than others in terms of wine characteristics and quality – it is the sum total of all the environmental factors that produces a good wine. Geographers are good at studying wine because geography is a synthesizing science that considers all physical and cultural elements in its analyses. An example is vineyard site selection research by John Boyer, a geographer at Virginia Tech University, who is a coauthor of articles on site selection of vineyards in order to minimize frost damage (Wolf and Boyer 2000) and a chapter on vineyard selection in the eastern United States (Wolf et al. 2009).

Landforms and Soil

A review of the literature shows only a few studies on the impact of landforms on viticulture and wine. Landforms are an important part of many regional articles, and landforms are an important factor in air drainage, elevation cooling, solar prospect, erosion rates, soil formation, and the importance of alluvial fans. A good example containing landforms is an article on the German Rhinegau by Schwartz (1997), which attempts to predict relative must quality of the Riesling from site factors by classification and regression tree (*CART*) analysis. While not arriving at a definitive conclusion on which factor was the most important, the study, nevertheless, reinforces the concept that it is a number of factors working together that contribute the special characteristic of the Rieslings from this region.

The landforms are especially important in the type of soil that forms in an area, or in the case of many great grape-growing areas, the lack of soil. Grapes are one of the few agricultural commodities that thrive in poor soils that are too rocky, too dry, and too nutrient deficient for other crops. Figure 1.5 illustrates a vineyard in Chateauneuf-du-Pape, France, where large rocks called *galettes* cover the surface of many vineyards, hardly an agricultural soil in anyone's perception.

Several studies of the soil and its impact on viticulture have appeared in the geographical literature. Elliott-Fisk and Noble (1992) investigate the environments of Napa Valley in California and how the landforms and soils influence the wine flavors of Cabernet Sauvignon. Elliott-Fisk (1993) provides an analysis of the soils and *terroirs* of vineyards in Napa Valley, CA by presenting an overview of the geological history, its effect on soil formation, a summary of the physical environment, and developing viticultural environments dividing Napa Valley into four geographical regions. The paper argues that the natural physical environment exerts important controls on viticulture and that subtle soil differences influence the wines of the region.

Soil erosion in the vineyards of the Ardeche region of France is the topic of research by Augustinus and Nieuwenhuyse (1986) who identify problems in the management and control of the vineyards and make recommendations to decrease the amount of erosion. In an article on gravel mulching in the vineyards of southern Switzerland, Nachtergaele et al. (1998), through field measurements, show the impact of gravel mulch on the thermal and hydrographic characteristic of the soil and its impact on vineyard productivity. Overton (2010) and Overton and Heitger (2008) explain how the Gimlett Gravels delimit a viticultural region in the Hawkes Bay area of New Zealand and result in its transformation from a virtual agricultural wasteland to one of the most sought-after and expensive wine-growing areas in the country. Several other studies of soil in vineyards appear in a later section on remote sensing.

Weather and Climate

Many scientific papers extolling the virtues of weather and climate in the creation of a great grape and wine-producing region have been written by geographers. An early paper by Aron (1976) provides an argument for why Oregon's climate is suitable for premium wine production, a forecast that has proven correct. Other early papers include a study of windbreaks in the Lower Rhone Valley by Gade (1978), wine production in a cool climate by Grant (1984), and location, scale, and climate by de Blij and Peters (1988). Geographer Gregory Jones is a coauthor in an article by Nemani et al. (2001) in which they assess asymmetric warming over coastal California and its impact on the premium wine industry. They find that warming trends are diurnally and seasonally asymmetric, with greatest warming during nights and in spring, and that warming is due to rising CO_2 levels with enhancement by increases in atmospheric water vapor.

Several other papers by Jones (2003a, b, 2004, 2006) study climate as one of several variables used to assess the Oregon wine potential. Jones and Storchmann (2001) use climate as one of the variables in analyzing wine market prices and investment under uncertainty using an econometric model for Bordeaux Crus Classés. They conclude that the price of Merlot wine is more sensitive to climate fluctuations, while Cabernet Sauvignon is more susceptible to fluctuations in Parker scoring. Jones et al. (2010) spatially map the climate in American Viticultural Areas (AVAs) in California, Oregon, Washington, and Idaho using the 1971–2000 PRISM 400-m resolution climate grids. They assess the statistical properties of four climate indices used to characterize suitability for viticulture – growing-degree days (GDD, or Winkler index, WI), the Huglin index (HI), the biologically effective degree-day index (BEDD), and average growing season temperatures (GST) – the result being the GST is the best calculation because of its ease of calculation.

California has been the site of weather and climate in papers by geographers, including a study by Eysberg (1987) on "cool air-conditioned valleys as the equivalent of warm sheltered *cotes*," an early comparison of California wines to those of Burgundy. Peters and Gosette (1990) present an analysis of the types of grapes grown in California and surmise that "viticultural mesoclimates appear to predominate as determinants of the geographic distribution of Californian cultivars," with special emphasis on Pinot Noir, Barbera, and Zinfandel.

Fitzharris and Endlucher (1996) study New Zealand as the site of their paper on climatic conditions for wine grape growing and show how the cool, marine conditions favor certain varieties. Jones and Davis (2000) analyze the importance of the weather and climate on the variety and the impact of climate on grapevine phenology, grape composition, and wine production in Bordeaux, France.

Canadian wine districts have been the location of several papers, including one in which Shaw (1999) analyzes the cool climate wine regions of eastern Canada and the ability to support quality vineyards. Shaw (2002) uses the Niagara region of eastern Canada to assess the use of wind machines for frost protection in the vineyards. In another analysis, Shaw cites Pelee Island and the Lake Erie North Shore as good locations for vineyards based on their location in Canada's warmest climatic region (Shaw 2001). Upon investigating the impact of the lake effect and soils on the grape varieties grown in the Niagara vineyards, Shaw (2005) concludes that the area's potential as a grape and wine producer compares favorably to Bordeaux and Burgundy climatically.

Climate Change

Geographers have published extensively on climate change, so it is not unusual to find several publications on how climate change influences grape cultivation and wine production. In his research on Oregon grape-growing regions, Jones (2003b) found changes in the trends in frost occurrence and the length of the frost-free growing season. Jones (2004) published another article in *Geotimes* about making wine in a changing climate and the potential impact on wine production. The impact of climate change on global wine quality is the topic of a study by Jones et al. (2005). In a study of quality wine areas around the world from 1950 to 1990, Jones et al. (2006) found a general warming of growing seasons with an accompanying increase in vintage ratings with less variability from year to year. Unfortunately, many of the best wine regions of Europe are already at their optimal climatic condition and continued warming may cause a negative impact on their quality. In another volume on *terroir*, Jones (2006) investigates the impacts of climate variability and change on wine. White et al. (2009) state that competition from the New World, a changing climate, and technological advances, have threatened the Burgundian notion that the quality of wine depends on regional geography and culture, and only flexibility can keep the concept of *terroir* alive.

In a series of publications, geographers from the University of Guelph explore the impact of climate change on agriculture in Canada. Belliveau et al. (2006a, b) published an occasional paper on the impact of climatic change on agriculture in Canada with references to the grape-growing industry. In addition, Belliveau et al. (2006b) survey the grape industry in the Okanagan Valley in terms of the climatic and non-climatic risks that growers perceive. The result is fear that changing climatic conditions may result in a lack of irrigation water and that changing demographic may interfere with their ability to adjust to climate change. Another related study looks at the apple and grape industry in Canada and its vulnerability to climate variability and change (Belliveau et al. 2007).

A special issue of the *Journal of Wine Research* in 2010 contains several papers by geographers. Jones and Webb (2010) introduce the topic by discussing the challenges and opportunities facing climate change research in viticulture. Another article, in which Jones is the coauthor, discusses climate induced historic and future changes in viticulture (Schultz and Jones 2010). In a paper reviewing the state of knowledge on the implications of climate change for viticulture and viniculture, Guelph geographers Holland and Smit (2010) suggest the vulnerability approach as a direction for future research on climate change. Switching over to a regional theme, Hadarits et al. (2010) present a study on the producers of Maule, Chile, and their adapting to climate change. Another work, including the geographer Tony Shaw, is an investigation of the potential use of weather derivatives to hedge harvest rainfall risk in Niagara, Canada (Cyr et al. 2010). This is an expansion of their work on hedging adverse conditions employing a short condor position (Cyr et al. 2008).

Biogeography

Geographers have an interest in diffusion of people, ideas, and even insects. Eichel (1975) presents an article on the impact of the root louse Phylloxera, which wiped out many of the world's vineyards in the mid-1800 s and is still a threat today. In another article on Phylloxera, Stevenson (1980) writes about the devastation of the phylloxera outbreak and its diffusion in Herault in the Languedoc of southern France.

In an example of a biogeographical work from South Africa, Fairbanks et al. (2004) examine the impact the wine industry is having through its expansion into areas of the unique and threatened vegetation of the Cape Floristic Region. They use predictive land use modeling by logistic regression techniques to determine suitable areas for vineyard cultivation according to climatic, topographic, and soil/geology variables, and as a result, they make recommendations to preserve the ecosystem from further damage.

Cultural Geography Research

In this section and the following section, the subdiscipline of human geography is broken down into two components – cultural geography and economic geography. Human geographers by the sheer size of their numbers have published a larger body of work than physical geographers. Under cultural geography, topics of discussion range from cultural landscapes, diffusion, and religion to historical, tourism, and related topics. Although the approach to the study of tourism can be an economic topic, it falls under the cultural heading because much of the work by geographers has been in terms of people's perception of the industry.

It is readily apparent that the physical environment influences the vineyard and that it has an indelible imprint on the resulting wine, but it is less likely for one to think about the involvement of cultural geographers with wine. When we think about a specific wine region, we often perceive the cultural landscape, or in this case, the winescape. The geographical works by Stanislawski (1970) in his classic work on the cultural landscape of Portugal and Peters (1997) in his book on wine landscapes in America provide good examples of how geographers can covey the essence of a wine region through well-crafted prose. Both works follow in the footsteps of Carl Sauer and the landscape tradition in geography.

Some cultural landscapes where wine is important are easy to visualize. Look at the photos in Fig. 1.6 that portray several classic cultural wine landscapes and identify which is the Douro Valley of Portugal, the Mosel Valley of Germany, and the Chianti region of Tuscany. Figure 1.6a is the classic Tuscan landscape made famous in the movie *Under the Tuscan Sun*, characterized by old stone farm buildings on the hillside, surrounded by vineyards and olive trees, and cypress trees growing along the driveways. Nor can one be mistaken that Fig. 1.6b is a vineyard along the Mosel River of Germany where the vineyard rows run up and down the steep, slate hillsides, above the Mosel River and the charming towns, e.g., Zell with its Germanic style architecture punctuated by the church towers piercing the sky, along its banks. Another example in Fig. 1.6c shows the World Heritage Site of the Douro River Valley in Portugal, with its horizontal terraces climbing up the steep hillsides, dotted with farms and quintas in the typical Iberian style of tile roofs and whitewashed walls. One can also visualize other wine cultural regions including the vineyards and stately chateau of the Loire of France, Burgundy with its small walled plots, Greece with its white buildings and blue domed architecture, or even Napa with its California faux Spanish style architecture and strip development.

An important part of culture is history, and as a result, geographers are making a significant contribution to wine literature in the realm of historical geography. Stanislawski enlightens us about the contributions of the Etruscans to Greek and Roman civilizations and their contribution to what eventually became the site for the Renaissance in Tuscany. He relates information on grape and wine production, including the invention of the pergola trellis system, by reviewing the literature of Virgil, Columella, Pliny, Strabo, and other writers of classical literature (Stanislawski 1977).

Stanislawski (1970) in a book entitled *Landscapes of Bacchus*, and in several journal articles (1973b, 1975), makes important contributions on how viticulture and wine making fueled the expansion of the Phoenician and Greek civilizations. He states that wine was used as a bartering agent for metals and other goods in far flung reaches as far away as Cadiz, Spain (Stanislawski 1973b:400). The Dionysus cult helped to expand the Greek influence and helped wealthy Greeks set up a colonial viticultural expansion and wine trade.

> Colonization based on grains and the vine, I suggest, represented the emergence of the earliest, specialized, monocultural, maritime, commercial agriculture of the Mediterranean basin (p408). That the vine was of early importance is not only a reasonable assumption, but may be shown by the fact that Naxos, the earliest colony in Sicily, stamped a depiction of the head of Dionysus on one side and a bunch of grapes on the other side of its earliest coins (p408). Early Phoenicians traded wine for metal--Hahn said it was wine; wherever they landed, they would tempt the barbarians to trade with them by offering it. Deodars Sickles is the author of the statement that wine was one of their chief trading products, even being carried as far as Spain. (Stanislawski 1973b:400)

Van Eetvelde and Antrop (2004) analyze structural and functional changes of traditional wine landscapes with two examples of urbanizing regions from southern France. In a study of terraced landscapes in Slovenia, Azman and Drago (2009) look at the origin and spatial distribution of terraces in various cultural regions of Slovenia and how they result in a characteristic Mediterranean landscape element. Older fruit-growing terraces are being abandoned, while newer grape-growing terraces are the product of modern, mechanized farming.

Diffusion is a topic of interest to geographers doing research in viticulture and wine. As spatial scientists, geographers are interested in how objects or phenomena change over time. Early work by Carl Sauer looks at the origin and dispersal or species and names the center of grape diffusion as the mountain valleys of the Caucasus-Turkish-Iranian borderlands (Sauer 1952:39). Stanislawski discusses the diffusion of the grape throughout the Mediterranean by the Phoenicians, Greek, and Romans (Stanislawski 1975). Additional work on the diffusion of Vitis vinifera occurs in the

Fig. 1.6 (**a**) Cultural winescape – Tuscan landscape with steep hillsides, vineyard, and villa. (**b**) Cultural winescape – Zell on the Mosel (Source: Mosel Tourism). (**c**) Cultural winescape – Duoro River Valley with river and terraces

works of Harm de Blij (1983b), Jonathan Sauer (1993), and Tim Unwin (1996). Also related to diffusion, Dickenson (1990) downplays the importance of the French in the diffusion of wine and gives credit to others by saying:

> It is noteworthy that despite France's pre-eminent reputation as a wine producer, the French have not been major diffusers of viticulture. With the exception of North Africa, where the strong wines of the baking south were used to give body to those of la Belle France in less favourable years, the French empire did not become a zone of viticulture. Instead, it was the Spaniards in California and South America, the Dutch in South Africa and the British in Australia and New Zealand who fostered non-European viticulture. In addition to pioneers such as Fr. Serra, Simon Van der Stel and James Busby, the role of migrants has been significant in the diffusion and maintenance of viticulture--Italians and Spaniards in Argentina (Morris 1969; Hansis 1977), Italians, Germans and Spaniards in the United States (Morrison 1950; Hewes 1953; Walker 1979); and Greeks and Italians in Australia (Stevenson 1988a).

Geography is a spatial science, but it also has a temporal element in that processes occur over time. Research in historical geography looks back in time and analyzes spatial aspects of viticulture. Saeidi and Unwin (2004) explore the symbolism of wine in medieval Persian poetry, focusing particularly on the works of Hafiz whose poetry provides a brief overview of the archeological and historical evidence pertaining to wine in Persia. Agnew (1945) reports on the historical evolution of the Bas-Languedoc of southern France and its transformation from the granary of the Roman world to a grape monoculture at the mercy of vagaries in the weather.

In another historical account, Dickenson and Unwin (1992) published a series of essays on the vine and wine in Spanish America and Brazil, documenting preconquest non-grape alcohol production, colonialization and its impact on grape and wine production, and in-depth analyses of the diffusion of wine production in Mexico and Brazil. Dickenson (1995) further explains the historical importance of Portugal's colonization on grape cultivation and wine production in pre-independence Brazil. Also of a historical geographic theme is Carter's (1987) work on Cracow's wine trade in the fourteenth to eighteenth centuries while Schenk (1992) investigates the viticulture in Franconia and the human and natural influences since AD 700.

The epitome of an historical geography work is the book by Tim Unwin, *Wine and the vine: An historical geography of viticulture and the wine trade* (Unwin 1996). Starting with the earliest evidence of grape growing and wine making, Unwin follows the diffusion of Vitis vinifera and wine making through the major periods of history, from the Greek and Romans to the impact of modern corporations. His treatment of the socioeconomic variables adds a dimension lacking in many other books, making it one of the most cited works in the geography of wine.

On a more contemporary note, Fredericks (1969) looks at the nineteenth century stonework in California's Napa Valley and gives an historical account for many of the grand, old winery buildings that form an important part of the cultural landscape. The buildings are a union of physical and cultural factors, including the availability of local building rock and the arrival of immigrants who were expert stone workers. Crowley (2000) gives an historical account of the wine industry in Chile by detailing how the industry has changed spatially and in terms of the quality of the product.

Many geographers work in education and teach introductory cultural geography, so it is no surprise their research is oriented towards the spatial distribution of people and how cultures vary. Kaminske (2005) reports on a viticultural project in which students and teachers at four wine region universities in France, Italy, Slovenia, and Germany work together to demonstrate the importance of viticulture across these countries and study how viticulture has played an important part in their history and culture. In another cultural study, Holmes and Hartig (2007) examine the emergence of a sharply delineated socioeconomic and sociocultural fault line between Cessnock's former coal towns and the immediately adjacent economically healthy Hunter Valley Wine Country in Australia and the resulting class structure.

South Africa is a country that has seen cultural upheaval and problems with class structure. Geographers have written several studies with McEwan and Bek (2006) stating how the wine industry has been important to the political empowerment of the minorities. Bek et al. (2007) continue to develop the theme in a work on ethical trading and socioeconomic transformations created by the wine industry in South Africa. McEwan and Bek (2009a) investigate the impact of social and environmental certification in the South African wine industry, and in another paper, analyze approaches to ethical trade in the South African wine industry through a case study of the Wine and

Agricultural Ethical Trade Association (WIETA) (McEwan and Bek 2009b).

Alcohol is one of the most regulated substances, and wine is no exception. Therefore, it is no surprise that geographers are involved in research about political controls, both by government action and by the industry. Government decisions can both help and hinder the wine industry. Lewis (2007) examines local and state-level liquor laws from 1907 to 1919 and the relationship between cultural groups (urban, religious, and immigrant populations) and the adoption of prohibition legislation based on census data and county-level voting data. His results show that cultural differences are important on the local level but less relevant in larger political divisions. In another study, Grigg (2004) looks at world patterns of consumption of wine, spirits, and beer and comments on why each is favored in different parts of the world. Sechrist (2004) analyzes wine sales in the state-controlled system in Pennsylvania and finds interesting differences in preferences between urban and rural areas as well as between cultural groups.

Tourism is another area of research that has caught the interest of geographers. Carmichael (2005) investigates the visitor perceptions of the physical and built environment in the wine industry in Niagara, Ontario, and the juxtaposition of the agriculture and tourism trend. In the Catalan region of Priorat, Spain, where wine has become the major tourist draw, an example of "quality agro-food products" stimulating the economy in a traditional agricultural community is the theme of a chapter by López et al. (2006). Another Spanish tourism study is the article on new forms of tourism in Spain that looks at wine, food, and rural tourism (Canoves and Suhut de Morais 2011).

Wargenau and Che (2006) present a wine tourism study on the motivations, expectations, and successes of Southwest Michigan Wine Trail member wineries in developing horizontal and vertical alliances. Also in Michigan, Veeck et al. (2006) examine agricultural tourism. Che and Wargenau (2011) use visits to winery tasting rooms in a study of education potential and direct marketing. Tomljenovic and Getz (2009) examine the emergence of wine tourism in two Croatian wine regions and assess related life-cycle implications in order to make recommendations for the growth of the wine industry in Croatia. In the Napa Valley of California, Skinner (2010) looks at the impact of sustainable development and wine legislation on tourism.

A population study on the workforce in the Marlborough, New Zealand wine industry by Beer and Lewis (2006), shows that wine tourists make up a significant number of the "casual" vineyard workers, along with transnationals and illegals to supplement the local workforce.

Economic Geography

The contribution of economic geographers to the geography of wine is substantial. Grapes and wine are important products of trade from ancient times to the present. Geographers are interested in the economic health of production regions, agriculture, foreign trade, trade agreements, marketing, globalization, governance, political impact, and value chains. The economic geographer differs in the approach to the geography of wine from the economists in that the geographical approach is spatial and takes into consideration the impact of the interaction of the physical and human environments.

The economic importance of grapes and wine production to *agricultural productivity* has been of interest to geographers throughout much of the history of the discipline, from the time of Strabo and Pliny the Elder. To substantiate that importance, even in modern times, Stanislawski states the following about the importance of grapes and wine to the economy of Mediterranean countries:

> Yet in Mediterranean Europe it is a staple, economically more important than are most industrial products given extended attention in textbooks. If the production of wine were to be stopped, the Greek economy would suffer; the economies of Italy and Portugal would collapse; and those of France and Spain would be put in serious disarray. (Dan Stanislawski 1975:427)

Many early studies of viticulture by geographers are descriptive regional studies that slowly evolved into economic geography as we know it today. Much of the early research by economic geographers pertained to production figures for orchards and vineyards (Meigs 1941) and to why vineyards were so spatially concentrated (Olmstead 1956). Other articles were case studies of a particular aspect in a wine region such as Dobby (1936) on the economic geography of the Port region, Weigend (1955) on the changing port functions of Bordeaux, Dahlberg (1961) on the Concord grape industry of the Chautaqua-Erie area, and Morris (1969) about the development of irrigation and its impact on the Mendoza wine region. In a modern look at water, Bulman (2004) looks at water management implications to viticultural practices.

Foreign trade based on wine has been a topic in geography with Kohn (1986) writing about American wines in foreign and domestic trade. On an international level, Jones and Storchmann (2001) analyze wine market prices and investment under uncertainty and develop an econometric model for Bordeaux Crus Classés. Another example includes a study by Gwynne (2006a) that applies new forms of political economy to a comparative analysis of land, production, and exports in the wine sectors of both Chile and New Zealand. Cusmano et al. (2010) compare Chile, Italy, and South Africa in terms of their international markets for wine and how they are supplying the demand.

International trade agreements are another topic of interest to geographers. Barton et al. (2007) use the signing of a strategic economic partnership (the Trans-Pacific SEP or P4) between Chile, New Zealand, Singapore, and Brunei Darussalam in 2005 to discuss the emergence of a new generation of trade agreements that seek to promote longer-term synergies and cooperation. In a Canadian study, Kingsbury and Hayter (2006) examine the impact of NAFTA and the response of the Okanagan wine region of British Columbia and note the resiliency of the industry to change from the outside.

Marketing is a key to selling the wine from any particular location, either by drawing attention to its environmental advantages or by highlighting a particular attribute. In a study of the language of branding, Pawson (1997) presents the effectiveness of the invention and use of place name branding in New Zealand. In a study of the Gimlett soils in Hawkes Bay, New Zealand, Overton and Heitger (2008) explore the physical basis for this area becoming an important wine region and, in addition, develop a case of how this area was transformed from a "peri-urban wasteland" into one of the most sought-after and expensive wine-growing areas in the country. A mix of physical, political, and economic factors have transformed the area over a period of 20 years. Banks et al. (2007) discuss the notion of "place" as an important marketing tool for the development of the New Zealand wine industry. On a micromarketing level is the study by Charters et al. (1999) on consumer responses to wine bottle back labels and Newman's article on "what's in a name?" (Newman 1988).

Globalization or the trend towards multinational corporations and uniformity in wine production, as opposed to the local *terroir* influence, has been the subject of several geographical studies. Jones (1989a), in an early study of globalization, reports on the attempt of Napa's Mondavi Corporation to enter the French wine market through acquisitions of vineyards and the resulting opposition. In a study of the impact of globalization on the American wine industry, Holly (1994) investigates the move towards the concentration of production and distribution of wine by large, vertically and horizontally integrated organizations, while Morris (2000) studies the impact of globalization on the regional differentiation of the Mendoza wine industry in Argentina. In an article on the "flying winemakers," Lagendijk (2004) reports on how they are producing global wines in interconnected locales and overcoming the local *terroir* oriented production. In his analysis of "geography versus brands in a global wine market," Schamel (2006) finds that New World wines have not caught up with European wines, with regional designations, in price and prestige; but large American corporations are able to narrow the gap differential substantially.

Hayward and Lewis (2008) investigate the impact of regional dynamics on the globalization of the wine trade in New Zealand, while Rosas (2008) does a comparative study of Chile and Argentina to show how globalization in Chile and the lack thereof in Argentina has a marked influence on the development of the wine industry in the two countries. Corby (2010) approaches the impact of globalization on wine cooperatives in Argentina by revealing how they are coping with the threat in order to stay in business. In a reply to those who cubbyhole wines into Old World and New World, Banks and Overton (2010:57) argue that we "use the lens of recent work on globalization to argue that such production requires us to reexamine the dichotomous Old/New distinction which structures much of the thinking around the global wine industry".

Governance is a subject that has attracted considerable attention by geographers because of its economic impact. It can take the form of government imposed regulations or it can be self-regulating agreements set up by the industry. Reporting on a misdirected government funded "vine pull scheme" in South Australia, Barrett (1989) blames the failure of the program on the lack of a long-term focus and the inability to identify grape varieties for the program. In their article on the

New Zealand wine industry, Barker, Lewis, and Moran (2001) give a detailed history of the New Zealand wine industry with special reference attributing its rapid rise to a number of factors including liberalization of government regulations and a free trade policy. Lewis et al. (2002) also use New Zealand as an example of territoriality playing a major role in governance. Barker (2004) provides a discussion of the definition of governance in different contexts and explores the impact on the New Zealand and French wine industry. In a California context, Guthey (2004) investigates the *terroir* and the influence of politics in the agro-industry of California's North Coast wine district.

Chile is the site of two governance studies with Visser and de Langen (2006) investigating the importance and quality of governance in the Chilean wine industry, and Gwynne (2006b) studying governance and the wine commodity chain strategies in New Zealand and Chilean wine firms. As an example of self-imposed governance, Patchell (2008) applies common pool resource theory (CPR) to assess the attempt by Bordeaux wine territories of St-Emilion and Blaye to construct self-governance in order to achieve a competitive position in global markets.

Related to governance is the impact of *land use regulations* and the impact of expanding urban areas on vineyards. In a paper of the metropolitan fringe, Peters (1998) points out the problems of vineyards being in the path of urban development. Closely related to land use regulation is the process of viticultural zoning, ranging from government action in creating appellations to government land use planning decisions, as reported by Vaudour and Shaw (2005). In a Canadian example of urbanization, Senese (2010) reports on the urban-rural fringe that is threatening the vineyards of the Okanagan Valley, Canada.

Geographers are also contributing to the study of *value chains*, *cluster analysis*, and *networking*. A study on the governance and wine commodity chain in New Zealand and Chilean wine firms by Gwynne (2006a, b) and a study on the selective nature of knowledge networks in clusters in the wine industry of Italy and Chile by Giuliani (2007) are examples. An issue of the *Journal of Economic Geography*, with Robert Gwynne (2008a) the editor, contains articles on Chilean wine production and the impact of UK retail supermarkets on the value chain (Gwynne 2008b), Chilean Fairtrade wine value chains (Kleine 2008), the New Zealand wine industry and its domination by investment from transnational wine corporations (Hayward and Lewis 2008), and a case study of the marketing of places in the global value chain in wine and the use of *terroir* as a marketing tool (Guthey 2008).

A more recent article by Turner (2010) on networks of learning within the British wine industry shows the impact of "firms" on the growth of the industry. Although written by economists, Larreina et a l. (2011) present an analysis of cluster analysis in Rioja, Spain in *The Open Geography Journal*.

Terroir

If any topic in the geography of wine is equivalent to opening Pandora's Box, it is discussing *terroir*. In the strictest definition, *terroir* is French for "land" or "soil," but in the language of wine, *terroir* is a construct that means much more. The French word *terroir* epitomizes the belief that each wine derives its characteristics from the environment in which the grapes are grown. Although many researchers in the past have highlighted the physical variables, the following material will also indicate the importance of cultural aspects. Thus, support for the concept of *terroir* is a belief that geography gives wine a sense of place and makes it unique from wine produced elsewhere. *Terroir* is therefore a powerful geographical concept because it brings together the spatial elements of the natural environment and synthesizes them with the socioeconomic factors, one of the fundamental tenets of the field of geography.

Over thousands of years of experimentation, European winemakers have identified the best grapes for wine production in a particular region. No grape variety is on the label; they just assume that the geographical name identifies what grape variety is in the bottle. Red wine from Burgundy comes from Pinot Noir, Tuscan Chianti from Sangiovese, Barolo from Nebbiolo, and red Bordeaux from Cabernet Sauvignon, Merlot, and Cabernet Franc. "New world" wineries have little history and are still experimenting to find what they can grow well; therefore, the grape variety is prominently displayed, not the geographic region.

Parker (2008) states, "One of France's most celebrated wine regions, Burgundy is often cited as the best place to search for the fullest expression of *terroir*." The same Pinot Noir grape grown in California, New York, New Zealand, or Washington will produce a wine with a different style, aroma, and taste.

Not only do the French think that a Pinot Noir from Burgundy tastes different from a Pinot Noir produced in California, they also believe that wines produced from Pinot Noir in different parts of a vineyard will differ because of the microgeographical conditions. Parker states the following about *terroir*:

> It embraces the soil itself, the subsoil and rocks beneath it, its physical and chemical properties and how they interact with the local climate, the microclimate of the region, to determine both the mesoclimate of a particular vineyard and the microclimate of a particular vine. This includes for example how quickly a patch of land drains, whether it reflects sunlight or absorbs the heat, its elevation, its degree of slope, its orientation to the sun, and whether it is close to a cooling or sheltering forest, or a warming lake, river or the sea. (Parker 2008:26)

The English wine authority Jancis Robinson states, "*Terroir* is a much discussed term for the total natural environment of any viticultural site. No precise English equivalent exists for this quintessentially French term and concept" (Robinson 1994:966). She further states that the components of *terroir* are a function of the soil, local topography, and climate. Both Parker and Robinson define *terroir* in terms of physical characteristics of the vineyard and wine region.

In scientific literature, geographers have come late to the table in terms of studying *terroir*. Much of the early research in the US is by geologists with James Wilson's book on *Terroir: The role of geology, climate, and culture in the making of French wines* being the groundbreaking work (Wilson 1998), Swinchatt and Howell (2004) with their work on Napa Valley, and MacQueen and Meinert (2006) publishing articles from several symposia on *terroir*. Fanet (2004) and Gladstones (2011) are both soil scientists who have published volumes with a healthy dose of *terroir* in them. The book by Sommers on the geography of wine is one of the few books by a geographer that unapologetically subscribes to a belief in *terroir* (Sommers 2008).

Some argue that *terroir* is the characteristic that gives a wine its unique olfactory and taste profile, while others are dubious and infer that this is a marketing ploy by the French to set their wines apart based on their special soils and landscape (Parker 2008:26). Geographers do not agree in their support of *terroir*. Dickenson and Salt take a dubious stand, as has Gade (2004) in the quotes that follow:

> …no methodology has yet been produced for satisfactorily correlating micro-environment characteristics with those of the end product. Everyone says there is a relationship, but no one understands why. (Dickenson and Salt 1982:169)

> Wine is too complex to be able to derive a simplistic causation from the climate or soils of a particular terroir. Elevation of environmental factors in the popular mindset to account for wine character and quality is testimony to the tenaciousness of deterministic thinking. Much terroir-based explanation reflects the partitioned mind-set that presupposes "human" factors as separate from "physical" factors. (Gade 2004:865)

On the other hand, Sommers in his book on the geography of wine takes a proactive stance and states:

> For those of you not familiar with the concept, terroir is French for "ground" or "soil," but it is more than that. It is used to describe all the local features of environment and society that have an effect on wine. Many people believe that all of the features of a place taken as a whole-its terroir-have a distinctive influence that you can taste in the wine. (Sommers 2008:19)

The views of Dickenson and Salt (1982) and Gade (2004) reflect a reaction against environmental determinism, a philosophy popular in the early half of the last century and abandoned by geographers. In fact, the field made a major swing towards cultural geography, urban geography, and techniques geography and away from the physical. Sommers, on the other hand, represents a new interest in environmental processes that characterizes many geographers today.

Moran (1988) questions whether *terroir* is an environmental description or an "economic device," a marketing tool. In a comparison of French and Californian approaches to appellation legislation and *terroir*, Moran (1993) states that the distribution of vine varieties in France and California illustrates the "tight control evident in France compared with the *laissez-faire* policies in California." Moran (2001) also forays into the *terroir* battle again in his work on *terroir* – the human factor – by stating:

> To attribute priority to the physical environment over the cultural is also a mistake… But let's be careful not to lose sight of the people who have their gumboots in these terroirs and on the floors of the cellars. The soil exists but the terroir arrives when somebody makes an expressive wine from grapes grown in it. Without people and wine the word terroir would not exist. (Moran 2001)

The idea of *terroir* is so acceptable that several countries have a legally binding set of rules defining specific regions. The French have the *Appellation d'Origine Contrôlée*, the Italians have the *Denominazione di Origine Controllata*, and the

Americans have the *American Viticultural Area* (AVA). There are 187 AVA's in the United States as of June 2011 (GPO Access 2011). Theoretically, you can taste the difference in a Cabernet Sauvignon produced in Napa Valley AVA versus one made in the Sonoma Valley AVA because of the uniqueness of the *terroir*.

There are many variables influencing *terroir* or the geography of wine. Of primary importance is the physical geography including climate, microclimate, air drainage, solar prospect, water drainage, geology, topography, altitude, and soil. There are also variations in local socioeconomic factors that have an impact on the regional characteristics of a wine. This is especially true in the more rustic producers on small family farms in France and Italy and less likely in the giant production facilities associated with Australia and Chile where globalization reigns. The following section highlights some of the cultural variables that may combine with the physical variables to give wine its resulting characteristics.

The type of grape planted in an area is a cultural decision because the grape grower has a choice of possibly hundreds of grapes that can grow in his or her area. Many grapes – other than Pinot Noir – can grow in Burgundy, but early in time, the growers in the region made a collective decision to grow Pinot Noir. In fact, by legislation, Pinot Noir is the only red grape that can be grown and carry the name of Burgundy. Even if the same grape type grows in several regions, the clone may be different, and the resulting wine may smell and taste different from another clone. An example is the Sangiovese Grosso, the Brunello di Montalcino grape, which through centuries of selective breeding results in a more disease resistant and more marketable grape than its simple cousin, the Sangiovese grape of neighboring Chianti. The Sangiovese Grosso is the result of human decisions made over many centuries, resulting in a grape and a wine that has a different flavor profile and demands a much higher price than Chianti.

Vineyard practices are another important cultural factor influencing the resulting wine. In the Medoc, the vines may be spaced one meter by one meter apart, while in Australia, the vines may be four meters apart in order to accommodate mechanical harvesting. The type of trellising, pruning methods, and orientation of rows, along with practices such as green harvesting and canopy management, make for great differences in the quality and quantity of the grape yield and influence the resulting wine. Figure 1.7 shows three types of grape training with systems ranging from the "head pruned" Zinfandel of Lodi or the Garnacha of Navarra and Catalonia, Spain (Fig. 1.7a); to the wire trained trellis systems common in Napa and Sonoma (Fig. 1.7b); to the pergola systems found in areas of Tuscany, Argentina, and the Val d'Aosta (Fig. 1.7c). Vineyard practices not only have an impact on the quantity of grapes harvested but may also have an impact on ripeness and the brix, or sugar level of the grape.

Although most discussions of *terroir* do not include wine making techniques as part of *terroir*, the following material makes an argument for its inclusion. Many of the tools and methods available to the wine maker are characteristic of a region and become part of the cultural contribution to the *terroir*. The choice to use oak aging or not, or even other wood varieties, is often a regional characteristic of many wines. The selection of French, American, or Hungarian oak is a wine maker's decision as well as the amount of "toast" or charring desired in the barrel and the amount of new oak barrels. In addition, the amount of time in oak and the amount of new oak is a human decision that has a great impact on the regional identity, tannin content, and aging potential of many wines.

Examples of other decisions that influence the local wine include whole berry carbonic maceration, batonage and punching down the cap, amount of aging in barrel or bottle, amount of residual sugar, *sur lees* aging, and the percent of grapes in a blend. Another interesting local decision or cultural imprint on a wine is the purposeful use of oxidation to give a wine a regional characteristic that is readily identifiable. The solara system used for making Sherry, Madeira, and Marsala and the Vin Santo process are examples in which the winemaker induces oxidation. All of these techniques have a profound impact on the smell and taste of wines, and many are regional in adoption and, hence, give an identifiable taste to the wine from a particular area.

Style is also important in determining the type of wine that is characteristic of a region. Sherry and Port are high-alcohol styles of wine that have a distinct regional imprint rather than portraying the characteristics of the grape variety. The Touriga Nacional grape is the major ingredient of Port and Douro red still wines, but the Port is a sweet, high-alcohol dessert wine and the Douro is a dry, tannic dinner wine – a good example of human action causing a major dichotomy in the

Fig. 1.7 (**a**) Head pruned vineyard system, Apulia, Italy. (**b**) Trellis vineyard system, Grinzane Cavour, Italy. (**c**) Pergola vineyard system at Corteforte, Italy

resulting wines. Even with the vinification of the same grape variety in areas that are far apart, one can pick out traits in the smell and taste of the wine that may not be attributable to physical conditions alone. For example, the fruity nature of a Cabernet Sauvignon from California versus one from Bordeaux results from a style that is preferred in a producing region. Some wine producers in California even try to emulate the French style and do so successfully by producing a low fruit, mineral character to the wine (Balzer 1989). Some French and Italian wine producers are making a more ready to drink style that emphasizes the fruit instead of highly tannic wines made for long aging as has been common in the past.

In addition, local indigenous yeasts are important in giving a regional imprint to a wine – another human decision. The rustic and barnyard character of many "old world" wines is attributable to *Brettanomyces*, wild yeast that can give complexity to wines, but at the same time can be very obnoxious to many palates in the American market. Wines with the smell of dog hair, wool, and barnyard wastes do not sound enticing to many in the American market but are more acceptable and even considered desirable in European markets. Other yeasts that impart distinctiveness to a wine include the flor yeast in Sherries. The choice of using natural or wild yeasts and the choice of making a wine in the "organic" or "biodynamic" style are cultural decisions that may result in a markedly different finished product. Organisms other than yeasts, which are characteristic of one area and not another, can influence the taste of wine.

Botrytis cinerea, a gray fungus that infests grapes, is both a curse and a blessing to vineyards in various parts of the world. On most grapes, it leads to rot and ruination of the crop. On other grapes, the "noble rot" results in a highly desirable taste resulting in some of the most expensive dessert wines in the world including Tokaji in Hungary, Trockenbeerenauslese from Germany, and Sauternes from France. Of the Sauternes, Chateau d'Yquem from Bordeaux is the most expensive white wine in the world.

Food is a major part of a culture and has an impact on the wine produced in a region; hence, it must be taken into consideration as part of the cultural environment that influences the wine. For example, the dry, steely white wines of the Vinho Verde of northwestern coastal Portugal and the Alberinos of Rio Baixas of northwestern coastal Spain are a perfect accompaniment for the bounty of fish, crustaceans, and mollusks from the nearby sea. Is this a case where these grapes are the best choices because of the climate, or is this a case where the food helped dictate the type of wine and is therefore a cultural decision? This is characteristic of many grape-growing areas where a particular food is thought of as the perfect accompaniment for the wine – the pasta and sauces of Chianti with the Sangiovese grape, the plethora of beef in the Argentine diet with the big, bold Malbec wines, and the great variety of German and Alsatian foods with the whites made from Riesling and similar grapes. All of these areas can produce other grape varieties, but the collective cultural thought has been to specialize in a certain grape variety that complements the food.

Distinct local characteristics caused by physical and cultural factors can be easily destroyed by blending, a procedure that is increasingly being practiced by the large wine producers. Many consumers who have grown up with the McDonalds fast food model prize uniformity in food and wine. Blended wines often lack distinct characteristics and are similar from year to year. This is a criticism of Australian "critter wines" which are blended from many different sources that are widely separated.

Another threat to *terroir* driven wine is the influence of globalization. Many wine experts are critical of Robert Parker, Michel Rolland, and consultants called flying winemakers because of the belief that they are edging us towards more homogeneous international style wines lacking a conviction of place. The premise of the movie *Mondovino* infers there is a conspiracy to globalize the taste of wines by forcing winemakers to conform to the norms established by the people who rate wines for the major wine publications (Nossiter 2005). The globalization of wine has removed much of the rustic and local character of wine and has substituted certain prized varieties for local grape types. Although this may be true of the mass produced commercial blends, there are still many wines produced by smaller wineries that try to preserve the local imprint of the geography.

In terms of the total impact of *terroir*, considering all of the physical and cultural variables, scale is an important factor on the resulting wine as has been shown by Vaudour (2002). Grapes produced from a large geographic area will display little of the unique characteristics of *terroir*, while those from a single block in a vineyard will show the most characteristic *terroir*, especially physical variables such as soil, drainage, and solar prospect. The latter is the reason estate bottled wines and vineyard specific wines have such a distinct geographic personality and demand higher prices in the market.

Terroir is therefore the sum total of all of the environmental factors which make wine a unique agricultural product. The smaller the geographic unit, the more important is the *terroir*. The larger the area, the less important *terroir* is as a distinctive characteristic. Small plots in Burgundy have a geographic personality and illustrate the concept of *terroir* in that adjacent wineries may have great differences in quality even though they are separated only by a ditch (Pitte 2008:192). On the other hand, mass produced "critter wines" from Australia, although very tasty and receiving accolades from many, do not have uniqueness, but are consistent because they are blended from a wide variety of sources.

Figure 1.8 is the "*terroir* triangle" showing a regional hierarchy in which the effect of *terroir* is inversely proportional to the size of the source areas for the grapes from which a wine is made. Blocks within a vineyard, and the parcels called *lieux-dits* in Burgundian vineyards, are the smallest named designations of area and display the most pure sense of place or *terroir*. On the next higher source area, the vineyard produces grapes and wines that are marketed with a vineyard designation, e.g., the famous To Kalon vineyard in the Oakville District of Napa. The resulting wines still have a regional distinctiveness and display the *terroir*. A collection of vineyards called an estate form the next higher level and bottles that carry the "estate" label come from adjacent vineyards owned by one proprietor. At the next higher level is a subdivision of an American Viticultural Area, an example being the Oakville AVA in Napa Valley, where the grapes grow in close proximity but on different soil types with differing fertility and water characteristics; thus, there is more variability in the characteristic of the wine produced from one place to another. At the next step is the Napa Valley AVA or a county-level designation, where in many cases, the grapes grow on different soils, geological formations, and climatic conditions and result in a blend that is starting to lose its regional distinctiveness. Differences between Napa wines are often greater

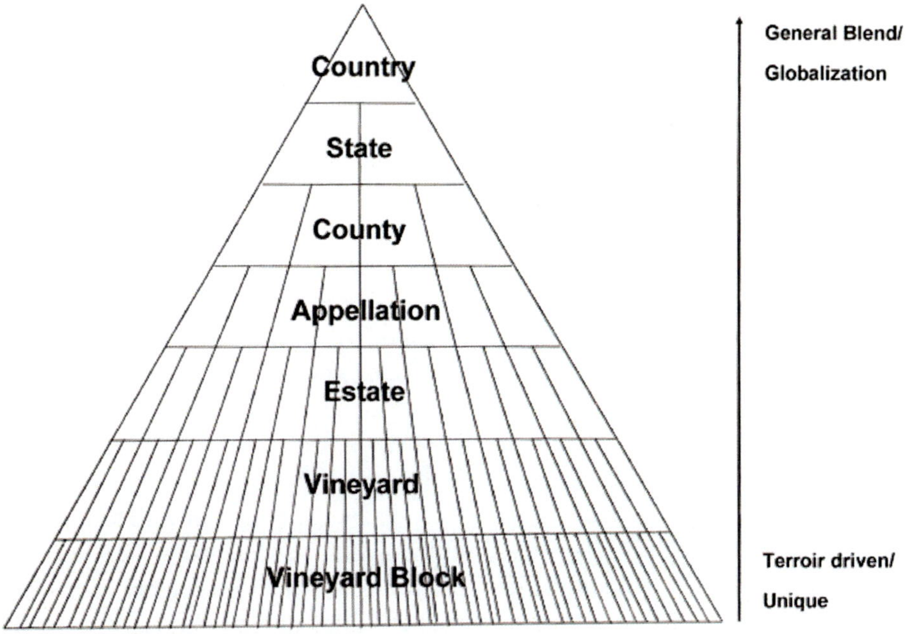

Fig. 1.8 *Terroir* triangle indicating the relative importance of *terroir* by size of the geographical area

than the difference between some Napa and Sonoma wines. Napa has such great diversity that some parts have soils derived from volcanic rock and other parts have soils derived from sedimentary or metamorphic rock. Other Napa vineyards are located on the flat valley floor as opposed to those found in the mountains where the solar prospect and microclimate varies greatly. By the time a wine has a county or large regional name, it has lost much of its regional distinctiveness. A wine labeled by a state name, such as California, is such a blend from different *terroirs* that it has lost virtually all of its regional distinctiveness.

The bottom line is there is no agreement among geographers as to the validity of the concept of *terroir*. It ranges from a belief that *terroir* is a marketing tool to the belief that the vineyard block is the reason for the true expression of wine. The idea is truly geographical in nature for it integrates the physical and cultural elements into the spatial analysis or wine regions.

Geographic Techniques

Just as carpenters need a hammer and a saw to do their job, the geographer needs tools of the trade to perform spatial analysis. The arsenal of tools includes math and statistics, maps, air photos, remote sensing imagery, and geographical information systems. This is a portion of the field that has mushroomed with the advent of satellite imagery, digital image analysis, and the popularity of geographical information systems.

Geographer John Nystuen is part of a team of researchers who investigated discrete mathematics and counting derangements in blind wine tastings and computed the probability of guessing the identity of the wines in a blind tasting (Arlingahaus et al. 1995). Reynolds and Outcalt (1975) evaluated the extent to which a double-dependence Markov process mirrors the autocorrelation function for a variety of different data sets involving weather, wine, and tree rings. Another quantitative study by Schwarz (1997) predicts wine quality from terrain characteristics by regression trees in a Riesling-producing area of Germany. In a principal components analysis and cluster analysis of 14 variables at 39 climate stations in Galicia, Blanco-Ward et al. (2007) set up spatial climate zones for the Mino River Valley of Spain.

An important geographical tool is the map, and it is used to great avail to plot and to study the spatial distribution of grapes and wine production. The epitome of this venue is the wine atlas with its collection of maps illustrating production areas as well as the environmental factors influencing the growth of grapes. The *World Atlas of Wine* by British wine experts Huge Johnson and Jancis Robinson (2008) is a compendium of 400 pages with color illustrations, photos, and 200 maps. It is widely used by anyone interested in wine. Other world atlases include those by Oz Clarke (2007) and the Hammond World Wine Atlas (2009). Several regional atlases are available including California by Halliday (1993), California and Pacific Northwest by Thompson (1993), Germany by Pigott (1995), Italy by Anderson and Pigott (1997), France by Duijker and Johnson (1997), Langhe by Petrini (2003), Canada by Aspler (2006), Australia by Halliday (2006), and New Zealand by Cooper and McDermott (2010).

Although the field of geography is the home to cartography, geographers are generally lacking from the list of producers of wine atlases and wine maps. The major exception in the United States is Donald Holtgrieve who published *The California Wine Atlas* (1978) and sheet maps with commentary on Napa-Sonoma (1995) and California (1997). In addition, de Blij (1986f) wrote a short article on viticultural areas and mental maps. Many journal articles and books by geographers contain excellent maps, but, in most cases, these do not appear separately or in an edited volume.

Geographers have always relied on air photos for a bird's eye view of the earth. Aerial reconnaissance has been a great way to look at spatial patterns and analyze why distributions are the way they are. This became even more apparent with the Landsat digital image revolution in 1982 when satellite imagery became an important tool in the geographer's arsenal of tools. Early satellite resolution at 79 m was not sufficient to assess vineyard conditions, but submeter resolution of today has opened a host of applications. It is surprising that more work is not forthcoming.

Excellent examples of work in this field include Vaudour et al. (2008) in which SPOT imagery is used to assess soil reflectance and Vaudour (2008) in which soil characteristics in the southern Rhone Valley are studied using SPOT imagery. Vadour et al. (2010) used 20-m SPOT imagery to map *terroir* by using bootstrapped decision trees on morphometric data and multitemporal SPOT images.

In a study of leaf area in two commercial vineyards in California's North Coast, Johnson et al. (2003) used IKONOS high spatial resolution imagery to perform a normalized difference vegetation index (NDVI) on a per pixel basis that was ground calibrated to convert the NDVI maps to a leaf area index. In another study, Johnson looked at temporal stability of an NDVI/LAI relationship in a Napa vineyard using satellite imagery. Johnson (2003) in yet another article analyzes the temporal stability of an NDVI–LAI relationship in a Napa Valley vineyard.

Not all of remote sensing imagery is spaceborne. Qin et al. (2010) led an interdisciplinary team, including geographers, using artificial intelligence to establish the relationship between the leaf thickness and red-edge/near-infrared (NIR) reflectance in different grape cultivars at different stages of growth. In another ground-based study, Keightley and Bawden (2010) use a tripod mounted laser scanner to generate high-resolution volumetric measures of vegetation structure and perennial woody tissue for the calculation of standing biomass.

Geographers have used remote sensing techniques to study land use patterns with Manandhar et al. (2009) utilizing Landsat data from 1985 to 2005 to study land use changes by change detection analysis in the Lower Hunter Valley of New South Wales, Australia. On the other hand, Van Eetvelde and Antrop (2004) digitized air photos from southern France, where landscapes are in transition between new residential urbanization and land abandonment, in order to study structural changes in land use, building and field patterns from 1960 to 1999 by using GIS analysis.

Boyer and Wolf (1998) use GIS and GPS to study viticultural potential in general, and in another article (Boyer and Wolf 2000a), they use the application specifically in Virginia. Also, Boyer and Wolf (2000b) use geographical information systems (GIS) in a vineyard suitability study of Virginia by using a combination of spatial modeling and geographical information system (GIS) data. Jones (2003b) utilizes GPS and GIS to analyze varietals and management characteristics in the Umpqua Valley, and in a related study, Jones et al. (2004) perform a GIS analysis of the *terroir* potential of the Umpqua Valley, Oregon. In a similar study, Jones (2001) and Jones et al. (2006) use GPS and GIS to assess the viticultural status and potential of the Rogue River Valley of Oregon. Chin (2011) analyzes nine GIS variables/layers, namely, growing-degree days, length of frost-free period, minimum winter temperature, aspect, slope, soil drainage, soil pH, organic matter, and land use to determine agricultural suitability in Nebraska for Edelweiss and Cynthiana/Norton cultivars.

Although cartography, GIS, and remote sensing are taught primarily in geography departments, there are not many books or journal articles published by geographers on these topics. More viticultural research originates from agriculture and engineering departments. This may result from the previous departments being applied in nature, whereas most geography departments are theoretical in the larger, more prestigious universities in the US, or education oriented in many of the state universities.

Conclusion

In conclusion, whether it is regional analysis, comparative studies, *terroir*, climate, GIS analysis, diffusion studies, economic analysis, or studying cultural winescapes, the geographer is uniquely equipped to study the geography of wine. The geographer is best qualified to study the complex interaction of the physical, cultural, and economic variables that influence the type of grape and the type of wine produced in each region. Whether you call it *terroir* or geography, there are great differences in the wine produced in each part of the world, and it is important to recognize these differences in order to get the most enjoyment from each bottle of wine.

References

Ackerman, E. A. 1937. The wine valley of north Portugal. *Journal of Geography* 36 (9): 333–353

Ackerman, William V. 1978. Testing central place concepts in western Argentina. *The Professional Geographer* 30 (4): 377–388

Agnew, S. (1945) The vine in Bas-Languedoc. *Geographical Review* 36: 67–79.

Albertson, Dean. 1950. Puritan liquor in the planting of New England. *The New England Quarterly* 23 (4): 477–490.

Anderson, Burton, and Stuart Pigott. 1997. *The wine atlas of Italy.* Octopus Publishing Group.

Arlingahaus, Sandra L., William C. Arlinghaus, and John D. Nystuen. 1995. Discrete mathematics and counting derangements in blind wine tastings. *Solstice: An Electronic Journal of Geography and Mathematics.* 6(1). http://deepblue.lib.umich.edu/bitstream/2027.42/58290/2/sols195.html. Last accessed June 9, 2011.

Aron, Robert. 1976. Oregon's climatic suitability for premium wine grapes. *California Geographer* 16: 53–51.

Aspler, Tony. 2006. *The wine atlas of Canada.* Toronto, Canada: Random House of Canada.

Augustinus· P.G.E.F., and P.J. Nieuwenhuyse. 1986. Soil erosion in vineyards in the Ardeche region. *Catena* 13 (3): 295–304.

Ažman, Lučka, and Drago Kladnik. 2009. Terraced landscapes in Slovenia. *Acta Geographica Slovenica* 49 (1): 7–37.

Balzer, Robert Lawrence. 1989. A French accent: Chateau Potelle: New California wines with a Gallic taste. *Los Angeles Times.* March 19, 1989.

Banks, G. and S. Sharpe. 2006. Wines, regions and geographic imperative: The Coonawarra example. *New Zealand Geographer* 62 (3): 173–184.

Banks, G., S. Kelly, N. Lewis, and S. Sharpe. 2007. Place "From one glance": The use of place in the marketing of New Zealand and Australian wines. *Australian Geographer* 38 (1): 15–35

Banks, Glenn, and John Overton. 2010. Old world, new world, Third world? Reconceptualising the worlds of wine. *Journal of Wine Research* 21(1): 57 – 75.

Barker, J., N. Lewis, and W. Moran. 2001. Reregulation and the development of the New Zealand wine industry. *Journal of Wine Research* 12 (3): 199–221

Barker, J.P.H. 2004. *Different worlds: Law and the changing geographies of wine in France and New Zealand.* Unpublished PhD thesis, University of Auckland.

Barnard, Hans, Alek N. Dooley, Gregory Areshian, Boris Gasparyan, and Kym F. Faull. 2011. Chemical evidence for wine production around 4000 BCE in the Late Chalcolithic Near Eastern highlands. *Journal of Archaeological Science* 38 (5): 977–984.

Barrett, Steven. 1989. An assessment of the vine pull scheme: A case study of the southern vales of South Australia. *Australian Geographer* 20 (2) 185 – 190.

Barton, Jonathan R.; Robert N. Gwynne; and Warrick E. Murray. 2007. Competition and co-operation in the semi-periphery: Closer economic partnership and sectoral transformations in Chile and New Zealand. *Geographical Journal* 173 (3): 224–241.

Baxevanis, John J. 1987a. *The wines of Bordeaux and western France.* Totowa, NJ: Rowman and Littlefield.

Baxevanis, John J. 1987b. *The wines of Champagne, Burgundy, eastern and southern France.* Totowa, NJ: Rowman & Littlefield.

Baxevanis, J.J. 1990. The Wines of the Sierra Foothills, *Vinifera Wine Growers Journal* 17(4): 216–224.

Baxevanis, John J. 1992. *The* wine *regions of America: Geographical reflections and appraisals.* Stroudsburg, PA: Vinifera Wine Growers Journal.

Beer, Chris, and Nicolas Lewis. 2006. Labouring in the vineyards of Marlborough: Experiences, meanings and policy. *Journal of Wine Research* 17(2): 95–106

Bek, David, Cheryl McEwan, and Karen Bek. 2007. Ethical trading and socioeconomic transformation: Critical reflections on the South African wine industry. *Environment & Planning A* 39 (2): 301–319.

Belliveau, Sue, Ben Bradshaw, Barry Smit, Susanna Reid, Doug Ramsey, Margaret Tarleton, and Bronwyn Sawyer. 2006a. *Farm-level adaptation to multiple risks: Climate change and other concerns.* Occasional Paper No. 27. Guelph, Canada: Department of Geography, University of Guelph.

Belliveau, Suzanne, Barry Smit· and Ben Bradshaw. 2006b. Multiple exposures and dynamic vulnerability: Evidence from the grape industry in the Okanagan Valley, Canada. *Global Environmental Change* 16 (4): 364–378.

Belliveau, Susanne, Ben Bradshaw, and Barry Smit. 2007. Comparing apples and grapes: Farm-level vulnerability to climate variability and change. In *Farming in a changing climate: Agricultural adaptation in Canada,* ed. Ellen Wall, Barry Smit, and Johanna Wandel, 157–172. Seattle, WA: University of Washington Press.

Bierly, Gregory D. and Jay D. Gatrell. 2004. Structural and compositional change **in** geography graduate programs in the United States: 1991–2001. *Professional Geographer* 56 (3): 337–344.

Blanchard, W.O. and E. R. Blanchard. 1929. The Grape Industry of Spain and Portugal. *Economic Geography* 5 (2): 183–193.

Blanco-Ward, D., J. M. Garcia Queueiero, and G. V. Jones. 2007. Spatial climate variability and viticulture in the Miño River Valley of Spain. *Vitis* 46 (2): 60–70.

Boyer, John. 2009. *Drink this now*. Hoboken, NJ: Wiley Custom Services.

Boyer, J., and T.K. Wolf. 1998. GIS and GPS identify viticulture potential. *Amer. J. Enol. Vitic.* 49: 449.

Boyer, J. and T.K. Wolf. 2000a. GIS and GPS aid the exploration of viticultural potential in Virginia. *Vineyard and Winery Management* Nov/Dec 2000: 48–54

Boyer, J., and T.K. Wolf. 2000b. Development and preliminary validation of a Geographical Information System approach to vineyard site suitability assessment in Virginia. In: *Proceeding of the 5th International Symposium on Cool Climate Viticulture and Oenology*, 16–20. Melbourne, Australia.

Bruce, W.C. 1917. *Benjamin Franklin self-revealed: A biographical and critical study based mainly on his own writings.* New York: G. Putnam's Sons.

Bulman, Teresa L. 1991. Development of the grape monoculture of Napa County. *Yearbook of the Association of Pacific Coast Geographers* 53: 61–86.

Bulman, Teresa L. 2003. The joy of geography. *Yearbook of the Association of Pacific Coast Geographers* 66: 9–26.

Bulman, T. L. 2004. Water management implications of modern viticultural practices. In *Viticulture in geographic perspective: Proceedings of the 1991 Miami AAG symposium*, ed. H. J. de Blij, 73–86. Miami: Miami Geographical Society.

Canoves, Gemma, and Raul Suhett de Morais. 2011. New forms of tourism in Spain: Wine, gastrometric and rural tourism. In *Tourism and agriculture: New Geographies of consumption, productions and rural structuring*, ed. Rebecca Maria Torres and Janet Henshall Momsen. New York: Routledge.

Carmichael, Barbara. 2005. Understanding the wine tourism experience for winery visitors in the Niagara Region, Ontario, Canada. *Tourism Geographies* 57 (2): 185–204.

Carter, F.W. 1987. Cracow's wine trade (fourteenth to eighteenth centuries). *The Slavonic and East European Review* 65 (4): 537–578

Charters· Steve, Larry Lockshin, and Tim Unwin. 1999. Consumer responses to wine bottle back labels. *Journal of Wine Research* 10 (3):183–195

Che, Deborah, and Astrid Wargenau. 2011. Visiting winery tasting rooms: Venues for education, differentiation and direct marketing. In *Tourism and agriculture: New Geographies of consumption, productions and rural structuring.* ed. Rebecca Maria Torres and Janet Henshall Momsen. New York: Routledge

Chin, Ting. 2011. *Using a geographic information system to define regions of grape-cultivar suitability in Nebraska.* Masters thesis: University of Nebraska-Lincoln.

Clarke, Oz. 1997. *The essential wine book: An indispensible guide to the wines of the world.* New York: Simon & Schuster.

Clarke, Oz. 2007. *Oz Clarke's wine atlas: Wines and wine regions of the world.* Brighton, UK: Pavilion.

Clinkerbread, Barbara G. 1982. Lesbian wine and storage amphoras: A progress report on identification. *Hesperia: The Journal of the American School of Classical Studies at Athens* 51 (3): 248–268.

Cooper, Michael, and John McDermott. 2010. *Wine atlas of New Zealand.* Andrews McMeel Publishing.

Corby, Julia H. Kentnor. 2010. For Members and markets: Neoliberalism and cooperativism in Mendoza's wine industry. *Journal of Latin American Geography* 9 (2): 27–47.

Crowley, William K. 1983. Changes in the French winescape. *Geographical Review* 83 (3): 252–268.

Crowley, William K. 2000. Chile's wine industry: Historical character and changing geography. *Conference of Latin Americanist Geographers Yearbook* 87 (15).

Crowley, William K. 2001. Is AOC really the best? *Focus* 46 (3): 41.

Cusmano, Lucia, Andrea Morrison and Roberta Rabellotti· 2010. Catching up trajectories in the wine sector: A comparative study of Chile, Italy, and South Africa. *World Development* 38 (11): 1588–1602.

Cyr, Donald, Martin Kusy, and A.B. Shaw. 2008. Hedging adverse bioclimatic conditions employing a short condor position. *Journal of Wine Economics* 3 (2): 149–171.

Cyr, Don, Martin Kusy, and Anthony B. Shaw. 2010. Climate change and the potential use of weather derivatives to hedge vineyard harvest rainfall risk in the Niagara region. *Journal of Wine Research* 21 (2–3): 207–227.

Dahlberg, R.E. 1961. The Concord grape industry of the Chautauqua-Erie area. *Economic Geography* 37: 150–169.

Davies, Sir Norman De Garis. 1917. *The tomb of Nakht at Thebes*. New York: Metropolitan Museum of Art.

de Blij, Harm J. 1981. Wine in China. *The Wine Report*, September, 1981

de Blij, Harm J. 1982. Wine in the Land of the Rising Sun. *Wines and Vines* 63 (5).

de Blij, Harm J. 1983a. Geography of viticulture: Rationale and resource. *Journal of Geography* 82 (3):112–121.

de Blij, Harm J. 1983b. Regional environments for viticulture in Argentina and Chile. *Chronicle of the Society of Wine Educators* 6: Winter, 1983.

de Blij, Harm J. 1983c. *Wine: A geographic appreciation.* Totowa, NJ: Rowman and Allanheld Publishers.

de Blij, Harm J. 1985a. From grape to wine. *Focus* 35 (4).

de Blij, Harm J. 1985b. The wine cellar. *Focus* 35 (3).

de Blij, Harm J. 1985c. The wine industry of Brazil. *Chronicle of the Society of Wine Educators* 8: Spring, 1985.

de Blij, Harm J. 1985d. *Wine regions of the Southern Hemisphere*. Totowa, NJ: Rowman & Allenheld.

de Blij, Harm J. 1986a. Devastating Phylloxera. *Focus* 36 (2).

de Blij, Harm J. 1986b. Malta. *Focus* 36 (1).

de Blij, Harm J. 1986c. Nine canons of the geography of viticulture. *East Lakes Geographer* 21 (1).

de Blij, Harm J. 1986d. Regional wine bottles. *Focus* 36 (1).

de Blij, Harm J. 1986e. Wine label geography. *Focus* 36 (3).

de Blij, Harm J. 1986 f. Viticultural areas and mental maps. *East Lakes Geographer* 21 (1).

de Blij, Harm J. 1987a. African wine frontiers. *Focus* 37 (3).

de Blij, Harm J. 1987b. Closure. *Focus* 37 (4).

de Blij, Harm J. 1987c. Vintage years and wine futures. *Focus* 37 (2).

de Blij, Harm J. 1987d. Viticulture and viniculture in the southeastern United States. *The Southeastern Geographer* 27 (2). 115–130.

de Blij, Harm J. 1987e. Wines of the Southeast. *Focus* 37 (1).

de Blij, Harm J. 1988a. The appeal of appellations. *Focus* 38 (2).

de Blij, Harm J. 1988b. Chile's maturing wine map. *Focus* 38 (1).

de Blij, Harm J. 1988c. In praise of older wines. *Focus* 38 (3).

de Blij, Harm J., and G. Peters 1988. Location, scale, and climate: Geographic dimensions of viticulture. *The Mid-South Geographer* 4 (1).

de Blij, Harm J. 1989a. Iberian regions. *Focus* 39 (4),

de Blij, Harm J. 1989b. Rise of the lowly blender. *Focus* 39 (3).

de Blij, Harm J. 1989c. Wine list geography. *Focus* 39 (2).

de Blij, Harm J. 1990a. Back labels. *Focus* 40 (1).

de Blij, Harm J. 1990b. The German question. *Focus* 40 (3).

de Blij, Harm J. 1990c. High-latitude challenge. *Focus* 40 (2).

de Blij, Harm J. 1991a. America's Zinfandel. *Focus* 41 (1).

de Blij, Harm J. 1991b. Enchanting Chateauneuf. *Focus* 41 (3).

de Blij, Harm J. 1991c. Hope at the Cape. *Focus* 41 (2).

de Blij, Harm J. 1991d. Rioja resurgent. *Focus* 41 (4).

de Blij, Harm J. 1992a. Wine in China? *Focus* 42 (1).

de Blij, Harm J. 1992b. Of wine and music. *Focus* 42 (3).

de Blij, Harm J. 1992c. Trouble in Beaujolais. *Focus* 42 (2).

de Blij, Harm J. 1992d. *Viticulture in geographic perspective: Proceedings of the 1991 Miami Symposium*. Editor. Coral Gables: Miami Geographical Society.

de Blij, Harm J. 2005. *Why geography matters: Three challenges facing America: Climate change, the rise of China, and global terrorism*. New York: Oxford University Press.

Dickenson, John. 1985. *Organizational structures in the Portuguese wine industry*. Liverpool papers in human geography, Working paper/University of Liverpool, Department of Geography -- no.19. Liverpool, U.K.: University of Liverpool, Dept. of Geography in Liverpool

Dickenson, J. P., 1989, Nine axioms for the geography of viticulture: A countervailing perspective, *The East Lakes Geographer* 24:1–12.

Dickenson, John. 1990. Viticultural geography: An introduction to the literature. *Journal of Wine Research* 1 (1):5–24.

Dickenson, John 1991. Vineyards under threat in Portugal. *The Vinifera Wine Growers Journal* 18 (2).

Dickenson, John. 1992a. Changes in the geography of wine: New wine in new bottles. *Interdisciplinary Science Reviews* 17 (2): 178–184

Dickenson, J. 1992b. Dimensions of a wine experience: The Pennsylvania experience. *Ontario Geographer* 39: 23–34.

Dickenson, John. 1995. Viticulture in pre-independence Brazil. *Journal of Wine Research* 6 (3): 195–200

Dickenson, J.P. and L. Salt. 1982. In vino veritas: An introduction to the geography of wine. *Progress in Human Geography* 6: 159–189.

Dickenson, J. and T. Unwin. 1992. *Viticulture in colonial Latin America: Essays on alcohol, the vine and wine in Spanish America and Brazil*. Liverpool Univ. (United Kingdom). Inst. of Latin American Studies 1992.

Dobby, E.H. 1936. Economic geography of the Port wine region. *Economic Geography* 12: 311–321.

Dougherty, P.H. 2010. Wine. In *Encyclopedia of Geography*, ed. Barney Warf. London: Sage Publications.

Duijker, Hubrecht, and Hugh Johnson. 1997. *The wine atlas of France: And traveller's guide to the vineyards*. Mitchell Beazley.

Egyptian Tourist Bureau 2011. The tomb of Nakht on the west bank at Luxor. http://www.touregypt.net/featurestories/nakht2.htm. Last accessed October 5, 2011.

Eichel, M.H. 1975. Phylloxera and its consequences. *Geographical Review* 65: 526

Elliott-Fisk, D., and A. Noble. 1992. Environments in Napa Valley, California and their influence on Cabernet Sauvignon wine flavors. In *Viticulture in Geographical Perspective*, ed. H. De Blij, 45–72. Miami: Miami Geographical Society.

Elliott-Fisk, D. 1993. Viticultural soils of California, with special reference to the Napa valley. *Journal of Wine Research* 4 (3): 67–77.

Estreicher, Stephen K. 2006. W*ine from neolithic times to the 21st century*. New York: Algora Publishing.

Eysberg, C.D. 1987. Viticulture in California: Cool air-conditioned valleys as the equivalent of warm sheltered 'cotes.' *Geo Journal* 15: 361–373.

Fairbanks, Dean H.K., Catherine J. Hughes, and Jane K. Turpie. 2004. Potential impact of viticulture expansion on habitat types in the Cape Floristic Region, South Africa. *Biodiversity and Conservation* 13 (6):1075–1100.

Fanet, Jaques. 2004. *Great wine terroirs*. Berkeley, CA: U of CA Press.

Fisher, Cathy. 2011. Review of the industry: U.S. wineries grow 9% to 6,785. *Wine Business Monthly*. February 15, 2011. http://www.winebusiness.com/wbm/?go=getArticle&dataId=85190. Last accessed May 31 2010.

Fisher, W.B. 1975. 100,000,000 bottles. *Geographical Magazine* 47: 320–326.

Fisher, W.B. 1978. Wine: The geographical elements. *Geographical Magazine* 51: 86.

Fisher, W.B. 1981. Wine: French wines with an English connection. *Geographical Magazine* 53: 753.

Fisher, W.B. 1984. Wine: Another flask of Rhenish. *Geographical Magazine* 56

Fitzharris, B.B., and W, Endlucher. 1996. Climatic conditions for wine grape growing. *New Zealand Geographer* 52 (1): 1–11.

Fredericks, R. 1969. Nineteenth century stonework in California's Napa Valley. *California Geographer* 9:39–48.

Gade, D.W. 1978. Windbreaks in the lower Rhone Valley. *Geographical Review* 68: 127–144

Gade, Daniel W. 2004. Tradition, territory and *terroir* in French vitaculture: Cassis, France, and Appellation Controlee. *Annals of the Association of American Geographers* 94: 848–867.

Giuliani E. 2007. The selective nature of knowledge networks in clusters: evidence from the wine industry, *Journal of Economic Geography* 7: 139–168.

Giuliani, Elisa, Andrea Morrison, Carlo Pietrobelli and Roberta Rabellotti. 2010. Who are the researchers that are collaborating with industry? An analysis of the wine sectors in Chile, South Africa and Italy. *Research Policy* 39 (6): 748–761

Gladstones, J. 2011. *Wine, terroir and climate change*. Kent Town, South Australia: Wakefield Press.

Gold, Susannah. 2007. Italian indigenous grape varieties: Albarola from Liguria. http://avvinare.com/2009/06/12/

italian-indigenous-grape-varieties-albarola-from-liguria/. Last accessed April 25, 2011.

Goldberg, Kevin Douglas. 2010. *German wine and the fermentation of modern taste, 1850–1914*. Ph.D. dissertation. University of California, Los Angeles.

GPO Access. 2011. Title 27: *Alcohol, Tobacco and Firearms. Part 9—American viticultural areas. Subpart C—Approved American Viticultural Areas*. http://ecfr.gpoaccess.gov/cgi/t/text/text-idx?c=ecfr&sid=057f99d792668247a3c45b4699417291&rgn=div5&view=text&node=27:1.0.1.1.7&idno=27. Last accessed June 9, 2011.

Grant, E. 1984. Wine from a cool climate. *Geographical Magazine* 56: 482–485.

Grigg, David. 2004. Wine, spirits and beer: World patterns of consumption. *Geography* 89 (2): 99–110.

Gruber, J.N. 1981. The northern Virginia wine industry. *Virginia Geographer* 14:1–20.

Guthey, Greig Tor. 2004. *Terroir and the politics of agro-industry in California's North Coast Wine District*. Ph.D. dissertation. University of California, Berkeley, 2004. 3165393.

Guthey, Greig Tor. 2008. Agro-industrial conventions: Some evidence from northern California's wine industry. *Geographical Journal* 174 (2): 138–148.

Gwynne, Robert N. 2006a. Export-orientation and enterprise development: A comparison of New Zealand and Chilean wine production. *Tijdschrift voor Economische en Sociale Geografie (Journal of Economic & Social Geography)* 97 (2): 138–156.

Gwynne, Robert N. 2006b. Governance and the wine commodity chain: Upstream and downstream strategies in New Zealand and Chilean wine firms. *Asia Pacific Viewpoint* 47 (3): 381–395

Gwynne, Robert N. 2008a. Value chains and the geographies of wine production and consumption. Editorial. *Geographical Journal* 74 (2): 95–96.

Gwynne, Robert N. 2008b. UK retail concentration, Chilean wine producers and value chains. *Geographical Journal* 74 (2): 97–108.

Hadarits, Monica, Barry Smit, and Harry Diaz. 2010. Adaptation in viticulture: A case study of producers in the Maule Region of Chile. *Journal of Wine Research* 21 (2–3): 167–178.

Halliday, J., 1993, *Wine atlas of California*. New York: Viking Adult.

Halliday, James. 2006. *Wine atlas of Australia*. University of California Press, 2006.

Hammond World Atlas Corporation. 2009. *The wine atlas*. Long Island City, NY: Hammond World Atlas Corporation.

Hansis, R.A. 1977. Land tenure, hazards and the economy: Viticulture in the Mendoza oasis, Argentina. *Economic Geography* 53: 368–371.

Hawkins, Anthony J. 2007. The super gigantic y2k winegrape glossary. http://www.wineloverspage.com/wineguest/wgg.html. Accessed 4/25/2011.

Hayward, D., Lewis, N., 2008. Regional dynamics in the globalising wine industry: the case of Marlborough, New Zealand. *The Geographical Journal* 174 (2): 124–137.

Hewes, L. 1953. Tontitown: Ozark vineyard center. *Economic Geography* 29:125–143.

Holland, Tara, and Barry Smit. 2010. Climate change and the wine industry: Current research themes and new directions. *Journal of Wine Research* 21 (2 & 3): 125–136.

Holly Brian P. 1994. Organisational structure of wine production in the US. *Journal of Wine Research* 5 (2): 91–101

Holmes, John, and Kate Hartig. 2007. Metropolitan colonisation and the reinvention of place: Class polarisation along the Cessnock-Pokolbin fault-line. *Geographical Research* 45 (1): 54–70.

Holtgrieve, Donald Gordon. 1972. Geography of the vine. *Oregon Geographer* 5 (6): 1–25.

Holtgrieve, Donald Gordon. 1978. *The California wine atlas*. Hayward, California: Ecumene Associates. 46 pages. cartographer, James Treavors. Folio of sheets in folder.

Holtgrieve, Donald Gordon. 1995. Map. *Vineyards and Wineries of the Napa-Sonoma Region*. Medford, OR: Raven Maps and Images.

Holtgrieve, Donald Gordon, and Stuart Allan. 1997. Map. *Wineries and vineyards of California*. Medford, OR: Raven Maps and Images.

Homer. Translated by Ian Johnson. 2006. *The Odyssey*. Arlington, VA: Richter Resources Publications.

Hort, Sir Arthur (translator). 1916. *Theophrastus: Enquiry into plants* (2 vol.). London: G.P. Putnam.

Hutton, Robert W. 1993. Tokaji. *Journal of the American Wine Society* 25 (2): 42–45.

Hutton, Robert W. 1997 Wines of Hungary. *Journal of the American Wine Society* 29 (4): 111–115.

Hutton, Robert W. 2004 Romanian wines. *Journal of the American Wine Society* 36 (1): 26–30.

Ilbery, B.W. 1983. The renaissance of viticulture in England and Wales. *Geograph.* 68: 341–344.

Ilbery, B.W. 1985. Factors affecting the structure of viticulture in England and Wales. *Area* 17: 147–154.

Johnson, Hugh and Jancis Robinson. 2008. *The world atlas of wine*. London: Octopus Publishing Group.

Johnson, L.F., 2003. Temporal stability of an NDVI–LAI relationship in a Napa Valley vineyard. *Australian Journal of Grape and Wine Research* 9: 96–101.

Johnson, L.F., D. E. Roczen, S. K. Youkhana, R. R. Nemani, and D. F. Bosch. 2003. Mapping vineyard leaf area with multispectral satellite imagery. *Computers and Electronics in Agriculture* 38 (1): 33–44.

Johnston, Patricia A. 1999. *Huc pater o Lenaee veni:* The cultivation of wine in Vergil's Georgics. *Journal of Wine Research* 10 (3): 207 – 221

Jones, A. 1989a. Reform of the European community's table wine sector: Agricultural despecialisation in the Languedoc. *Geography* 74: 29–37.

Jones, A. 1989b. The role of the SAFER in agricultural restructuring: The case of the Languedoc-Roussillon, France. *Land Use Policy* 6 (3): 249–261

Jones, Gregory V., and Robert E. Davis. 2000. Climate influences on grapevine phenology, grape composition, and wine production and quality for Bordeaux, France. *Am. J. Enol. Vitic.* 51 (3): 249–261.

Jones, Gregory V., and Karl-Heinz Storchmann. 2001. Wine market prices and investment under uncertainty: An econometric model for Bordeaux Crus Classés. *Agricultural Economics* 26 (2): 115–133.

Jones, G.V., 2001, Using GIS/GPS technology to assess viticultural status and potential in the Rogue Valley AVA: *Proceedings of the 115th Annual Meeting of the Oregon Horticultural Society,* Portland, Oregon.

Jones, G.V. 2003a. Trends in frost occurrence and the frost-free growing season length in Oregon grape-growing regions: *Proceedings of the 117th Annual Meeting of the Oregon Horticultural Society*, Portland, Oregon, p 223–231.

Jones, G. V. 2003b. Umqua Valley AVA: A GPS and GIS vineyards mapping and analysis of varietal, climate, landscape, and management characteristics. in *Report to the Oregon Wine Advisory Board and the Umpqua Chapter of the Oregon Winegrape Growers Association.*

Jones, G.V. and E. Hellman. 2003a. Oregon winegrowing regions. In *Oregon Viticulture,* ed. E. Hellman. 5th Edition., Corvallis, Oregon: Oregon State University Press.

Jones, G.V. and E. Hellman. 2003b. Site assessment. In *Oregon Viticulture,* ed. E. Hellman. 5th Edition., Corvallis, Oregon: Oregon State University Press.

Jones, Gregory V., Nicholas Snead, and Peder Nelson. 2004. Geology and wine 8. Modeling viticultural landscapes: A GIS analysis of the *terroir* potential in the Umpqua Valley of Oregon. *Geoscience Canada* 31 (4).

Jones, Gregory. 2004. Making wine in a changing climate. *Geotimes*. 2004 (8). http://www.geotimes.org/aug04/feature_wineclime.html

Jones, Gregory, Michael White, Owen Cooper, and Karl Storchmann. 2005. Climate change and global wine quality. *Climatic Change* 73 (3): 319–344.

Jones, G.V. 2006. Climate and *terroir*: Impacts of climate variability and change on wine. In *Fine wine and terroir - The geoscience perspective,* ed. R.W. Macqueen and L.D. Meinert. St. John's, Newfoundland: Geoscience Canada Reprint Series Number 9, Geological Association of Canada. St. John's, Newfoundland,

Jones, G.V., A.A. Duff, and J.M. Myers. 2006. Modeling viticultural landscapes: A GIS analysis of the viticultural potential in the Rogue Valley of Oregon. *Proceedings of the VIth terroir congress, Bordeaux and Montpellier, France. July 3–7, 2006.*

Gregory V. Jones, Andrew A. Duff, Andrew Hall, and Joseph W. Myers. 2010. Spatial analysis of climate in winegrape growing regions in the western United States. *American Journal of Enology and Viticulture* 61 (3): 313–326.

Jones, Gregory V., and Leanne B. Webb. 2010. Climate change, viticulture, and wine: Challenges and opportunities. *Journal of Wine Research* 21 (2–3): 103–106.

Jordan, David J. 2002. *An offering of wine: An introductory exploration of the role of wine in the Hebrew Bible and ancient Judaism through the examination of the semantics of some keywords.* A Thesis submitted for the degree Doctor of Philosophy (PhD). The Department of Semitic Studies, University of Sydney

Kaminske, Volker. 2005. Viticultural project. *International Research in Geographical & Environmental Education* 14 (1): 46–60.

Keightley, Keir E., and Gerald W. Bawden. 2010. 3D volumetric modeling of grapevine biomass using tripod LiDAR. *Computers and Electronics in Agriculture* 74 (2): 305–312.

Keys, David. Archaeology Correspondent. 2003. Now that's what you call a real vintage: Professor unearths 8,000-year-old wine. Sunday, 28 December 2003 London: *The Independent.*

Kingsbury, Aaron, and Roger Hayter. 2006. Business associations and local development: The Okanagan wine industry's response to NAFTA. *Geoforum* 37 (4): 596–609.

Kleeman, Grant. 2006. The Australian wine industry: From boom to gloom? *Geography Bulletin* 38 (4): 32–37.

Kleine, Dorothea. 2008. Negotiating partnerships, understanding power: Doing action research on Chilean fairtrade wine value chains. *Geographical Journal* 74 (2): 109–123.

Kohn, C.F. 1986. American wines in foreign and domestic trade. *Arkansas Journal of Geography* 2: 22–31.

Kolpan, Steven, Brian H. Smith, and Michael A. Weiss. 2010. *Exploring wine: The culinary institute of America's guide to wines.* New York: John Wiley and Sons.

Lagendijk, Arnoud. 2004. Global 'lifeworlds' versus local 'systemworlds': How flying winemakers produce global wines in interconnected locales. *Tijdschrift voor Economische en Sociale Geografie* 95 (5): 511–526

Larreina, Mikel, Fernando Gómez-Bezares, and Ricardo Aguado. 2011. Development rooted on Riojan soil: The wine cluster and beyond. *The Open Geography Journal* 4: 3–15.

Lewis, Michael. 2007. Cultural norms and political mobilization: Accounting for local and state-level liquor laws, 1907–1919. *Journal of Cultural Geography* 24 (2): 31–52.

Lewis, N., W. Moran, P. Perrier-Cornet, J. Barker, J. 2002. Territoriality, régles mentation in industry governance. *Progress in Human Geography* 26 (4): 433–462.;

López, Armesto, A. Xosé, and Belen Gomez Martín. 2006. Tourism and quality agro-food products: An opportunity for the Spanish countryside. *Tijdschrift voor Economische en Sociale Geografie* 97 (2): 166–177.

Macdonald, Gerald M. and Denyse Lemaire. 1995. American Viticultural Areas: A problem in regional geography. *Journal of Geography* 94 (2): 376–383.

MacQueen, R.W., and Meinert, L.D. eds. 2006. *Fine wines and terroir: The geoscience perspective.* Geological Association of Canada.

Manandhar, Ramita, Inakwu O.A. Odeh, and Robert Gilmore Pontius Jr. 2009. Analysis of twenty years of categorical land transitions in the Lower Hunter of New South Wales, Australia. *Agriculture, Ecosystems & Environment.* 135 (4): 336–346

McCarthy, Ed and Mary Ewing-Mulligan. 2006. *Wine for dummies.* Hoboken, NJ: Wiley Publishing Inc.

McEwan, Cheryl, and David Bek. 2006. (Re)politicizing empowerment: Lessons from the South African wine industry. *Geoforum* 37: 1021–1034

McEwan, Cheryl and David Bek. 2009a. The political economy of alternative trade: Social and environmental certification in the South African wine industry. *Journal of Rural Studies* 25 (3): 255–266.

McEwan, Cheryl; Bek, David. 2009b. Placing ethical trade in context: Wieta and the South African wine industry. *Third World Quarterly* 30 (4): 723–742.

McGovern, Patrick E. 2007. *Ancient wine: The search for the origins of viniculture.* Princeton, NJ: Princeton University Press.

Meigs, P. 1941. Current trends in California orchards and vineyards. *Economic Geography* 17: 275–286.

Moran, W. 1988. The wine appellation: environmental description or economic device? In: *Proceedings of the Second International Cool Climate Viticulture and Oenology Symposium New Zealand Society for Viticulture and Oenology*, ed. R.E. Smart, R.J. Thornton, S.B. Rodriguez, and J.E. Young, J.E., 356–360. Auckland, New Zealand.

Moran, Warren. 1993. The wine appellation as territory in France and California. *Annals of the Association of American Geographers* 83 (4): 694–717.

Moran, W. 2001. *Terroir* – The human factor. *Australian and New Zealand Wine Industry Journal* 16 (2): 32–51.

Morris, A.S. 1969. The development of the irrigation economy of Mendoza, Argentina. *Annals Association American Geographers* 59: 97–115.

Morris, Arthur. 2000. Globalisation and regional differentiation: The Mendoza wine region. *Journal of Wine Research* 11 (2): 145–153.

Morrison, P.C. 1936. Viticulture in Ohio. *Economic Geography* 12: 71–85.

Morrison, P.C. 1950. Kelley's Island: Ohio: An economy in transition. *Economic Geography* 26:105–124.

Nachtergaele, Jeroen, Jean Poesen, and Bas van Wesemael. 1998. Gravel mulching in vineyards of southern Switzerland. *Soil and Tillage Research* 46 (1–2): 51–59

Nemani, Ramakrishna R., Michael A. White, Daniel R. Cayan, Gregory V. Jones, Steven W. Running, and Joseph C. Coughlan. 2001. Asymmetric warming over coastal California and its impact on the premium wine industry. *Climate Research* 19: 25–34.

Newman, J.L. 1986. Vines, wines and regional identity in the Finger Lakes region. *Geographical Review* 76: 301–316.

Newman, James L. 1988. What's in a name? *Focus* 38 (4).

Newman, James L. 1992. Decline and development in the Finger Lakes wine region of New York State. *Journal of Wine Research* 3 (2): 79–95.

Newman, James L. 2000. Wine. In *The Cambridge History of Food*, ed. Kenneth F. Kiple and Kriemhilf Conee Ornelas. Cambridge: Cambridge University Press.

Nichols, Sandra Lucile. 2002. *Saints, peaches and wine: Mexican migrants and the transformation of Los Haro, Zacatecas and Napa, California.* Ph.D. dissertation. University of California, Berkeley. 3063502.

Nossiter, Jonathan. 2005. *Mondovino* Documentary film. New York: ThinkFilm.

Olmstead, C.W. 1956. American orchard and vineyard regions. *Economic Geography* 32:189–236.

Overton, J., and Heitger, J., 2008. Maps, markets and merlot: The making of an antipodean regional wine appellation. *Journal of Rural Studies* 24 (4), 440–449.

Overton, John. 2010. The consumption of space: Land, capital and place in the New Zealand wine industry. *Geoforum* 41(5):752–762.

Overton, John and Warwick E. Murray. 2011. Playing the scales: Regional transformations and the differentiation of rural space in the Chilean wine industry. *Journal of Rural Studies* 27(1):63–72.

Parker, Robert M. 2008. *Wine Buyers Guide.* Seventh edition. New York: Simon and Schuster.

Patchell, Jerry (2008) Collectivity and Differentiation: A tale of two wine territories. *Environment and Planning A* 40(10): 2364 – 2383.

Pawson, Eric. 1997. Branding strategies and languages of consumption. *New Zealand Geographer* 53(2): 16–21.

Peters, G.L. 1984. Trends in California viticulture. *Geographical Review* 74:455–467.

Peters, G.L. 1987. The emergence of regional cultivar specializations in California viticulture. *Professional Geographer* 39 (3): 287–297.

Peters, Gary L. 1997. *American winescapes: The cultural landscapes of America's wine country,* Boulder, CO: Westview Press.

Peters, Gary. 1998. Wines & vines on the metropolitan fringe. *Focus On Geography* 45 (2): 33

Peters, G. and F. Gosette. 1990. Geographic variations in cultivar distributions: Pinot Noir, Barbera and Zinfandel in California. *Journal of Wine Research* 1 (2):121–138.

Petrini, Carlo. 2003. A *wine atlas of the Langhe: The great Barolo and Barbaresco vineyards.* White River Junction, VT: Chelsea Green Publishing. 352 pages.

Pigott, Stuart. 1995. The *wine atlas of Germany: And traveller's guide to the vineyards.* London: Mitchell Beazley.

Pitte, Jean-Robert. 2008. *Bordeaux/Burgundy: A vintage rivalry.* Translated by M.B. DeBevoise. Berkeley, CA: University of California Press.

Pliny. Translated by John Bostok, and H. T. Riley. 1892. *The natural history of Pliney.* Volume 3. New York: George Bell and Sons.

Puleo, Thomas Joseph. 2008. *Vitis vinifera and other parasites of the Vallis tellina.* Ph.D. dissertation. University of California, Los Angeles. 3351608.

Qin, Jiang Lin, Donald Rundquist, Anatoly Gitelson, Mark Steele, Christopher Harkins, and Rebecca Briles. 2010. A non-linear model for measuring grapevine leaf thickness by means of red-edge/near-infrared spectral reflectance. *Acta Ecologica Sinica* 30 (6): 297–303.

Reynolds, Robert G and S.I. Outcalt. 1975. The double-dependence Markov chain as an analogue to the autocorrelation function for a temporal series—Or weather wine and tree rings. *Geographical Analysis* 7 (2): 205–212.

Robinson, Jancis. (1986). *Vines, grape and wines.* London: Mitchell Beasley.

Robinson, Jancis. (1994). *The Oxford companion to wine.* New York: Oxford University Press USA.

Rosas, Margarita Riffo. 2008. Impactos espaciales y socio-económicos de la vitivinicultura en Chile y Argentina. *Revista Geográfica* 143 (1):163–211.

Saeidi, Ali, and Tim Unwin. 2004. Persian wine tradition and symbolism: Evidence from the medieval poetry of Hafiz. *Journal of Wine Research* 15 (2): 97–114.

Sánchez-Hernández, José Luis, Javier Aparicio-Amador, and José Luis Alonso-Santos. 2010. The shift between worlds of production as an innovative process in the wine industry in Castile and Leon (Spain). *Geoforum* 41 (3): 469–478.

Sanderson, Bruce. 2010a. Cool Riesling: The 2008 vintage from Germany is a year for fresh crisp, elegant whites. *Wine Spectator* 35 (1): 81–84.

Sanderson, Bruce. 2010b. *Terroir* and transparency: At Domaine de la Romanée-Conti, wine character is rooted in the earth. *Wine Spectator* 35 (2): 63–69.

Sanderson, Bruce. 2011. The jewel of northwest Italy: Piedmont's many appellations offer styles from rustic to refined. *Wine Spectator* 36 (1): 47–62.

Sauer, Carl Ortwin. 1952. *Agricultural origins and dispersals: Domestications of animals and foodstuffs.* New York: American Geographical Society.

Sauer, Jonathan D. 1993. *Historical geography of crop plants: A select roster.* Boca Raton, FL: CRC Press.
Schamel, G., 2006. Geography versus brands in a global wine market. *Agribusiness* 22 (3): 363–374.
Schenk, Winfried. 1992. Viticulture in Franconia along the River Main: Human and natural influences since AD 700. *Journal of Wine Research* 3 (3):185–203
Schultz, Hans R. and Gregory V. Jones. 2010. Climate induced historic and future changes in viticulture. *Journal of Wine Research* 21 (2–3): 137–145.
Schwarz, Reiner. 1997. Predicting wine quality from terrain characteristics by regression trees. 10th European Colloquium on Quantitative and Theoretical Geography. *Cybergeo: European Journal of Geography*. http://cybergeo.revues.org/361.
Sechrist, Robert. 2004. Pennsylvania wine sales September 2000 to August 2002. *Pennsylvania Geographer.*
Senese, Donna. 2010. Amenity resources and rural change in the Okanagan Valley of British Columbia. In *The rural-urban fringe in Canada: Conflict & controversy*. ed. Kenneth B. Beesley. Rural Development Institute, Canada: Brandon University Press.
Shaw, A.B. 1999. The emerging cool climate wine regions of eastern Canada. *Journal of Wine Research* 10 (2): 79–94.
Shaw, A. B. 2001. Pelee Island and Lake Erie North Shore, Ontario: A climatic analysis of Canada's warmest wine region. *Journal of Wine Research* 12 (1): 19–37.
Shaw, A.B. 2002. A climatic assessment of the Niagara peninsula viticulture area for the application of wind machines. *Journal of Wine Research* 13 (2): 143–164.
Shaw, Anthony B. 2005. A climatic analysis of Canada's largest wine region. *Journal of Wine Research* 16 (2): 85–103.
Skinner, Angela M. 2010. Napa Valley, California: A model for wine region development. In *Tourism around the world: Development, management and markets.* ed. C. Michael Hall, Liz Sharples, and Niki Macionis, 283–297. Oxford, UK: Butterworth-Heinemann.
Smith, Brian. 2005. Solving the puzzle of Barolo and Barbaresco. *Gastronomica* 15 (1): 88–94.
Smith, D.L. 1970. Viticulture in the Barossa region: Prospects and costs. *Australian Geographical Studies* 8 (2): 101–120.
Sommers, Brian J. 2008. The *geography of wine: How landscapes, cultures, terroir, and the weather make a good drop.* New York: Penguin Group.
Spellman, Greg, and Ken Field. 2002. The changed fortunes of United Kingdom viticulture? *Geography* 87 (4): 324–330.
Stanislawski, D. 1970. *Landscapes of Bacchus*. Austin: Texas University Press.
Stanislawski, D. 1973b. Dark Age contributions to the Mediterranean way of life. *Annals Association of American Geographers* 63 (4): 397–410.
Stanislawski, Dan. 1975. Dionysus westward: Early religion and the economic geography of wine. *The Geographical Review* 65 (1): 427–444.
Stanislawski, D. 1977. Seeds for the flowers of Tuscany. *Geographical Review* 67 (4): 412–42.
Stevenson, I. 1976. Sour grapes from rich harvests. *Geographical Magazine* 48: 262–264.
Stevenson, I. 1980. The diffusion of disaster: The phylloxera outbreak in the department of the Herault, 1862–80. *Journal of Historical Geography* 6: 47–63.
Swinchatt, Jonathan and David G. Howell. 2004. *The winemaker's dance: Exploring terroir in the Napa Valley.* Berkeley: U of CA Press.
Templer, O.W. 1986. Wine growing in arid and semi-arid west Texas. *Forum of the Association for Arid Land Studies* 99–105.
The Wine Institute. 2010. Statistics. http://www.wineinstitute.org/resources/statistics. Last accessed October 7, 2011.
Thompson, Bob. 1993. *The wine atlas of California and the Pacific Northwest: A traveler's guide to the vineyards.* New York: Simon & Schuster.
Tomljenović, Renata, and Donald Getz. 2009. Life-cycle stages in wine tourism development: A comparison of wine regions in Croatia. *Tourism Review International* 13 (1):31–49.
Turner, Simon. 2010. Networks of learning within the English wine industry. *Journal of Economic Geography* 10 (5): 685–715.
Unwin, T. 1994. European wine sector policy and the UK wine industry. *Journal of Wine Research* 5 (2) p. 135–146.
Unwin, Tim. 1996. *Wine and the vine: An historical geography of viticulture and the wine trade.* London: Routledge.
Van Eetvelde, Veerle, and Marc Antrop. 2004. Analyzing structural and functional changes of traditional landscapes—Two examples from Southern France. *Landscape and Urban Planning* 67 (1–4): 79–95.
Vaudour, E. 2002. The quality of grapes and wine in relation to geography: Notions of *terroir* at various scales. *Journal of Wine Research* 13 (2):117–141.
Vaudour E. 2008. Remote sensing of Red Mediterranean soils: A case study in the viticultural Southern Rhone Valley (France) using SPOT satellite imagery. *Geocarto International* 23 (3):197–216.
Vaudour, E. and A.B. Shaw. 2005. A worldwide perspective on viticultural zoning. *South African Journal of Enolology and Viticulture* 26 (2): 106–115.
Vaudour E., J. Moeys, J.M. Gilliot, Y. Coquet. 2008. Spatial retrieval of soil reflectance from SPOT multispectral data using the empirical line method. *International Journal of Remote Sensing* 29 (19): 5571–5584.
Vaudour, E., V.A. Carey, and J.M. Gilliot. 2010. Digital zoning of South African viticultural *terroirs* using bootstrapped decision trees on morphometric data and multitemporal SPOT images. *Remote Sensing of Environment* 114 (12): 2940–2950.
Veeck, Gregory, and Deborah Che, and Ann Veeck. 2006. America's changing farmscape: A study of agricultural tourism in Michigan. *The Professional Geographer* 58 (3): 235–248.
Velluzzi, Nicholas Daniel. 2007. *Fermenting growth: Institutions, agency and the competitive foundations of localized learning in the Walla Walla wine industry.* Ph.D. dissertation. University of Washington. 0819364.
Vins de Provence France. 2011. US wine market: Facts and figures. http://provencewineusa.com/resources/Market_Facts_v5.pdf. Last accessed May 31, 2011.
Visser, Evert-Jan, and Peter de Langen. 2006. The importance and quality of governance in the Chilean wine industry. *GeoJournal* 65 (3): 177–197.
Walker, G. 1979. Vineyardists in the Sonoma Valley, 1870–80: Elements of an ethnic geography. In: Vicero, R.D. ed. *Proceedings of Cukanzas* 1979.
Wargenau, Astrid, and Deborah Che. 2006. Wine tourism development and marketing strategies in Southwest Michigan. *International Journal of Wine Marketing* 18 (1): 45–60.

Weigend, G.G. 1954. The basis and significance of viticulture in south-west France. *Annals Association American Geographer* 44: 75–101.

Weigend, G.G. 1955. Bordeaux: an example of changing port functions. *Geographical Review* 45: 217–433.

White, Michael A., Philip Whalen and Gregory V. Jones. 2009. Land and wine. *Nature Geoscience* 2**:** 82 – 84.

Wilson, James E. 1998. *Terroir: The role of geology, climate, and culture in the making of French wines.* Berkeley: University of California Press.

Wolf, Tony K. and John D. Boyer. 2000. Site selection and other vine management principle and practices to minimize the threat of cold injury. *Proceedings of the Cold Hardiness Workshop,* American Society for Enology and Viticulture Annual Meeting, June 2000, Seattle, WA. *Amer. J. Enol. Vitic.* 51.

Wolf, Tony, John Boyer, Tim Martinson, and Fritz Westover. 2009. Chapter 2: Vineyard site selection. In *The wine grape production guide for Eastern North America.* Publication Number: NRAES-145. Blacksburg, VA: Virginia Tech.

Terroir: At the Heart of Geography

2

Tim Unwin

Geography is a flavour

(Starbucks 2005)

Abstract

This chapter provides an overview of the ways in which particular assemblages of vines and environments have emerged since prehistory, thereby giving rise to our notions of terroir. It reviews debates relating to the importance of terroir in the contemporary world of wine and develops two main arguments: first, that the land in which grapes are grown can, indeed, have a distinctive influence on the flavours of the wines made from them; and second, that different groups of people, be they grape growers or winemakers in various parts of the world, all have particular interests around which debates over terroir can be constructed.

Introduction

Few words have caused such debate and antagonism between winemakers, journalists and indeed academics as the French word *terroir*. In part, these debates reflect different usages and meanings attributed to the notion. All too often, it has been taken simply to imply the soil upon which vineyards are located. In the 1960s, for example, Winkler (1962: 64) commented: 'In most of the famous grape-growing and wine-producing regions of Europe and elsewhere, the belief is firm that a particular soil has much to do with local success. Perhaps it does in some instances, but in others it is largely irrelevant'. This was an over-simplification. Few grape growers and winemakers had ever made such claims about the soil alone; it is the complex physical environment, of which soil is but one part, that they have suggested lends character to a wine. Despite claims to the contrary, such as Hinkle's (1991) assertion that 'The literal French definition of terroir as we all know, is 'soil'', *terroir* actually means something much more subtle and complex.

French winemakers have long used the term *terroir* to refer to the complex interaction between all of the physical aspects of geology, soils, climate, geomorphology and vegetation that combine to create a particular 'place' where grapes are grown. The notion that soil and *terroir* are one and the same has nevertheless had lasting effect on popular and academic discourses on the relationships between the environment and wine. This has been beautifully captured by the cartoonist Blachon (Humoristes Associés 1980) in Fig. 2.1, which shows a wine bottle with roots going down into the ground beneath it, thereby implying the origin of the grapes that made it. This cartoon, though, also shows so much more. In the background is the village where the *vignerons* live, dominated by the church spire, emphasising also the deep significance that religion has played in shaping the history of wine. Likewise, this wine clearly 'grew up' in a rolling

T. Unwin (✉)
Department of Geography, Royal Holloway, University of London, Egham, Surrey TW20 0EX, UK
e-mail: tim.unwin@rhul.ac.uk

P.H. Dougherty (ed.), *The Geography of Wine: Regions, Terroir and Techniques*,
DOI 10.1007/978-94-007-0464-0_2, © Springer Science+Business Media B.V. 2012

Fig. 2.1 Cartoon by Blachon (Source: Humoristes Associés 1980)

landscape of hills – once more indicating a key element of the *terroir* that helped to craft it.

The debate over terroir was particularly highlighted in a series of articles in the popular wine magazine *Decanter* in the early 1980s between Bill Jekel (1982, 1983a, b) and Bruno Pats (1983). Jekel had established the winery that bore his name in 1972 in Monterey County, California, whereas Prats was the owner of Château Cos D'Estournel in Saint-Estèphe, Bordeaux. For Jekel, soils, and particularly their mineral content, had little influence on the quality of a wine; for Prats, in contrast, soils were an important part of the *terroir* that gave France's great wines their characteristics. In practice, Prats and Jekel were not actually that far apart, because Jekel did not say that soils had no role, and Prats placed his emphasis on *terroir* rather than soils alone. Nevertheless, as the geographer John Dickenson (1990: 9) concluded from this debate, 'despite Prats's catholic interpretation of terroir, it is evident from the ensuing correspondence, popular wine books and advertisements for claret, that most protagonists *do* equate soil with terroir and believe that it is the determining factor in wine quality'.

Since the 1980s, these debates have continued to flourish. As Dickenson noted (1990: 9), 'Many contributors to the debate have little notion of environmental processes, and use terms such as soil, sub-soil and geology indiscriminately, while their reference to 'microclimate' is similarly cavalier. Most wine writers pass rapidly over environmental issues and express diffidence and confusion'. Despite the publication of many books on the effects of the environment on grape growing (Coombe and Dry 1988; Gladstone 1992), and the plethora of high quality research on the ways in which different aspects of the physical environment shape grape and wine quality (Winkler et al. 1992; Goode 2006), Dickinson's generalisation still remains largely true. Thus, instead of presenting yet another overview of the ways in which different aspects of the physical environment influence the aromas and flavours of wines, I wish here to direct attention to the issue of why these debates over *terroir* persist. Three interrelated binary oppositions are highlighted: the distinction between winemaking and grape growing, the different attitudes and approaches between Old World and New World wine producers and the tensions that exist between protectionism and innovation. Before I do this, however, it is necessary to situate the debate over *terroir* in its historical context.

The Differences that Physical Environments Make

One need only taste wines grown in different places but made from the same grape variety, in the same vintage, and vinified as far as possible in exactly the same way to know that 'place' matters. As discussed below, this effect can vary enormously, depending upon winemaking techniques. Anyone who doubts the importance of place, though, needs only to sit down with a winegrower such as Patrice Rion in the Côte de Nuits in Burgundy (http://www.patricerion.com/) and taste the different barrels of wine that he makes from the Pinot Noir grapes grown in nearby plots of land. Each has its own distinctive flavour. Or talk to one of the outstanding Australian winemakers who, when putting together a commercial blend of wine will, year after year, select fruit from a particular vineyard because they know that it gives them the distinctive characteristics for which they are looking. The final wine will not have definable place-based characteristics that one might associate with the notion of *terroir*, but its component elements almost certainly will. As Brian Walsh, chief winemaker for Yalumba, commented in 1999, 'In recent years, Australia has discovered its terroir.... We've moved from an emphasis on technology in the winery out to the vineyard' (Boyd 1999). Even the Jekel Vineyards website now reflects the importance of *terroir* (for a wider discussion of environment and taste in California, see Elliott-Fisk and Noble 1992):

> *The Terroir*
> Northern Monterey County, Monterey and Arroyo Seco appellations are located on an ancient alluvial plain, carved from an antediluvian river bed. The primary attribute of this region is the afternoon winds that bring inland the cooling air of Monterey Bay—allowing the vine to retain greater acidity and vibrancy in its fruit.
>
> This region's unique soil composition (cobblestone, depleted soils) challenges and yet nurtures the vine to yield grapes of depth, character and concentrated flavors. Its mild climate encourages bud break two weeks earlier than other regions, as well as extends the season to nearly two weeks longer-resulting in greater intensity of flavor. (http://www.jekel.com/ourvineyards.aspx, accessed 16th September 2009)

Defining Terroir

In order to make sense of debates over *terroir*, it is essential to reflect a little more on the precise definition of the term. Amy Trubek (2008: 18) has emphasised: 'In France, food and drink from a certain place are thought to possess unique tastes. Thus, more than words, terroir and goût du terroir are *categories* that frame perceptions and practices'. She goes on to argue that these categories are fundamentally important for 'framing and explaining people's relationship to the land, be it sensual, practical, or habitual. The connection is considered essential, as timeless as the earth itself' (Trubek 2008: 18). Such an understanding of the relationships between people and the physical environments in which they lived were of fundamental importance in shaping French conceptualisations of their differing *pays* in the nineteenth century, and thus of the regional geographies that were written about them during the twentieth century (Demangeon 1905; Cion 1908; see also de la Blache 1913, 1922; Brunhes 1925).

If geography is seen at least in part as being about 'place' and about the various interactions between humans and the physical environment that shape different places (Unwin 1992), then such a notion of *terroir* is of central importance to the discipline. Of all the products humans have shaped and crafted over the millennia, it is possibly wine that most reflects this distinctive relationship (Unwin 1991). The importance of this link between geography and taste is featured, for example, in Starbucks' slogan 'Geography is a flavour'. Starbucks (2005) go on to claim: 'Much like a fine wine, the aroma and flavour of different coffee varies depending on where it is grown and how dark the roast is ... Indeed, also like wine, coffee is tasted in a certain way to discover its aroma, flavour, acidity and body'. Wine's *terroir* has become archetypical for product placements far beyond its original meaning.

The original French word *terroir* has always meant something very different from simply 'soil', which in French is '*sol*'. Indeed, it is much more closely related to the notion of territory (French *territoire*), derived from Latin *territorium*, than it is simply to soil. A *terroir* is thus a bounded place where generations of people have crafted a distinctive way of life during which they have created distinct products (Asselin et al. 1996). A recent gathering under the auspices of UNESCO (2005) has emphasised that it is extremely difficult to translate *terroir* into other languages, but they suggest that the definition proposed by a working group of INRA/INAO could serve to permit a dialogue based on the following common ground:

> Un terroir est un espace géographique délimité défini à partir d'une communauté humaine qui construit au cours de son histoire un ensemble de traits culturels distinctifs, de savoirs, et de pratiques fondés sur un système d'interactions entre le milieu naturel et les facteurs humains. Les savoir-faire mis en jeu révèlent une originalité, confèrent une typicité et permettent une reconnaissance pour les produits ou services originaires de cet espace et donc pour les hommes qui y vivent. Les terroirs sont des espaces vivants et innovants qui ne peuvent être assimilés à la seule tradition.[1]

[1]My translation: A terroir is a delimited geographical space, defined from a human community which in the course of its history constructs an assemblage of distinctive cultural traits, knowledge and practices founded on a system of interaction between the natural environment and human factors. The skill set involved reveals originality, conferring a typicity and permitting recognition for the products or services originating from this space and thus for the men who live there. Terroirs are living and innovative spaces which cannot be assimilated into a single tradition.

Such a definition goes far beyond the usual more restricted notion that wine *terroirs* are primarily the result of physical environmental factors that contribute to a wine's distinctive character and flavour. James Wilson (1998: 55) captures some of the essence of this when he comments: 'The true concept is not easily grasped but includes physical elements of the vineyard habitat – the vine, subsoil, siting, drainage, and microclimate. Beyond the measurable ecosystem, there is an additional dimension – the spiritual aspect that recognizes the joys, the heartbreaks, the pride, the sweat, and the frustrations of history'. While precisely what this statement means is somewhat opaque, it does emphasise that for many people the notion of *terroir* goes well beyond mere physical dimensions.

Even if one tries to define a vineyard simply in terms of its physical characteristics, it remains extremely difficult to decide what should be included within the notion of *terroir*. For example, many vineyard soils have been so worked by previous generations of grape growers that they can hardly be seen as being 'natural'. Even rigorously organic or biodynamic grape growers interfere with the 'natural' soils by planting and cultivating vines in them. To adopt an extreme position, it could be argued that the notion of *terroir* should include the specific assemblage of 'natural yeasts' that have evolved alongside the grape growing and winemaking traditions in a particular place, and that these should be used to make a product truly characteristic of the *terroir*.

One important dimension that needs to be considered in shaping the definitions of *terroir* is the distinctive interests that underlie the various definitions. Thus, those who consider factors other than the soil as being of most importance in shaping the characteristics of wine can benefit by devaluing the term *terroir* to mean simply 'soil'. By doing so, their alternative explanations appear to gain strength. In contrast, those who want to delimit particular plots of land so that they can develop monopoly advantages (Unwin 1991), as with the French *Appellation d'origine contrôlée* system, are eager to attribute rather different characteristics to the word *terroir*. These various approaches are teased out in more detail in the sections that follow, but beforehand it is important to explore briefly the historical evolution of the notion that particular grapes grown in certain places can lead to distinct and different aromas and flavours in the resultant wines.

Terroir in Antiquity

The wild grapevine, *Vitis vinifera* subsp. *sylvestris*, was widely distributed in southwest Asia and around the coasts of the Mediterranean in pre-historic times. Considerable genetic diversity already existed within this range before humans started selecting specific vines that produced grapes with particular characteristics (Zohary 1995). Gradually, from the Neolithic period onwards, people began to transform this 'natural' distribution, seeking to identify vines that produced fruit with the characteristics that they wanted in particular places. As McGovern (2003: 12–13) has commented, 'The sheer number of cultivars or clonal types, which has been estimated to be as many as 10,000 worldwide … testifies to the plant's pliable, almost chimeric nature'.

By the Roman period, farmers were well aware that different types of vines grew particularly well in certain locations. Writing in the second century BC, Cato thus commented: 'On the ground that is called best for wine, and is exposed to the sun plant the small Aminean grape, the twin Eugenean and the little yellowish grape' (Cato 1933: 17). Herein is clear recognition of the value of *terroir*. A century later, Varro (1912) repeated many of Cato's observations with respect to the locations where vines should be cultivated, further emphasising that different types of grape ripened at different times and that at vintage time, good farmers selected their grapes for different purposes. However, it is in Columella's (1941, 1954, 1955) *de Re Rustica* that we find one of the best early accounts of the relationships between different types of vine and the environments in which they grew:

> The wise farmer will have discovered by test that the kind of vine proper for level country is one which endures mists and frosts without injury; for a hillside, one which withstands drought and wind. He will assign to fat and fertile land a vine that is slender and not too productive by nature; to lean land, a prolific vine; to heavy soil, a vigorous vine that puts forth much wood and foliage; to loose and rich soil, one that has few canes. He will know that is it not proper to commit to a moist place a vine with thin-skinned fruit and unusually large grapes, but one whose fruit is tough skinned, small, and full of seed. (Columella 1941, III.i.5: 229)

Over the millennia, grape growers and winemakers had therefore gradually been able to select particular vines that produced the fruit characteristics that they required for making wine in specific environments. Moreover, according to Strabo (1949–1954) writing in

Fig. 2.2 The slopes of Vesuvius from Pompeii, the wines from which Pliny would have known well. He died while rescuing friends from the eruption of the volcano in 79 AD

the early first century AD, vines were by then being cultivated in most parts of the Roman world and beyond. He cites Aristobulus who claimed that wine was produced in India and also comments that the vines of Hyrcania, southeast of the Caspian Sea, were very productive. To the north of Italy, the Ligurians made a small amount of poor quality wine, and at the western end of the Mediterranean, vines were cultivated in Iberia. Half a century later, Pliny (1945–1950) commented on the expansion of viticulture as the tentacles of Roman power spread further into Gaul and Iberia. Vineyards were being laid out in Narbonensis and at Vienne, just south of Lyon. He noted that a new type of vine, known as *carbonica*, was being planted extensively, giving a distinctive flavour of pitch to the wines made from it. In western France, the *biturica*, which may have originated in Spain, came to dominate production, and, in the area around Lyon, the indigenous variety grown by the Allobroges, known as the *allobrogica*, had come into prominence.

All of these Roman writers noted gradations of quality. Certain vines growing in particular places had gained distinctive reputations. For example, Strabo observed that the wines of the Aegean, especially those of Cos, Chios and Lesbos, were particularly praiseworthy. In Italy, he claimed that the top three wines were the Falernian, Satanian and Calenian, with the Surrentine wines coming close to them in terms of quality. Caecuban, Fundian, Setian and Alban wines were also famed. Writing in the first century AD, Pliny listed some 91 varieties of vine, 50 kinds of quality wine and 38 varieties of foreign wines, along with further salted, sweet and artificial wines (Fig. 2.2) (Unwin 1991). Of particular significance with regard to *terroir*, Pliny described some vines, such as the Nomentian and Apiana, as being particularly suited to cold climates, as in the mountainous terrain in the north of Italy.

These examples emphasise that two important aspects of *terroir* were already well established by the time of the expansion of the Roman Empire in Europe. First, grape growers and winemakers were already familiar with the notion that a close relationship existed between the vines grown in particular environments and the wines that could be produced from them. They also knew that certain places had gained a reputation for producing wines claimed to be of especially high quality. The fundamental tension within the idea of *terroir* was, therefore, already born at least two millennia ago: although *terroir* merely refers to the relationships between wines and the environments from which they originated, some places with distinctive *terroirs* were nevertheless also identified as producing higher quality wines than others. It is this conflation between notions of *terroir* and 'quality' that has been at the heart of most subsequent debates over the significance and meaning of *terroir*.

The Medieval and Early Modern Worlds

The spread of viticulture northwards during the Roman Empire led to the discovery of new cultivars of *Vitis vinifera*. Naturally occurring crosses between varieties led to even greater complexity, with grape growers throughout the subsequent centuries gradually selecting the varieties that best suited the environments in which they were grown.

By the fourteenth century, there is clear evidence that some of the distinctive names of vines that are still associated with particular winemaking regions had come into being. In 1395, for example, Philip the Bold denounced the introduction 'd'un trés mauvais et disloyal plant nommé gaamez'[2] and legislated that all Gamay vines in Burgundy were to be cut down and uprooted so as to enhance the quality of wine production in his realm (Berlow 1982; Lachiver 1988). A couple of centuries later, Andrea Bacci's (1596) agricultural treatise *De naturali vinorum historia, de vinis Italia,* provides detailed lists of 50 noble vines known in sixteenth century Italy, as well as much evidence about the wines then being produced in other parts of Europe. It is from this period that we have clear information about the Trebbiano in Tuscany and the Marche, the Vernaccia in Liguria and Schiava in the Po Valley (Jones 1966). Lachiver (1988) likewise notes that it was around the middle of the sixteenth century that regional specialties had become established in France, with the Chenin Blanc, known locally as the Pineau de la Loire, prominent in Anjou, and the Cabernet Franc, known locally as the Breton, popular around Tours and Orléans.

Recent research based on the analysis of nuclear DNA microsatellites has refined our knowledge of the actual origins of many of the crosses that probably occurred in medieval times and gave rise to the familiar modern grape varieties (McGovern 2003). Thus, Bowers and Meredith (1997) have shown that the classic grapevine of Bordeaux, the Cabernet Sauvignon, is a cross between the red grape producing Cabernet Franc and the white Sauvignon Blanc vines. Likewise, Chardonnay, the most important white wine grape of Burgundy, as well as Gamay Noir, Aligoté and Melon, were apparently the result of a cross between a single pair of parents, a Pinot and the little known and oft-maligned Gouais Blanc (Bowers and Meredith 1999).

Gradually, as certain grape varieties came to be grown in particular locations around the world, so too did local traditions of vine cultivation emerge. These were judged to be practices that produced the best quality of grapes for the purposes of winemaking, and they gradually became institutionalised in the regulatory processes of the late nineteenth and early twentieth centuries. This set in place delimited areas from which grapes could be grown to produce certain named wines, the grape varieties to be used, the methods of vine training and pruning and the ways in which the wines were made. Whilst such demarcation systems undoubtedly confer opportunities for monopoly rent, as so perceptively identified by Marx (1981: 910), it must be recalled that most of them were initially introduced at a time of crisis in order to help guarantee the quality and progeny of the wines.

Tensions over Terroir

Grape Growing and Winemaking

Across Europe in the early twentieth century, most wines were made by the people who grew the grapes. They shaped the terraces, cultivated the soil, planted and pruned the vines, kept an eye open for the weather, decided when to pick the grapes, selected the bunches that would be used for different *cuvées* and then fermented and matured the wine. However, during the second half of the twentieth century, considerable changes took place in the science of winemaking that shifted the balance of interest away from the cultivation of grapes to the making of wine. Driving this change was the application of research emanating from institutions such as the University of California Davis and Roseworthy College in Australia during the 1950s and 1960s (Winkler 1973; Bishop 1980; Unwin 1991; Lapsley 1997) which led to a very different kind of wine industry. Much of this research initially focused

[2]My translation: 'of a very bad and disloyal grape called Gamay'

on overcoming the faults that were all too often encountered in wines, thereby ensuring that the overall quality of wines improved. Although many developments had taken place in winemaking between the seventeenth and twentieth centuries (Unwin 1991), this new scientific approach to winemaking gradually led to a belief that the most important factor differentiating wines was the skills of the winemaker rather than the environment in which the grapes were grown, or indeed, the skills of the person growing the vines.

This emphasis on the importance of wine science over and above the practice of agriculture was not only felt in the wine industry but was also experienced more widely in terms of the status of agricultural practices and institutions across Europe and North America. All too often, scientists and students in prestigious academic research institutes and universities looked down their noses and made disparaging comments about people teaching and studying in agricultural colleges. Although much excellent agricultural research had been undertaken across a range of academic institutions, this division between the high respect given to laboratory sciences and the much lower status attributed to agriculture research is one that has persisted for a considerable time (Chamot 1926; Board on Agriculture, National Research Council 1992). To this day, and in many different parts of the world, it is still all too often said that the bright students go to study science at universities, whereas their less intelligent peers finish up learning how to farm (Lipton 1977). The net effect of this bias against agriculture was to create an over-emphasis on the science of winemaking that persists to this day. To be sure, considerable research has been done in the field of grape science and viticulture, but it is the winemaking that continues to have the prestige. This is well exemplified by the number of bottles of wine that now have details of the winemaker on their back labels, often with a quotation about the attributes that he, or more rarely she, has tried to give to the wine (Charters et al. 1999).

Interestingly, this division between wine science and grape growing is also one that resonated with debates in nineteenth century France as the country was coming to grips with the arrival of phylloxera. As Pouget (1990) has illustrated, attempts to save Europe's vineyards from the onslaught of the phylloxera aphid generally fell into two camps: the chemists and the plant specialists. Among the former, sprays and concoctions were developed to try to counter the infestation, whereas the latter experimented with the somewhat less fashionable, but ultimately more successful, solution of grafting European vine scions onto the more resistant North American rootstocks. Two aspects of this scenario are important for a discussion of *terroir*. First, the 'scientific' chemical approach was generally seen as being more glamorous and higher in status than was the less exciting solution of vine grafting (for the origins of scientific wine research see Pasteur 1866). However, second, and in retrospect, it is noteworthy that it was the grape growers rather than the chemists who eventually held sway. Indeed, it was they who were subsequently able to use their connections with the land to argue effectively for the basis of much European legislation over wine to be grounded in the uniqueness of place that is expressed in *terroir*.

In trying to overcome the chemical abuses of wine producers who mixed wine from various different sources and 'stretched' it with additives so that it went further, the legislatures in countries such as France sought to go back to the natural purity of wines made in distinctively different traditions located in specific places (Loubère 1990). Hence, the use of *terroir* in demarcation schemes such as the French *Appellation d'origine contrôlée* (AOC) system was explicitly designed as a way of trying to guarantee purity of production and characteristics that were seen as being typical of products made in particular places. The fact that these appellation systems have now often tended to become restrictive, and have been widely criticised for limiting the innovation that could lead to even higher quality wines (Barham 2003), is not so much the result of their emphasis on *terroir* as that of the wider commercial interests associated with the marketing and sale of wines made in other parts of the world. Although the AOC system has the notion of *terroir* at its heart, criticism of the former does not necessarily invalidate the importance of the latter.

Winemaking in the Old and New World

Much of the debate over *terroir* has been positioned as one between winemakers and writers in the New World and those in the Old World. In part, this derives from the contrasting views of Prats and Jekel noted earlier in this chapter and in part because many of the new scientific approaches to winemaking in the 1960s were pioneered in the New World. However, it was also a result of attempts to compare wines made in the Old and the New Worlds with the aim of judging which were superior in quality, typified by the series of challenges between Californian wines and those of Bordeaux. The first and most famous of these was the 1976 Challenge organised by Steven Spurrier in Paris in which French judges tasted Chardonnay and Cabernet Sauvignon wines from France and California and, to their surprise, rated Californian wines as the best in each category (Taber 2005). Although such tastings were about the qualities of wine that the judges preferred, the results also came to be used as a critique of the French notion of *terroir*. The Bordelais were seen as claiming that their *terroir* made the best wines, and by winning the competition, the Californians claimed to show that they could not only produce equally good wine but also invalidate the notion of *terroir*. Wines made in Californian environments could be just as good as those produced from the hallowed gravels of the Médoc (Fig. 2.3).

Tracing the complexity of these arguments is not easy, but at least three factors are of importance: the nineteenth century evolution of winemaking in the New World, the role of blending and the scale attributed to the concept of *terroir*. From the first introduction of vines to Mexico in the sixteenth century, grape growers were most concerned about finding the locations where vines could grow at all, rather than necessarily identifying the optimal locations for producing quality grapes for winemaking. It was thus the climatic conditions for vineyards that were foremost in their minds, rather than the specific geology or soils upon which the vines were planted. In Australia, for example, the first vines planted at Farm Cove near modern Sydney only produced a limited amount of fruit as a result of the high humidity. Captain Arthur Phillip thus ordered a new vineyard to be planted further inland in a site by the Partamatta River in 1791 (Laffer 1949). By the early twentieth century, this emphasis on climate lay behind the pioneering work of Winkler (1938, 1962), who used heat summation indices to identify areas in the world, and particularly in California, that were most suited to growing particular grapes for wine. He noted, for example, that the northern French vineyards of Champagne had approximately 1,050°C days a year of heat summation, whereas the Paarl district of South Africa had between 1,440°C and 1,450°C days a year, and the northern regions of Chile between 1,600°C and 1,700°C days a year. Although the impact

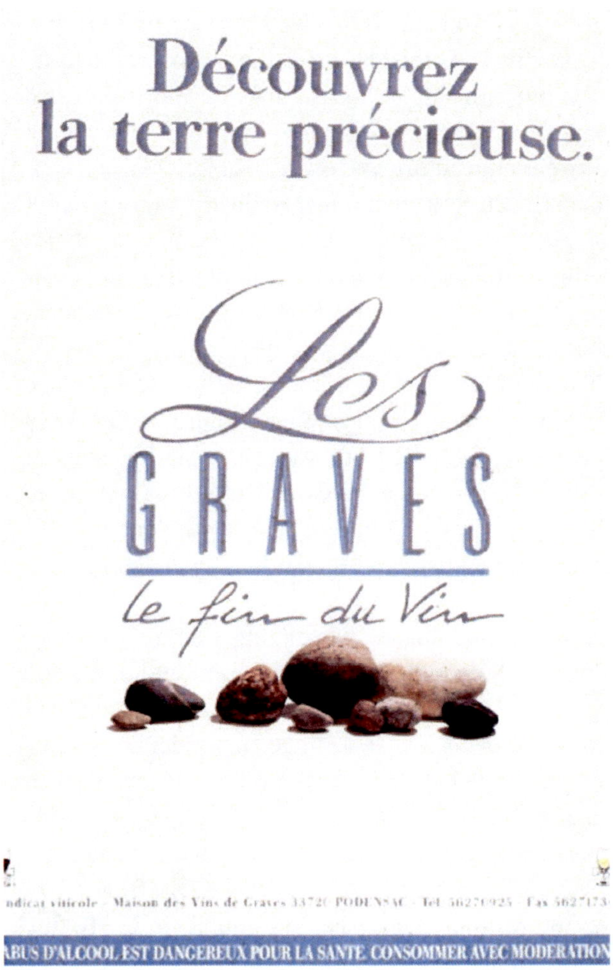

Fig. 2.3 Advertisement for wines from the Graves, showing the perceived importance of the gravels

of climate at a local scale is a critical feature of *terroir*, it was this overall focus on identifying general climate zones where grape growing was going to be possible that dominated much New World thinking. When this was combined with an emphasis on winemaking over grape growing, it can readily be seen how a concern with the impact of more local environmental factors was likely to have been given lesser importance. This was particularly evident in the wine policies that evolved in Australia in the 1990s in which much emphasis was placed on increasing the area under vines so as to serve the apparently never-ending increase in demand for Australian wines. Insufficient attention was therefore paid to the identification of the best environments in which to grow particular vines.

In part, this reflects the importance of blending in the production of bulk wines for international markets that now dominates much New World production. It

Fig. 2.4 Stephen Doyle in the early days at Bloodwood. Stephen began exploring the best physical sites for wine in the mid-1970s and eventually chose this land at Orange in New South Wales, where he now produces Bloodwood wines (Source: http://bloodwood.biz/our-history/130-chapter-two.html)

is important, though, to emphasise that there are significant exceptions to this generalisation, typified by winegrowers such as Stephen Doyle of Bloodwood in Orange, Australia (http://www.bloodwood.com.au), who have deliberately sought out the best environments in which to grow their vines (Fig. 2.4). The overwhelming emphasis, nevertheless, remains on the production of the best quality/value ratio for blended wines which are, in large part, destined for supermarkets in countries that do not produce much wine themselves. Again, it is an emphasis on the winemaker's skills that seems to predominate in decisions over the production of such wines. To be sure, huge strides have been made in grape growing in the New World (Smart and Robinson 1991), but when winemakers are bringing together batches of grapes, must or wine, sometimes from sources hundreds of miles apart, it becomes very difficult to claim that the resultant wines themselves have any distinctive *terroir* characteristics. Whilst the winemakers may well seek to identify fruit with particular characteristics to contribute to their blends, this is something very different from claiming that it is possible to identify the *terroir* from which the blended wine has emanated.

This brings into consideration the question of the scale at which the notion of *terroir* applies. When winegrowers turned the grapes from their small plots of land into a locally made wine, it could indeed be claimed that the resultant wines closely reflected the environments in which they grew up. Such a conceptualisation underlies the French AOC system, but the application of the notion of *terroir* to regional wines is much more problematic. While there may be some justification that grapes from different Premier and Grand Cru vineyards in Burgundy (Fig. 2.5) do have distinctively different characteristics, it is very different to claim that all Burgundian wines have a specific *terroir* that can be attributed to them generically. Nevertheless, the observation that experienced wine tasters are indeed able to identify wines made from grapes grown in different regions of the world does lend some credence to the argument that the idea of regional typicity could indeed have value. The situation is, nevertheless, made much more complex when, for example, a wine made in the New World has deliberately been made in an Old World style. Once again, the skills of the winemaker rise to the fore. Ultimately, the scale at which people wish to identify the influence of *terroir* on a wine is probably largely a personal choice. For purists, *terroir* is based on the notion that distinct environments produce wines of identifiable character, but its interpretation does depend considerably on the scale of similarity. Whereas in Burgundy, people can detect subtle

Fig. 2.5 La Romanée-Conti in the Côte de Nuits in Burgundy, producing one of the most expensive wines in the world

differences in the tastes of wines made from grapes grown on either side of a road, in places such as Coonawarra, Australia, there are those who describe whole regions as having a distinct *terroir*. As the Coonawarra website notes, 'Its secret lies in a magical marriage of rich red terra rossa soil, limestone, pure underground water and a long cool ripening season for the grapes' (http://www.coonawarra.org/).

Protectionism and Innovation

Many European vineyard demarcation systems are fundamentally based on the notion of *terroir* (Unwin 1991), with particular locations being seen as providing the basis for wines of particular quality. Once again, to use the example of Burgundy, the vineyards that are seen as producing the better quality wines (*premier crus* and *grands crus*) are usually on the slightly steeper slopes rising up above the *appellations communales* on the flatter ground towards the river Saône (Fig. 2.5). In origin, as discussed previously, these demarcation systems were designed to try to guarantee the authenticity of traditional products and the appellation regulations, therefore, specified such criteria as the methods of training used to grow the grapes, the yields permitted and the methods of vinification. Critics of such appellation systems argue that they can lead to complacency with some producers making wines of mediocre quality from grapes grown in prestigious vineyards (Gladstones 1992; Barham 2003). Moreover, by placing restrictive regulations on viticulture and vinification, they can also limit innovation. Such arguments are, nevertheless, primarily about faults with the appellation system rather than with the notion that different physical environments produce grapes that can be made into wines with different characteristics.

It is important to stress, though, that some innovation is actually possible within the constraints provided by systems such as the French AOC system or the German wine classification system where grape growers can, indeed, adopt very different vineyard practices and winemakers can use new technologies to improve the qualities of their products. Moreover, although they can be slow to change, such demarcation systems are themselves by no means immutable, as debates over the German classification system in the early 2000s (Robinson 2003) and the introduction of the Italian *Indicazione di Geografica Tipica* system in the early 1990s indicated. For blended wines designed for the mass market, *terroir* may have little value as a concept, but for winegrowers' intent on making the best possible wines from the grapes grown on particular plots of land, the distinctive characteristics that can be found in the notion of *terroir* are crucial.

It may well be that the importance of *terroir* does vary geographically as well. In more marginal areas, where obtaining fruit of a consistently high quality year after year is a real challenge, certain favoured plots of land, sheltered from the winds or with slightly better aspect and higher insolation levels, may consistently produce grapes with better acid–sugar balance and fruit characteristics than other areas. Thus, *terroir* is probably of more importance to the quality of the wines in Burgundy where vintages are much less consistent than they are, for example, in central Italy. This might also go some way to explaining differences between Old World and New World attitudes to *terroir*. In many of the New World's winemaking regions, the climate conditions are generally more benign for grape growing than they are in the more northerly vineyards of Europe. Of most importance, the sunshine that is crucial for ripening the grapes is much more consistent in most of the New World's wine regions than it is in northern Europe. Winemakers there tend to have less difficulty in obtaining fruit of consistent quality and, therefore, produce more consistent vintages. Here, *terroir* will become a less important factor in determining the overall quality of a wine than it would in places where in a damp, wet year only certain vineyards might produce grapes of sufficient ripeness to make high quality wines.

Conclusions

This chapter has provided both an overview of how notions of *terroir* have emerged and a review of some of the more important debates relating to the importance of *terroir* in the modern wine world. Its essential arguments have been twofold: first, that the land in which grapes are grown does, indeed, have a distinctive influence on the characteristics of the wines made from them, and second, that particular groups of people, be they grape growers or winemakers in different parts of the world, all have particular interests around which debates over *terroir* can be constructed.

Much of the debate about *terroir* has, though, hinged on a conflation of the term with ideas about wine quality. To take but one example, Gergaud and Ginsburgh (2008: F149–50) have argued: 'The French terroir legend obviously does not hold, at least in the Haut-Médoc region which is probably one of the most famous in the world. Nowadays, high quality wines are produced in many different environments, including Languedoc in the South of France, a region which was supposed to be able to grow table wines only'. This observation completely misses the point about *terroir*. In its origin, *terroir* has little to do with wine quality. It merely means that wines made from grapes grown in different places vary in aroma and flavour characteristics. The difficulties have arisen when certain groups of producers have tried to use the notion of *terroir* to claim that their particular environments are necessarily better than others for making the highest quality wines. To be sure, in some parts of the world, certain plots of land do, indeed, consistently produce wines that tasters say are of higher quality than others, but this need not always be so. Making wines is about turning grapes into a beverage that has particular characteristics. One needs both grape growing and winemaking knowledge to do this effectively. In the past, the relative balance of importance between these two sciences has varied, but to make the best wines from any particular plot of land, it is essential to know the attributes that the physical environment can contribute and to know how these can best be manipulated to make the sorts of wine that a particular winegrower wishes to place on the market. Geography remains as essential to an understanding of wine today as it was in the days of Roman writers such as Columella and Pliny.

References

Asselin, C., Flanzy, C., Sapis, J.C. and Martin, J.D. (eds) 1996 *Les terroirs viticoles – concept, produit, valorisation.* Angers: Institut National de Recherches Agronomiques.

Bacci, A. 1596 *De naturali vinorum historia, de vinis Italiae...*, Rome: N. Mutii.

Barham, E. 2003 Translating terroir: the global challenge of French AOC labeling, *Journal of Rural Studies* 19(1): 127–38.

Berlow, R.K. 1982 The "disloyal grape": the agrarian crisis of late fourteenth-century Burgundy, *Agricultural History* 56: 426–38.

Bishop, G.C. 1980 *Australian Winemaking: the Roseworthy Influence*. Hawthordene, South Australia: Investigator Press.

Board on Agriculture, National Research Council (1992) *Agriculture and the Undergraduate*. Washington DC: National Academy Press.

Bowers, J.E. and Meredith, C.P. 1997 The parentage of a classic wine grape, Cabernet Sauvignon, *Nature Genetics* 16: 721–24.

Bowers, J.E. and Meredith, C.P. 1999 Historical genetics: the parentage of Chardonnay, Gamay and other wine grapes of northeastern France, *Science* 285: 1562–65.

Boyd, G.D. 1999 The Barossa Valley, *Wine News* February/March (http://www.thewinenews.com/febmar99/barossa.html. accessed 16th September 2009).

Brunhes, J. 1925 Human geography. In *The history and prospects of the social sciences*, ed. H.E. Barnes, 55–105. New York: Alfred A. Knopf.

Cato, M.P. 1933 *On Farming*, trans. E. Brehaut. New York: Columbia University Press.

Chamot, E.M. 1926 Applied science in the provincial universities in France, *The Scientific Monthly* 22(5): 377–99.

Charters, S., Lockshin, L. and Unwin, T. 1999 Consumer responses to wine bottle back labels. *Journal of Wine Research* 10(3): 183–95.

Cion, J. 1908 *Les paysans de la Normandie orientale*. Paris: Colin.

Columella, L.J.M. 1941, 1954, 1955 *On Agriculture*, trans. H.B. Ash, E.S. Forster and E.H. Heffner. London: Heinemann, 3 vols.

Coombe, B.G. and Dry, P.R. (eds) 1988 *Viticulture, Volume 1: Resources in Australia*, Adelaide: Australian Industrial Publishers.

Demangeon, A. 1905 *La plaine picardie*. Paris: Colin.

Dickenson, J. 1990 Viticultural geography: an introduction to the literature in English, *Journal of Wine Research* 1(1): 5–24.

Elliott-Fisk, D. and Noble, A. 1992 Environments in Napa Valley, California and their influence on Cabernet Sauvignon wine flavors. In *Viticulture in Geographical Perspective*, ed. H. De Blij, 45–72. Miami: Miami Geographical Society.

Gergaud, O. and Ginsburgh, V. 2008 Natural endowments, production technologies and the quality of wines in Bordeaux. Does terroir matter, *The Economic Journal* 118: F142-F157.

Gladstones, J. 1992 *Viticulture and Environment*, Adelaide: Wine titles.

Goode, J. 2006 *The Science of Wine: From vine to glass*. Berkeley, CA: University of California Press.

Hinkle, R.P. 1991 Just what is terroir anyway? *Wine & Vines*, March, http://findarticles.com/p/articles/mi_m3488/is_n3_v72/ai_10645037/?tag=content;col1. accessed 16 September 2009.

Humoristes Associés (1980) *Le Vin*. Humoristes Associés, imprimé en Espagne par Tonsa.

Jekel, B. 1982 California wines: the quality factors, *Decanter* 7(12): 67–8.

Jekel, B. 1983a Soil and the taste of wine, *Decanter* 8(5): 59

Jekel, B. 1983b Soil and wine – the debate continues, *Decanter* 8(10): 15

Jones, P. 1966 Italy. In *The Cambridge Economic History of Europe, Volume 1: the agrarian life of the Middle Ages*, ed. M.M. Postan, 340–431. Cambridge: Cambridge University Press.

Lachiver, M. 1988 *Vins, Vignes et Vignerons: histoire des vignobles français*, Paris: Fayard.

Laffer, H.F. 1949 *The Wine Industry of Australia*. Adelaide: Australian Wine Board.

Lapsley, J. 1997 *Bottled Poetry: Napa Winemaking from Prohibition to the Modern Era*. Berkeley, CA: University of California Press.

Lipton, M. 1977 *Why Poor People Stay Poor*. Boston: Harvard University Press

Loubère, L.A. 1990 *The Wine Revolution in France*, Princeton: Princeton University Press.

Marx, K. 1981 *Capital: a critique of political economy, Volume 3*. Harmondsworth: Penguin (first published 1894).

McGovern, P.E. 2003 *Ancient Wine: the search for the origins of viniculture*, Princeton, NJ: Princeton University Press.

Pasteur, L. 1866 *Etude sur le Vin, ses Maladies, Causes qui les Provoquent, Procédés Nouveaux pour le Conserver et pour le Vieiller*. Paris: Imprimerie Impériale.

Pliny 1945–50 *Natural History*, trans. H. Rackham. London: Heinemann.

Pouget, R. 1990 *Histoire de la Lutte Contre le Phylloxéra de la Vigne en France*. Paris: INRA.

Prats, B. 1983 The terroir is important, *Decanter* 8(7): 16.

Robinson, J. (2003) New German classification systems sees sense, http://www.jancisrobinson.com/articles/jr7044.html, accessed 12 October 2009.

Smart, R. and Robinson, M. 1991 *Sunlight into Wine. A Handbook for Winegrape Canopy Management*. Adelaide: Wine titles.

Starbucks 2005 Geography is a flavour, http://www.starbucks.co.uk/en-GB/_Worlds+Best+Coffee/. accessed 16 September 2009.

Strabo 1949–54 *The Geography of Strabo*, trans. H.L. Jones. London: William Heinemann.

Taber, G.M. 2005 *Judgement of Paris: California vs. France and the Historic 1976 Paris Tasting that Revolutionized Wine*. New York: Scribner.

Trubek, A.B. 2008 *The Taste of Place: a cultural journey into terroir*, Berkeley, CA: University of California Press.

UNESCO 2005 *Rencontres Internationales Planete Terroirs UNESCO 2005: Actes*, Montpellier: UNESCO, SC-2007/WS/41.

Unwin, T. 1991 *Wine and the Vine: an historical geography of viticulture and the wine trade*. London: Routledge.

Unwin, T. 1992 *The Place of Geography*. Harlow: Longman.

Varro 1912 *On Farming*, trans. L. Storr-Best. London: G. Bell & Sons.

Vidal de la Blache 1913 Les caractères distinctifs de la géographie, *Annales de Géographie*, 5, 129–42.

Vidal de la Blache 1922 *Principes de géographie humaine*. Paris: Colin

Wilson, J.E. 1998 *Terroir: the role of geology, climate and culture in the making of French wines*, Berkeley, CA: University of California Press.

Winkler, A.J. 1938 The effect of climatic regions, *Wine Review* 6, 14–16, 32.

Winkler, A.J. 1962 *General Viticulture*, Berkeley, CA: University of California Press.

Winkler, A.J. 1973 *Viticultural research at University of California Davis, 1921–1971*. Davis, CA: University of California, Davis.

Winkler, A.J., Cook, J.A., Kliewer, W.M. and Lider, L.A. 1992 *General Viticulture*, 2nd edition, Berkeley, CA: University of California Press.

Zohary, D. 1995 The domestication of the grapevine. In *The Origins and Ancient History of Wine*, eds P.E. McGovern, S.J. Fleming and S.H. Katz, 23–30. Luxembourg: Gordon and Beach.

3 Geography and the American Viticultural Areas Process, Including a Case Study of Lodi, California

Deborah L. Elliott-Fisk

Abstract

The United States has a legal system for the formal designation of wine appellations of origin overseen by the Department of the Treasury. By regulating the labeling of wine to ascertain the area of origin of the wine for the consumer, the Department's Alcohol and Tobacco Tax and Trade Bureau controls place names that can be used on wine labels and, indirectly then, throughout the wine trade. Approved places are known as American Viticultural Areas (AVAs) and are established (or rejected) through a formal public petition, review, comment, and sometimes, a hearing process. The geographical distinctiveness of the place drives its approval as a formal AVA. This is primarily based on physical geographic criteria such as climate, topography, soils, and geology, and also on a function of place history and name as cultural elements of the landscape. Geographers are well engaged in the AVA establishment process and through this, contribute to the education of winegrowers, the trade, the government, and the public on place-based distinctiveness of wines. Over 200 AVAs have been established in the United States since the rule-making process began in 1978, with over 100 of these AVAs in the State of California. Over the last 30 years, the federal agency has become stricter in following its criteria to delimit geographically distinctive areas with historical name recognition, less sympathetic to letting in wine-growers outside the proposed AVA boundaries without cause, and more systematic in how AVAs nest within one another. The best examples of the AVA establishment process now are also about building a sense of community and place through contestation and discourse and the give and take that shapes a community of place. The petition to divide the large Lodi, California American Viticultural Area into seven distinct AVAs is outlined as a case study in this process, where winegrowers and many other community members came together to support the establishment of finer-scale AVAs with the large Lodi AVA.

Introduction

This chapter discusses the United States federal government's public process, as outlined in the Code of Federal Regulations ATF-53, to establish place-based

D.L. Elliott-Fisk (✉)
Graduate Group in Geography, University of California Davis, One Shields Ave., Davis, CA 95616, USA
e-mail: debbieef@wildblue.net

P.H. Dougherty (ed.), *The Geography of Wine: Regions, Terroir and Techniques*,
DOI 10.1007/978-94-007-0464-0_3,

wine appellations carrying the appropriate place names and defined as geographically distinctive. These appellations are called American Viticultural Areas (AVAs). Through time, they have become better defined as both the public, who submit petitions to establish the AVAs, and the federal agency staff gain knowledge to improve the AVA establishment process, thus creating AVAs with meaning viticulturally. The process is also about building knowledge, sharing information, and forming a place-based community with common goals, focused on the commodities of wine and wine grapes, and supporting a diverse, agriculturally based economy that extends from production to consumption, farming to tourism, and the local to the global economy and trade.

Where is my food from? Where was that lovely bottle of wine made? Across the world, it is common to label agricultural products such as cheese, coffee, fruit, and wine as to their place-based area of origin, which may be a country, state, or more local region. Many nations and cultures have practiced the use of place-based area of origin names for centuries. With premium grape wines receiving much scrutiny and acclaim from winemakers, the press, and the public, many countries and some states have established rules, i.e., laws and regulations, that oversee how wines are labeled that derive from those places. Some countries, such as France, go further in regulating agricultural practices within areas of wine origin.

The French *Appellation d'Origine Contrôlée* (AOC) is the most famous of the area of origin regulations and oversees the certification of not only wine but also cheese, butter, and other agricultural products as to their geographic origin. The system is managed by the French Ministry of Agriculture as their Institut National des Appellations d'Origine (INAO), established in 1935. The Côtes du Rhône appellation of origin was the first AOC legally approved in 1937. In this system, not only is geographic origin ascertained but also the varietals of grapes grown in any appellation, maximum yields, alcohol content, and other wine characters are regulated.

In the United States, where private land's rights and choice are so ingrained in the culture of most American farmers and entrepreneurs, movements to regulate agriculture processes are not viewed with much favor. However, misuse or inappropriate use of "famous" place names in the marketing of agricultural products has occurred over the last century in the United States, e.g., use of the name Champagne for American sparkling wines and the Napa Ridge winery making primarily non-Napa Valley wines. This led to the formulation of a specific US federal regulation to oversee the proper use of geographic areas of origin on wine labels and the establishment of an appellation system. This system gives consumers confidence that a wine carrying a certain appellation of origin designation is actually from that particular place. And if they like a particular wine from that place, they might buy another variety of wine from that same place. Appellation of origin systems has important educational and marketing value and can result in economic gain for winegrowers and winemakers, thus making the production of high-quality wine a "win–win system" for wine lovers and producers.

The American Viticultural Areas system is overseen by the US Department of the Treasury, Alcohol and Tobacco Tax and Trade Bureau (TTB; www.ttb.gov). Regulations contained in Title 27, Chapter 1, Part 9 of the Code of Federal Regulations (27 CFR Part 9) oversee the establishment of appellations of origin as American Viticultural Areas. Subpart A of the regulations presents the general provisions of the program, subpart B the formal definitions, and subpart C the approved American Viticultural Areas (AVAs), with the latter consistently updated as new AVAs are approved. As of 2010, 217 AVAs have been approved via a petition and rule-making process, with many applications currently pending. The program began in 1978, with the first AVA approved in 1980 (Augusta in Missouri) and the second and perhaps most internationally renowned (Napa Valley in California) in 1981.

Wine labels – which must be pre-approved by TTB – are an important source of information for consumers. Eighteen regulations under 27 CFR control the information that can be placed on a wine label, with the American Viticultural Areas 27 CFR Part 9 being one of these. Wine labels may contain one or more AVA names and must be approved by the TTB as Certificates of Label Approval (COLAs). For example, the appellation of origin on a wine label is the name for the place in which the dominant grapes used in the wine were grown. It can be the name of an approved political unit (country, state, or county) or a geographical region called a viticultural area. An American Viticultural Area is a defined grape-growing region in the US with geographic features, such as soil and climate that set it apart from the surrounding areas. An American Viticultural Area appellation on the label indicates that 85% or more of the wine was produced from grapes grown in the named area (TTB P 5190.1, 2008).

The US Department of the Treasury defines wine appellations of origin under US regulations (specifically 27 CFR 4.25) as:

- A country.
- A US state or the foreign equivalent.
- For US wine, a listing of up to three states (multi-state appellation).
- A US county or the foreign equivalent.
- For US wine, a listing of up to three counties (multi-county appellation).
- A US or foreign government recognized delimited grape-growing area (referred to as a viticultural area under US regulations).

With a viticultural area appellation of origin, the wine is:

- Made from grapes grown in the area labeled an American Viticultural Area approved under US regulations (specifically 27 CFR Part 9).
- Not less than 85% of the volume of the wine is derived from grapes grown in the labeled American Viticultural Area.
- The wine is fully finished (except for cellar treatment and/or blending which does not alter the class and type of the wine) in the state or one of the states where the viticultural area is located.

In this regulatory process, a viticultural area is defined as a delimited, grape-growing region distinguishable by geographical features, the boundaries of which have been delineated in subpart C during the process. There are no size limits, either small or large, to these areas, and as such they vary greatly in size from 150 acres to 16 million acres. The largest AVA at over 16 million acres (16,640,000 ac) is the Ohio River Valley located in four states: Indiana, Kentucky, Ohio, and West Virginia. The smallest AVA is the Cole Ranch at 150 acres in Northern California. Many of the AVAs are between 10,000 and 50,000 acres in size. While the size is quite variable, it, along with the function of the landscape, the area planted to wine grapes, and the knowledge of experts, including the winemakers and winegrowers, makes up the geographical factors that the region conveys to wine and grape distinctiveness.

Along with this, smaller AVAs may be nested inside larger AVAs, and in the past, the Department has approved AVAs which actually overlap. Flaws like the latter are being addressed petition by petition and through requests to consider changing the code.

The Process to Establish a New American Viticultural Area

Although the American Viticultural Areas system is based on law, policy, and regulation, anyone may present a written petition to the Treasury, Alcohol and Tobacco Tax and Trade Bureau (TTB) to establish a new American Viticultural Area (AVA). Once the properly prepared petition is submitted, it is reviewed by the TTB staff and administrators for completeness, and it may or may not go back to the petitioners for further information or clarification. Once TTB is satisfied that the petition meets all criteria and presents a reasonable argument for the uniqueness of the new AVA, they publish a Notice of Proposed Rule-Making (NPRM) in the *Federal Register*. This gives all members of both the wine trade and the public a chance to read the condensed petition, as summarized by TTB staff, and to provide comments to assist with the decision-making process. Comments either in support of or against the petition, are requested by a given date, can be made via the internet or postal services. The TTB staff and administrators review comments and then decide to (a) no longer consider the petition, (b) request additional information from the petitioners, or (c) approve the AVA petition via a Final Rule published in the *Federal Register*.

As this is a public process, it can take considerable time to go through the procedure to establish a new AVA. Whoever submits the petition has to have geographically defined and bounded the area and put together historical and contemporary research that supports the new AVAs name, geographical distinctiveness, and boundaries. Cartography and GIS play key roles in accurately mapping and describing the proposed boundaries and in compiling map-based data about the area both inside and outside of the proposed AVA. The initial research and involvement of winemakers, grape growers, other industry representatives, political leaders, and the public in the petition typically take from 1 to 3 years, with the review, comment, and approval taking another 1–4 years. Although early in the process (1980s), petitions may have been in the form of a two- to four-page letter with the required US Geological Survey map depicting the proposed AVA boundaries, e.g., the petition for the Howell Mountain, California AVA approved in 1983, more recent petitions are replete with scientific data, historical documents, evidence of name maps and related sources, and letters of support.

Considerable expertise, time, and money are required to gain approval of a new AVA under almost all circumstances today. There is inherent tension in a process where one party may gain financially in being allowed legal use of an appellation of origin name while someone across the road, river, or ridge who is outside the boundary of the AVA may lose financially or not benefit from that marketing tool. The process poses, then, the potential to build community and community-capacity or to fracture the community during the public petition and comment process.

The Role of Geography

Geography plays a key role in the establishment of new American Viticultural Areas, as each written petition requesting establishment of a new AVA must present clear information showing:

1. Evidence that the name of the viticultural area is locally and/or nationally known as referring to the area specified in the application.
2. Historical or current evidence that the boundaries of the viticultural area are as specified in the application.
3. Evidence relating to the geographical features, such as climate, soil, elevation, physical features, etc., which distinguish the viticultural features of the proposed area from surrounding areas.
4. The specific boundaries of the viticultural area, based on features which can be found on United States Geological Survey (USGS) maps of the largest applicable scale.
5. A copy of the appropriate USGS map(s) with the boundaries prominently marked.

[T.D. ATF–60, 44 FR 56692, Oct. 2, 1979, as amended by T.D. ATF–92, 46 FR 46913, Sept. 23, 1981; T.D. ATF–355, 59 FR 14553, Mar. 29, 1994; T.D. ATF–432, 65 FR 69253, Nov. 16, 2000].

Although an appropriate name for the place or region proposed as the new AVA must be requested, the geographical features of the place are key in establishing it, differentiating it from others areas, and in drawing appropriate boundaries. Many battles have been fought by neighbors for years over where the appropriate boundaries should be and who is in versus out. This is one of the most challenging parts of the geographical sciences used in designating AVAs, as boundaries are difficult to draw for features related to weather and climate but better demarcated where clear soil, geomorphological, and geological changes occur on the landscape. However, petitioners typically avoid splitting properties or legal parcels when drawing boundaries and use the nearest road, railroad, river, topographic contour line, or other linear feature to approximate the boundary.

Geographical Distinctiveness

Winegrowers, as all farmers, know their land and its capability for crops or in this case, different wine varietals, e.g., Pinot Noir versus Chardonnay versus Cabernet Sauvignon. Different vineyard blocks or parcels with different local climates, topography, and soils will be more or less optimal for different wine varietals, with grape yield and character also influenced by farming practices such as irrigation amounts, pruning regimes, trellis system, etc. There is a general belief now in the production of high-quality wines that the best wines are a direct expression of the vineyard or *terroir*, the place where the grapes come from, with as little manipulation in the winery as practical. As such, if a winegrower's property has various *terroirs* on it, with different soils, depth to bedrock, duration of cooling evening breezes, drainage, etc., the grower will plant, manage, and harvest these differently. Wines are produced and labeled perhaps to specific vineyards that are distinctive, but most often to viticultural area. Some people in the wine trade express this as "value added" to the product, with both grapes and wines from well-recognized American Viticultural Areas, and especially for specific varietals within those areas, selling for more money, and being collected and consumed by wine aficionados.

Yet, despite differences in how the grapes are grown, clear expressions of similarity in wine characters, such as flavor, aroma, tannins, phenols, etc., are seen in the wines across regions of various sizes. Winemakers and wine tasters have known this for decades, with well-made wines expressive of the vineyard and identifiable to it. These individuals will tell you that they can recognize a Cabernet Sauvignon from the AVAs of the Stag's Leap District versus Rutherford versus St. Helena in the Napa Valley and differentiate Chardonnay from the AVAs of Carneros (Napa–Sonoma) and one from the

Oak Knoll District or Howell Mountain in the Napa Valley. These AVAs were established as the winemakers and winegrowers recognized these differences and submitted and defended, as necessary, their petitions to the TTB. Research by Noble and Elliott-Fisk (1990), Elliott-Fisk et al. (1992) shows these wine distinctions for Cabernet Sauvignon from various AVAs in the Napa Valley and how this relates to terroir and specifically to soils (Elliott-Fisk 1993).

From the time of the establishment of the American Viticultural Areas program in 1978, winemakers and winegrowers have worked to gain approval of what many growers believed were clearly distinctive growing regions, namely, Los Carneros, Howell Mountain, Sonoma Valley, Napa Valley, Lodi, Paso Robles, etc. With grapes continuing to be planted across some of these regions, and growers and winemakers gaining more and more knowledge of their fruit and its expression in the wine, a movement took hold to create smaller, more distinctive viticultural areas within some of the larger areas. These discussions started in the 1980s and have resulted in the Napa Valley AVA, approved in 1981 and comprised of 225,280 acres, now having 15 AVAs and two others partially within it. Petitions are still pending and will be written in the future undoubtedly for additional AVAs within the Napa Valley AVA.

In 2006, TTB approved establishing seven new AVAs totally within the Lodi AVA, which was approved in 1986 and is made up of 450,000 acres. In 2007, petitioners submitted AVA petitions to establish 11 new AVAs within the Paso Robles AVA which was originally approved in 1983 and contains 666,618 acres. The author participated as the geographer, helping to establish AVA boundaries, characterizing their geographical distinctiveness, and writing parts of the petitions for the new Paso Robles and Lodi AVAs, as well as for more than half of the Napa Valley AVAs, and doing research since 1986 across these and other viticultural areas of California. The process used to identify and establish the seven new AVAs with the large Lodi AVA is presented as a case study, lending details and insights into the AVA designation procedure.

An Example: Lodi, California and Seven New AVAs Within It

In 2001, a small group of winegrowers and winemakers in Lodi, California, approached the author to gain her insights and expertise on establishing a new AVA within the large Lodi AVA around the town of Clements in the lower foothills to the east of Lodi. In meetings to review a series of maps and the process, then visiting various vineyards in the field, it became apparent that this group believed that there were several distinct areas within the region that could have appellation status. We undertook research across the region and decided to pursue an ambitious, but cost effective, effort to petition TTB for seven new AVAs. A project like this had never been presented to TTB before, instead with "sub-AVAs" being submitted one by one, usually by different parties, across a decade or more.

Thus, in 2002, after more than a year of discourse with the public in open meetings and much thought, the Lodi American Viticultural Areas Steering Committee, where the author was the lead scientist, submitted a petition, proposing the designation of seven regions as new American Viticultural Areas. They are fully within and inclusive of the entire established Lodi American Viticultural Area, with no overlapping boundaries or omissions. These AVAs are within both Sacramento and San Joaquin Counties, California, as is the Lodi AVA.

These proposed AVAs can be seen on the map in Fig. 3.1 and are the:

1. Cosumnes River – approximately 54,700 acres
2. Alta Mesa – approximately 55,400 acres
3. Sloughhouse (original name of Deer Creek Hills replaced as the best name during the process) – approximately 78,800 acres
4. Borden Ranch – approximately 70,000 acres
5. Jahant – approximately 28,000 acres
6. Clements Hills – approximately 85,400 acres
7. Mokelumne River – approximately 85,700 acres

It should also be noted that two other petitions were in the drafting stages by individuals who were not members of the Steering Committee to both expand the existing Lodi AVA to the west of the Delta and to the south of the Calaveras River into Linden to Farmington, east of Stockton. The two proposed expansion areas were not part of the approved Lodi AVA in 2002 and thus were not part of this AVA petition process to establish the seven new AVAs.

Lodi is a region with a very rich agricultural history Martin (1904). Wine grapes have been grown, and wine produced here for well over 150 years. Many changes have occurred through time in the grape

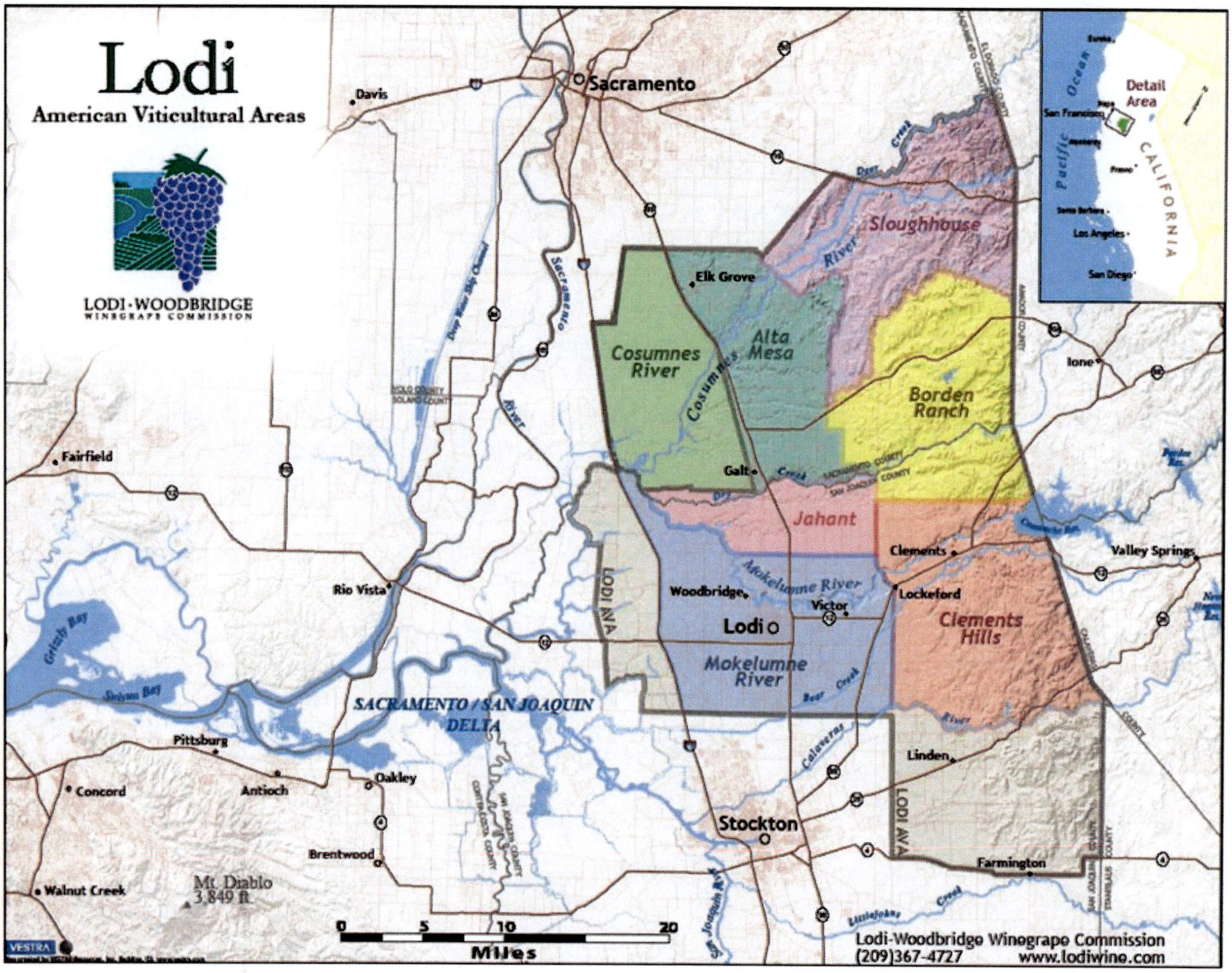

Fig. 3.1 The seven new appellations resulting from the division of the Lodi AVA (Source: Lodi-Woodbridge Winegrape Commission)

varietals planted, the style of wines made, and in both viticultural and winemaking practices. Through decades of farming, knowledge of site potential and the diversity of sites within the Lodi AVA have grown, especially since the Lodi AVA was established in 1986 and with the organization of the Lodi–Woodbridge Wine Grape Commission in 1991. Winemakers across California had recognized the Lodi appellation as an excellent source of high-quality fruit for wines carrying the "California" label, for Lodi AVA wines, and for select vineyard designates. Definition of smaller viticultural areas with geographically distinctive landforms, soils, climates, and topographies contributing to *terroir* would provide more accurate information to consumers on grape origin, which we know has long been recognized in both the literature and in tasting as a key control of the characteristics of wine.

It was the LAVA Steering Committee's belief that the designation of smaller, geographically distinct viticultural areas within the large Lodi AVA would:

1. Further promote the grapes and wines of the Lodi viticultural area, the new and smaller viticultural areas within it, and the efforts of all groups.
2. Provide a better understanding of the viticultural geography of the Lodi appellation and the new AVAs, aiding winegrowers and winemakers in their selection of vineyard sites, grapes, viticultural practices, and winemaking techniques.
3. Provide the grape buyer, consumer, and media with more information as to the diversity and distinctiveness of the viticultural areas, their grapes, and wines.
4. Follow the successful history and approach of other viticultural regions, e.g., Sonoma Coast, Napa Valley, Sierra Foothills, in defining and promoting

small viticultural areas that produce distinctive fruit and wines.

This vision was sold to the public, other winegrowers who were not part of the LAVA Steering Committee, and the Lodi–Woodbridge Wine Grape Commission, who were part of the discussions but not part of the LAVA Steering Committee, as the Commission represented a very large constituency of winegrowers.

Lodi is, from a physical geographic perspective, a fascinating part of California. With the low relief of the landscape in this valley floor setting, across distances exceeding 25 miles as the vista, the viticultural diversity of the region is best recognized by the winegrowers themselves who are cognizant of important differences in local climates, surface soils, and subsurface sediments. They recognize that the differences strongly influence the performance of rootstocks, clones, yields, and fruit characteristics.

Scale is important here. What appear to be very subtle environmental changes across long gradients to the casual observer are important, local, or site specific conditions for wine grapes. It is this diversity and distinctiveness that contributes unique character to Lodi wines across the region and the several unique viticultural regions which are now being showcased to increase the reputation and consumer recognition of the area.

Like the Napa Valley floor, the Lodi region is composed of a set of geologically old to young alluvial fans with the incursion of bay/delta deposits on its western fringe and the underlying bedrock of the Sierra Nevada to the northeast (Jennings 1977), with distinct climatic gradients across the region. Soils, as a function of geology, climate, and time, are a strong influence on vineyard site potential. The alluvial geomorphic surfaces in the Lodi region are much larger than those for the Napa Valley, Alexander Valley, or other valleys of the Coast Ranges of California due to the sheer size of the Sierra Nevada mountain range and its large watersheds from which the alluvial fans issue. The surfaces represent old Pleistocene, glacial age alluvial fans, and river terraces with the older soils on higher river terraces and fan sand buried at depth by more recent deposits, especially to the west.

Geographically, the Lodi area is characterized by a cooler, maritime climate with smaller daily and monthly temperature ranges than to the north (Sacramento), the south (Stockton), or east (Camp Pardee east of Clements Hills and Folsom Dam east of Sloughhouse). This has long been recognized by botanists and agronomists as relevant to both the native vegetation of the region and the regional potential for agriculture. With the passage of winds off the Pacific Ocean through the Golden Gate topographic gap which continues up the Carquinez Straits, these coastal winds encounter the landmass at Lodi – effectively, the toe of the Mokelumne River alluvial fan – and decrease in intensity (speed) and duration (time) toward the east due to friction with the ground as elevations increase. With low temperatures and elevations, precipitation in the Lodi AVA is the least in the lower Mokelumne River near the town of Lodi, e.g., in the southwest corner of the Lodi AVA and the proposed Mokelumne River and Jahant AVAs, with precipitation increasing upslope, in particular from southwest to northeast, associated with the increase in elevation, thereby creating a gradient in local climates. The climate of the entire Lodi AVA using the global-scale climate classification system of Koppen (see Geiger and Pohl 1953) is Mediterranean warm summer (Csa), although the town of Lodi itself is almost a Mediterranean cool summer climate (Csb). [Note: Climate classification systems are really based on large geographical regions for the globe, similar to the classification of soils to orders for the globe.] Climates are not classified to the fine level, e.g., series, as are soils.

Distinct climatic gradients exist across the large Lodi AVA (Climatological Data, National Climate Data Center, NOAA). The combination of climatological parameters, e.g., temperature, precipitation, wind, and evaporation, provides the basis for distinctive local or topographic climates that characterize each of our proposed seven AVAs (Climatological Data, Western Regional Climate Center, Desert Research Institute, University of Nevada, Reno). As shown by using the standard climate parameters of temperature, precipitation, growing degree days, actual evaporation, and wind, and as depicted graphically from the City of Lodi both inland to the east and north toward Sacramento, the annual precipitation increases; the mean annual temperature increases; the mean seasonal temperatures for winter, spring, summer, and fall all increase; the growing degree days over a base of 50°F, the standard used in viticulture, (See Amerine and Winkler 1944) increase, yet evaporation and wind duration/speed decrease. What these data show is that the city of Lodi and the Mokelumne River AVA is cooler and windier than the areas to the north – Cosumnes River AVA, Alta Mesa AVA, Sloughhouse AVA, and Sacramento and Folsom – and to the east – Borden Ranch AVA, Clements Hills AVA, Camp

Pardee, and the Sierra Foothills AVA. The climate of the Jahant AVA is similar to that of the Mokelumne River AVA where the city of Lodi is located.

The region is distinguished geologically by its tectonic history and complex of distinctive alluvial fans, with the current surface landforms spanning the last 1 Ma (million years; see Pavich et al. 1986). The region links the Sierra Nevada and the valley floor, with isostatic and tectonic processes leading to continued deposition by rivers to the south of the current river channels. The rivers themselves serve as good boundaries for viticultural areas. It is also evident that the subsurface environment for the vines is even more diverse than the surface climates, with this "*terroir*" significant to grape composition and wine flavors and aromas. Although some soils are very old as either surface or buried paleosols that formed in the past under different climates, such as the San Joaquin series, most soil profiles are largely immature, as the active erosion and alluviation (deposit) by streams has controlled soil profile development.

Diagnostic characteristics for soils in the region which attest to their various ages are (a) a dark mollic epipedon at the surface, with organic material accumulating and decomposing through time, (b) cambic and argillic B horizons where clays have formed as secondary minerals with pedogenesis and accumulated over time, eventually leading to the formation of a claypan, and (c) the formation of duripans, with soil horizons cemented by silica through weathering of granitic minerals and volcanic ash (Soil Conservation Service 1992, 1993). Pedogenic accumulation of clays and reddening, both denoting leaching and time, are seen in the San Joaquin and Redding series where land surfaces are more stable.

Soil pH is near neutral for most soils, and there are no strong chemical imbalances in the soils except for the very oldest soils on the eastern margin. Viticultural challenges are mostly due to impervious subsoils and in places, to very fine, sandy, permeable surface horizons. The Storie Index (Huntington 1992), which is used to rate soil potential for intensive agricultural uses based solely on the soil characteristics (from 100 highly suitable to 0 unsuitable), shows the diversity of soil types for viticulture in the Lodi AVA, with the Tokay/Hanford soils in the proposed Mokelumne River AVA near the town of Lodi having the highest rating (80–95) and being the location where grapes were first grown. The San Joaquin series and Jahant clay loams and loams are next with ratings around 25–40 – Cosumnes River, Alta Mesa, and parts of the Sloughhouse and Jahant AVAs – and then, the other more upland soils of the Sloughhouse, Borden Ranch, and Clements Hills AVA are rated as moderate to low suitability, 15–30 (Cosby and Carpenter 1937; Huntington 1992), which often means they are excellent for contemporary viticulture where reduced yields often lead to enhanced quality.

From both geographical and viticultural viewpoints, it is thus apparent that there are large differences seen by winegrowers from north to south and west to east across the 458,000-acre Lodi viticultural area, from the deep, sandy loam soils around cool Lodi, to the rocky, clay rich soils in the sunny hills above Clements, to the moderate sites on clay loam soils north near Galt. This has resulted in new varietals being planted in the last several years by subregion, for example, more white varietals in the cooler, western areas and red varietals in the warmer, eastern areas, careful rootstock selection by site/soil, new canopy management, and trellising techniques, and all the other contemporary cultural practices that are dependent on the *terroir*–varietal interaction. It should also be noted here that the Lodi growers are well recognized in California and elsewhere for their Lodi Rules for Sustainable Winegrowing program, under the leadership of the Lodi–Woodbridge Wine Grape Commission (www.lodiwine.com).

The 458,000-acre Lodi American Viticultural Area is characterized by:

1. A relatively maritime climate for its inland location in the Central Valley.
2. By alluvial valley floor soils derived from Sierra Nevada glaciofluvial and river deposits.
3. By a long history of grape-growing and wine production.

This large region has considerable geographical and historical diversity, supporting the establishment of seven new American Viticultural Areas. These proposed new AVAs are the:

1. Cosumnes River
2. Alta Mesa
3. Sloughhouse
4. Borden Ranch
5. Jahant
6. Clements Hills
7. Mokelumne River

The names of these AVAs are derived from well-recognized, locally significant, geographic place names and historical family names as reflected in the communities, roads, and ranches of the region. The elevations, climate, as reflected by growing degree days, temperature seasonality, annual precipitation, duration of fog, persistence of wind, and evapotranspiration demands, soil capabilities for viticulture, geomorphic surface type and age, soil types, and viticultural conditions are quite diverse. The framework of fully subdividing, with no exclusions or omissions, the Lodi AVA into these distinctive viticultural areas aids consumers, winemakers and viticulturalists in their pursuit of crafting and enjoying fine wines from this historic grape-growing region of California.

Following the publication of the Notice of Proposed Rule-making (Notice 50, 27 CFR Part 9, TTB) for the seven AVAs in the *Federal Register* (www.gpoaccess.gov/fr/) in 2005, there was overwhelming support from the winegrowers, winemakers, and local citizens to approve these new AVAs. TTB established the seven new AVAs in 2006, as published in the *Federal Register* (Treasury Decision TTB-50). The concept and effort was grasped and supported by the community, resulting in increased wine sales and visitation to the Lodi region. It has also intrigued other winegrowers and winemakers on how they may go about a similar effort to establish multiple AVAs supported by the local community in their region.

The Future

Undoubtedly, many more AVAs will be petitioned for by the winegrowing community over the next decade and beyond. TTB has sought input from the public the last several years on more detailed guidelines for what needs to be included in petitions to establish new AVAs (TTB 2007, Draft AVA Manual for Petitioners), and whether certain aspects of the process should be changed. In 2007, they proposed a revision of the AVA regulations (Notice No. 78 [Proposed Revision of American Viticultural Area Regulations (2006R-325P)]), which is still undergoing review. Hundreds of comments, mostly in opposition to changes in the regulations, were received by the TTB in 2007–2008. No ruling on the process has been made as of June, 2010.

For geographers, viticultural areas are fascinating to both define and study. They require detailed field sampling and original data collection and analysis, analysis of map/GIS/remote sensing data, discussions with the winegrowers and winemakers, careful definition of scientifically supported area boundaries, and accurate description of the boundaries. You cannot define and describe an AVA and what makes it geographically distinctive without doing this detailed, intimate work. And of course, getting to know the people, the place, and their wine is a great benefit, as is working to build the community and the economic-base of this place-based industry. It is all about a sense of place and an awareness of place – and as we state in geography – place matters!

References

Amerine, M.A., and A.J. Winkler. 1944. Composition and quality of musts and wines of California grapes. *Hilgardia* (15) 493–673.

Climatological Data, National Climate Data Center, NOAA.

Climatological Data, Western Regional Climate Center, Desert Research Institute, University of Nevada, Reno.

Cosby, S.W., and E.J. Carpenter. 1937. Soil Survey of the Lodi Area, California. USDA, Bureau of Chemistry and Soils, Series 1932, No. 14.

Elliott-Fisk, Deborah L. 1993. Viticultural Soils of California, with Special Reference to the Napa Valley. *Journal of Wine Research* 4 (2): 67–77.

Elliott-Fisk, Deborah L., and Ann C. Noble. 1992. The Diversity of Soils and Environments in Napa Valley, California and Their Influence on Cabernet Sauvignon Wine Flavors. pp. 45–71 In: H. J. de Blij ed. Viticulture in Geographic Perspective: Proceedings of the 1991 Miami AAG Symposium. Coral Gables: Miami Geographical Society. 111 pp.

Geiger, R., and W. Pohl. 1953. Revision of the Koppen-Geiger Klimakarte der Erde Erdkunde, Vol. 8, pp. 58–61.

Huntington, G. L. 1992. Storie Index Rating. pp. 157–159, in: Soil Conservation Service. 1992. Soil Survey of San Joaquin County. USDA, Soil Conservation Service.

Jennings, C.W. 1977. Geologic Map of California. California Division of Mines and Geology, Sacramento.

Martin, A.A. 1904. The Delta Lands of California: the Environs of Lodi, Woodbridge, Acampo and Lockeford, their Resources, Climate, Advantages and Opportunities; facts for the homeseeker, farmer, vineyardists, orchardist or investor. Realty Co., Lodi. [Printed by the Lodi Herald.]

Noble, A. C., and D.L. Elliott-Fisk. 1990. Evaluation of the effects of soil and other geographical parameters on wine composition and flavor: Napa Valley, California. pp. 37–45. Actualities Oenologiques 89, 4 Symposium International d'Oenologie. Paris: Dunod. 567 pp.

Pavich, M.J., L. Brown, J. Harden, J. Klein, and R.A. Middleton. 1986. 10Be distribution in soils from Merced River terraces, California. *Geochemica et Cosmochimica Acta* 50: 1727–1735.

Soil Conservation Service. 1992. Soil Survey of San Joaquin County. USDA, Soil Conservation Service.

Soil Conservation Service. 1993. Soil Survey of Sacramento County. USDA, Soil Conservation Service.

TTB. 2007. American Viticultural Areas Manual for Petitioners (draft 2). 28 pp.

Part II

Regional

4 Bordeaux and Burgundy: A Comparison of Two French Wine Regions in Transition

Denyse Lemaire and David Kasserman

Abstract

The vineyards of Burgundy and Bordeaux constitute one of the greatest cultural and culinary assets of France, producing some of the finest wines in the world. Exquisite Gevrey-Chambertin, Aloxe-Corton, Pouilly-Fuissé, or Nuits-Saint-Georges from Burgundy, and Petrus from Pomerol, Lafite-Rothschild of Pauillac, Léoville-Las-Cases of Saint-Julien, Château Cheval Blanc of Saint-Emilion, or other examples from Bordeaux are the envy of vintners from every other region and often the standard by which their products are judged. This predominance is partly the fortuitous product of a remarkable combination of climate and soil that maximizes the regions' wine grape production, but it is also derived from a wine-producing technology that is as old as the Middle Ages and as young as the computer age. By reviewing and contrasting both the physical and the cultural *terroir* of Burgundy and Bordeaux, this chapter provides a glimpse into the complex interplay of natural resource and creative genius that has made these regions into two of the world's greatest centers of fine wine production.

Introduction

France produced 42.5 million hl of wine in 2008, which represents a decrease of about 10% from the average production during the period of 2003–2007 (Production du vignoble 2009). Though second to Italy in total production, France remains the leader in terms of revenue from the export of wine, earning 9.4 billion Euros in 2008 compared to Italy's 4.4 billion in the same period (Annequin 2009). This exceptional success is primarily due to the export of highly valued wines from Bordeaux, Burgundy, and Champagne. The wines are the product of a complex interaction of physical and human *terroir* and are best understood through an analysis of both.

Recognizing the complexity of *terroir*, the *Institut National des Appellations d'Origine* (INAO or French Institute of Appellations and Origin) has both described and regulated the quality and value of all French wines through a strict system of categorization. Created in 1935, the *Institut* traditionally separated French wine into three categories: the *Appellation d'Origine Contrôlée* (AOC) or the Appellation of Origin, *Vins Délimités de Qualité Supérieure* (VDQS) or Quality Wines, and *Vins de Table* or Table Wines.

The AOC refers to a wine originating from a specific area or region of France and guarantees the consumer

D. Lemaire (✉) • D. Kasserman
Department of Anthropology and Geography, Rowan University, Glassboro, NJ 08028, USA
e-mail: Lemaire@rowan.edu; Kasserman@rowan.edu

P.H. Dougherty (ed.), *The Geography of Wine: Regions, Terroir and Techniques*,
DOI 10.1007/978-94-007-0464-0_4,

the authenticity of the region, the variety of grapevines used, the vinification process, the maximum yield per hectare, and a defined degree of alcohol. Bordeaux is the largest AOC wine-growing area of France with 118,900 ha in 2008. Burgundy ranks sixth in France in the AOC with 27,600 ha (Conseil Interprofessionnel du Vin de Bordeaux 2009, 2).

The second category, VDQS, is barely less stringent than the AOC and generally focuses on the same issues of production area, maximum yield in hectoliters per unit area, percent of alcohol, and typical vinification. Less than 1% of the wine produced in France falls in this category, usually originating in vineyards intending to apply for, or vineyards that have applied and are waiting to be granted, the AOC status.

The lowest category is *Vins de Table* or table wine. Beginning in 1973, and formally adopted in 1979, a large proportion of this table wine, because it originates in a well-defined geographic location, has been placed in a superior subcategory called *Vins de Pays* (country wines). These wines are divided into 13 regions and are sold under a regional appellation as well. These *Vins de Pays* reflect the variety of the French vineyards. They are subdivided into 93 *Vins de Pays*, to which wines of départements (the administrative divisions of France) are added to represent a hefty 150 different *Vins de Pays*. The area of each *Vins de Pays* is therefore a tiny fraction of the country of France (Vins de Pays 2006).

In 1999, the European Union began reorganizing the classification, production, and vinification practices in the 27 countries that are members of the union. Both AOC and VDQS still appear on labels, but since July 24, 2009, these two categories have been officially combined as *Vins Délimités de Qualité Supérieure* (VQPRD). The final step of this reorganization became official on August 1, 2009, after which all previously AOC wines must be recategorized as AOP for *Appellations d' Origine Protégée* or Appellations of Protected Origin. Most surviving table wines, if they come from a well-defined geographic region, are now labeled *Indication Géographique Protégée* (IGP) (Guide du demandeur d'une appellation d'origine 2009) (Fig. 4.1).

The region of Bordeaux, with its 144,333 ha, encompasses more than the vineyards classified as the AOC category (118,900 ha). Approximately one seventh of its vineyards planted produce the lower valued quality wines classified as table wine and country wine (Fig. 4.2).

Wine produced in the Bordeaux region, but outside of its AOC areas, may be labeled as *Vin de Pays de l'Atlantique* when it is produced in the départements of Charente-Maritime, Charente, Dordogne, Gironde, or Lot-et-Garonne. It may also be labeled as *Vin de Pays Charentais* if it is produced to the northeast of the Gironde estuary, *Vin de Pays du Périgord* if it is produced in Dordogne, or *Vin de Pays de l'Agenais* if it comes from the western portion of the département of Lot-et-Garonne.

In Burgundy, there are two *Vins de Pays,* labeled *Vins de Pays des Côteaux de l'Auxois* and *Vins de Pays de Sainte Marie La Blanche*. Wines from the Burgundy region may also be labeled by the départements in which they are produced: Saône-et-Loire and Côte-d'Or.

The five regions of Burgundy are, from north to south, the Chablis/Grand Auxerrois/Châtillonnais, Côte de Nuits, Côte de Beaune, Côte Chalonnaise, and Mâconnais. The Mâconnais and the Chablis/Grand Auxerrois/Châtillonnais each represent about 25% of the Burgundy vineyards, while the Côte de Beaune has 21%, the Côte Chalonnaise 16%, and the famous Côte de Nuits and Hautes Côtes de Nuits 14% of the area. The Chablis/Grand Auxerrois/Châtillonnais is west of the *Côtes* in a landscape of rolling hills where east- and southeast-facing slopes are planted in vines (Fig. 4.3).

The AOC of Burgundy covers 27,600 ha (over 68,000 acres) and produced about 193,000 bottles in 2008. There are 100 AOC in Burgundy. The AOC is subdivided into four categories. The lowest is *appellations régionales* which are scattered in the whole Burgundy region. They are allowed to place the word *Bourgogne* on their label. There are 23 of them, accounting for 52% of the Burgundy wine production in 2008. There are 44 *appellations communales* producing 36.6% of the Burgundy wine in 2008. The next higher category comprises the *appellations premier cru* (first growth appellations). The 635 first growth appellations produce 10% of the total production. Finally, the best of what Burgundy produces is labeled *appellations grand cru.* Only 33 *appellations grand cru* exist, and their production only represents 1.4% of the wine produced in Burgundy (Pitiot and Servant 2008, 1–3).

Nationwide, the last 10 years have seen a slight decrease in the area planted in vines while at the same time the average size of vineyards has increased. Both trends are encouraged by the European Union's desire

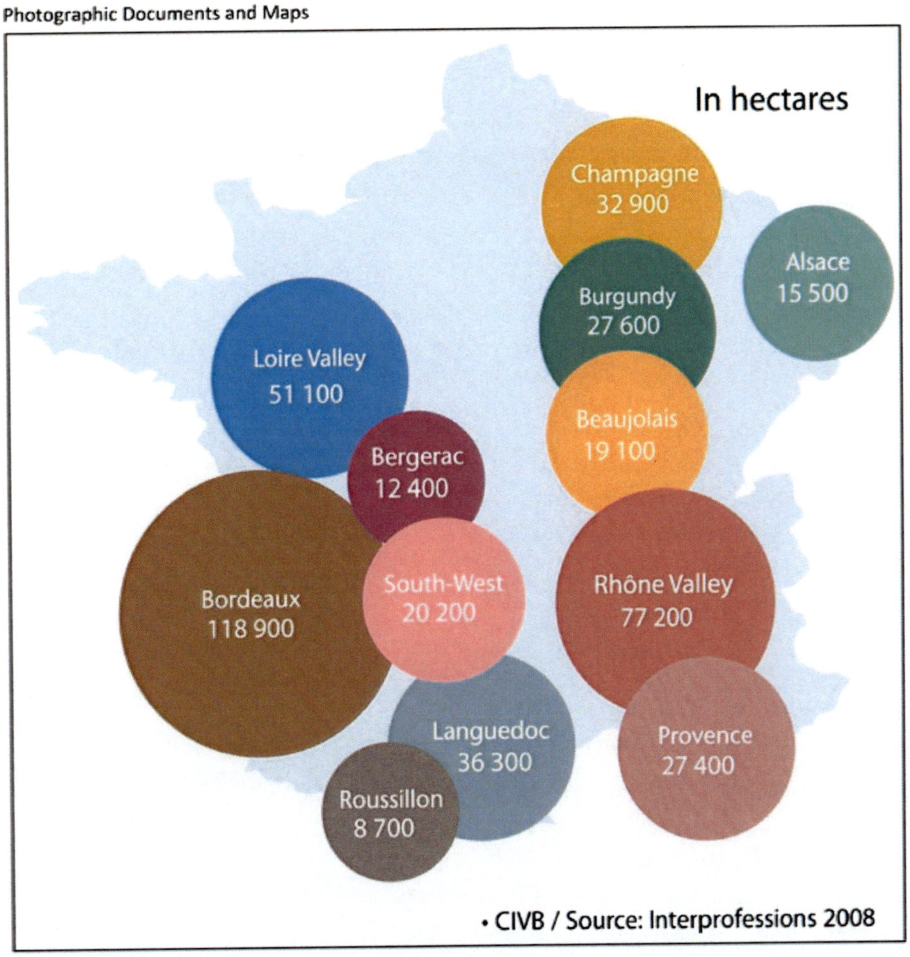

Fig. 4.1 Appellations of origin in France (AOC) with their areas of production in 2008 (in hectares)

to see low-quality vines removed and land use patterns restructured for greater efficiency. From 1997 to 2007, the area planted in vines in France lost 48,800 ha (over 120,000 acres). These changes predominantly affected the table-wine-producing regions. During this same period, Bordeaux and Burgundy actually experienced an expansion of 4,000 and 1,900 ha respectively. In the Burgundy region, the area of production has doubled between 1981 and 2008. In Bordeaux, the same period saw a decrease in the production of table wine from 20% to 1% (similar to the national trend), a decrease of the production of the white wine AOC from 22% to 11%, and an increase of the red wine AOC from 58% to 88% (Eléments clés de la Bourgogne viticole 2009) (Fig. 4.4).

Although Bordeaux and Burgundy are both AOC regions, their climates differ slightly because of their variation in latitude, altitude, and their distance from the moderating influence of the Atlantic Ocean. The northernmost extent of the Bordeaux region is located in *Médoc* at a latitude of 45°30′N while the southernmost point of the region is situated about one degree of latitude to the south at 44°35′N. The vast majority of the Bordeaux region is located in the department of Gironde where the city of Bordeaux is located. Because of its location, the climate of the area is temperate. Protected from the Atlantic winds by the forest of the *Landes* region paralleling the coast of the Atlantic Ocean, the area is still under the influence of the Gulf Stream, the ocean current that brings warm water from the Caribbean region to Western Europe. Winters are mild; springs are humid and seldom bring late frost. Summers are dry and sunny.

The total insolation (unobstructed sunlight) may vary substantially from one year to the next. The minimum number of hours of insolation is about 1,800 h per year, as occurred in 2000. However, it can rise to up to 2,400 h, as in 2003, which produced an excep-

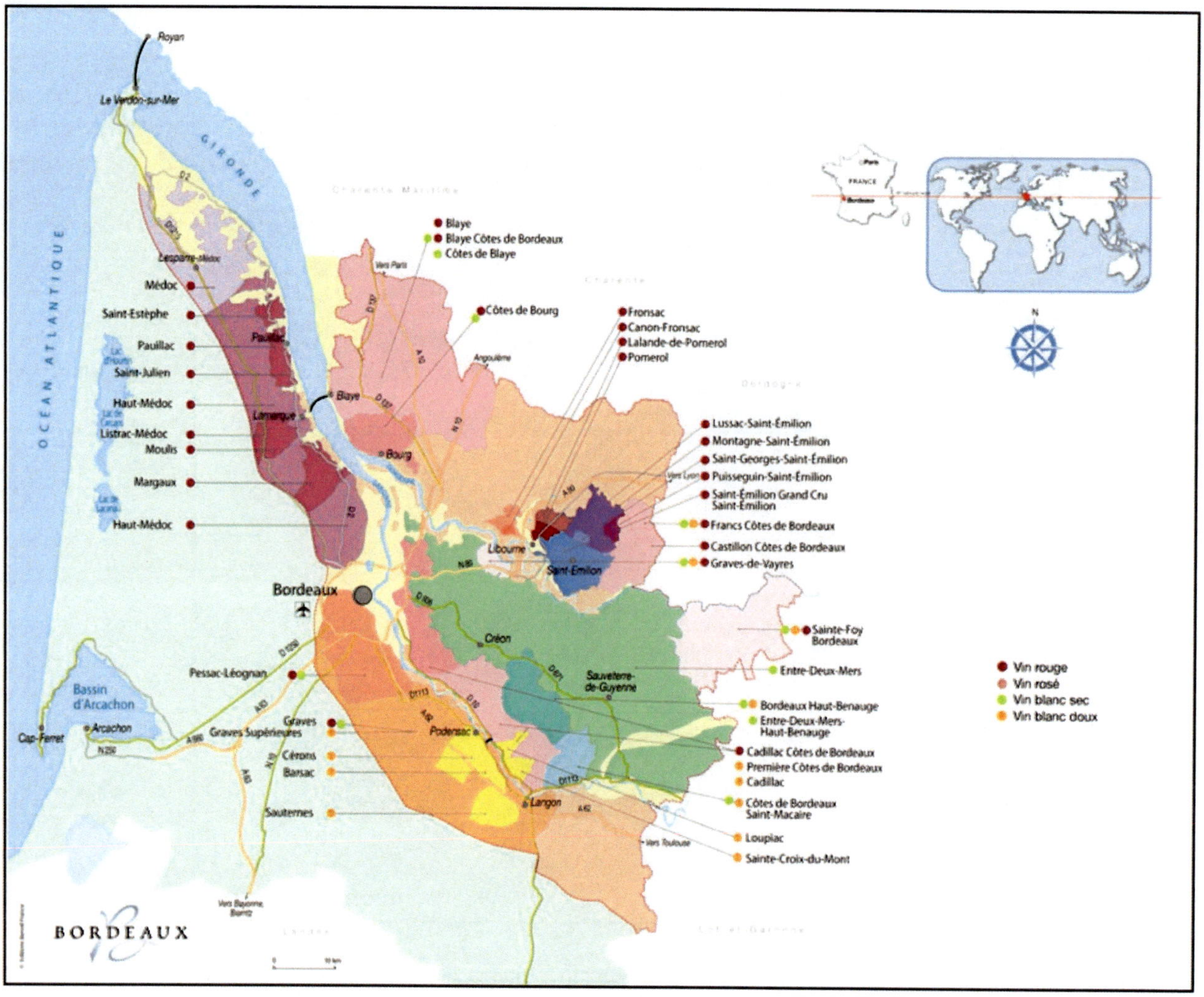

Fig. 4.2 The vineyards of Bordeaux (Source: CIVB Bordeaux 2008)

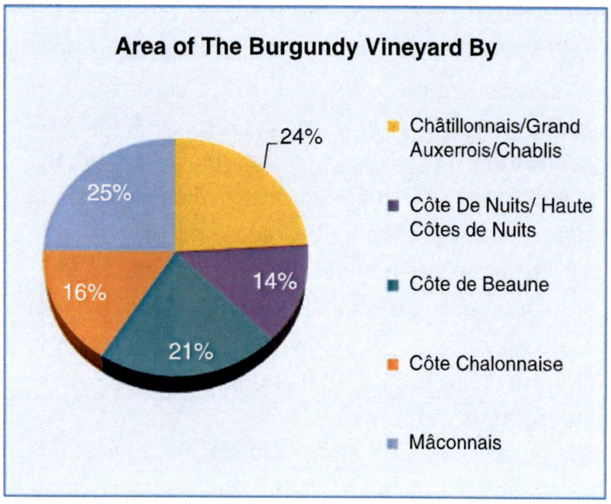

Fig. 4.3 Proportion of Burgundy AOC falling into the five wine regions of Burgundy (Bourgogne Viticole 2008)

tional vintage in the region. The average yearly precipitation of the Bordeaux region, computed from data collected between 1842 and 1989, is 851.3 mm (33.5 in.). This amount of precipitation is sufficient and not excessive in August and September during the ripening of the grapes. Although the climate is generally mild, the region may on occasions be hit by heavy hailstorms in the spring. Between May 11 and May 13, 2009, widespread hailstorms destroyed 30–100% of the crop in the Côtes de Blayes, Saint-Emilion, and Graves AOC appellations (Nouveaux dégâts dans les vignes du Bordelais 2009). On May 25, 2009, another hailstorm hit the area of Entre-deux-Mers in Gironde. These severe hailstorms damaged 15,000 ha (over 37,000 acres) of excellent vineyards (Cassagnes 2009). This became a hardship for growers since most do not buy weather-related insurance.

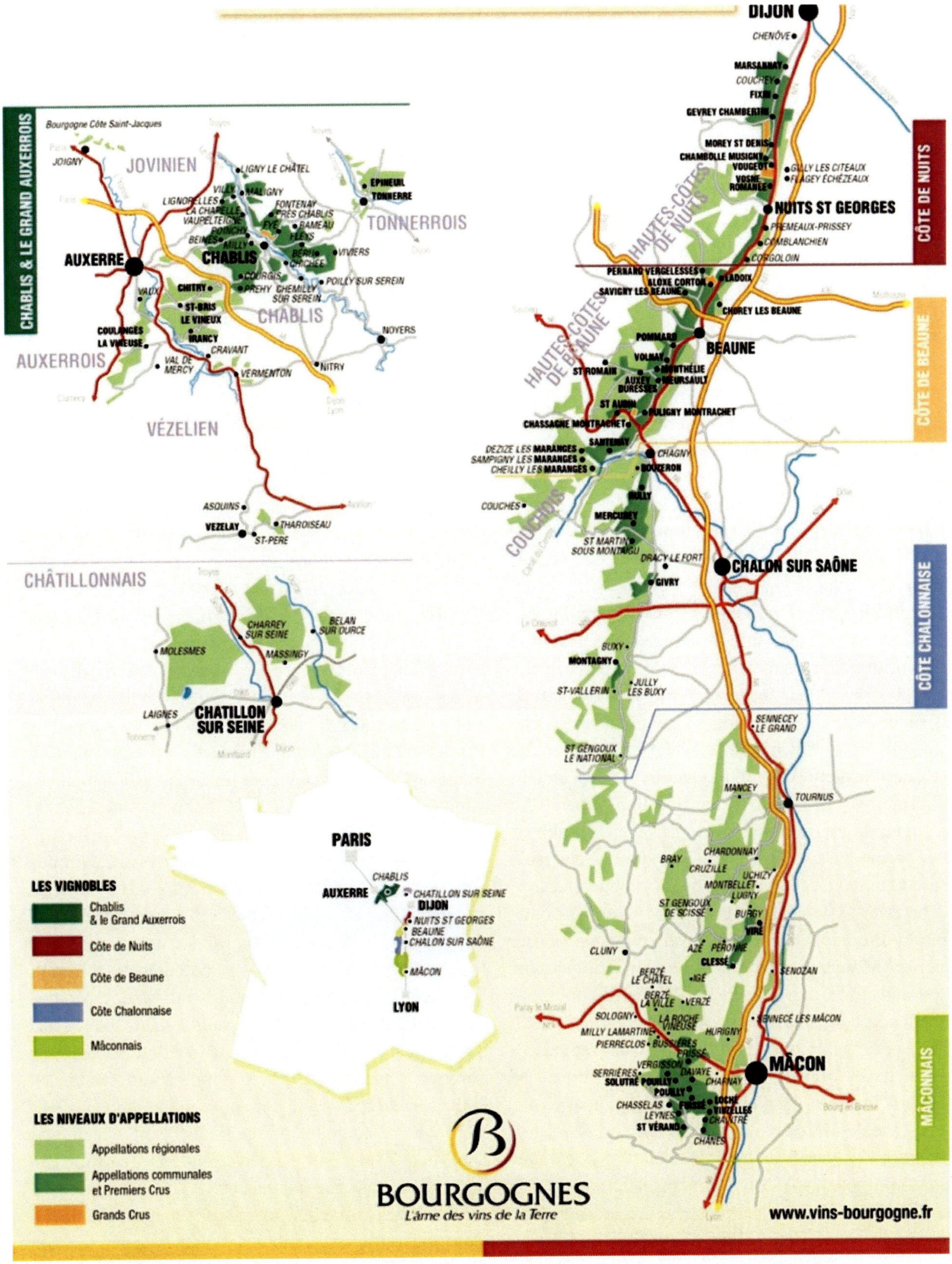

Fig. 4.4 The five regions of the vineyard of Burgundy AOC

Burgundy is located between 47°19′N in Dijon in the north of the region and 46°16′N in Mâcon in the south. Because this region is located in the center of France, the maritime influence produced by the westerly winds delivering moisture from the Atlantic Ocean is diminished. The climate is instead affected by Mediterranean influences when weather systems move north along the Rhône and Saône Rivers corridor, bringing high temperatures (the highest recorded is 38.1°C). Continental influences during the winter cause the temperature to plunge (the lowest recorded is −21.3°C).

Dijon receives an average of 705.1 mm of precipitation per year (1831–1989). This is about 17% less than in Bordeaux. The average yearly insolation is 1,831 h, much less than in the Bordeaux region. However, a hilly terrain provides some compensation when vineyards are located on steep, east-facing slopes. Like Bordeaux, Burgundy can be subjected to severe storms. On July 13, 2009, a hailstorm destroyed 30% of the vineyard of Saint Véran located about 20 km west of Mâcon (Grêle sur le vignoble 2009).

The Bordeaux region is drained by two major rivers, the Garonne and the Dordogne. The Garonne River, originating in the Pyrénées Mountains in the south, has deposited thick layers of pebbles, gravel, and sand over a long period of geologic time. The Dordogne River flows from the highest peak of the vast and old Massif Central mountain range located in the center of France. At the location where the Dordogne flows into the Garonne, their combined water forms an estuary called Gironde that channels their combined discharge toward the Atlantic Ocean.

The river system of the Burgundy region is dominated by the Saône River. The river originates on the Faucilles Mount located about 120 km northeast of Dijon and flows into the Rhône River in Lyon about 130 km south of Chalon-sur-Saône.

The geographic boundaries of the wine regions in France have been drawn according to precisely bounded "*terroirs*." However, while the boundaries are precise, the definition of the term *terroir* varies greatly. It is a broad concept that includes both physical and human components. Most *terroir* specialists agree that the physical *terroir* encompasses climate, soil type, geologic bedrock, slope, slope orientation, and drainage. The human component includes the type of vine grown, the vinification process employed, the production of single-varietal wines (as in Burgundy) or multi-varietal blends (as in Bordeaux), and the use of manual or mechanical harvesting. According to Cornelis van Leeuwen, professor of viticulture at E.N.I.T.A. (Ecole Nationale d'Ingénieurs des Travaux Agricoles) in Bordeaux, "*terroir* is concerned with the relationship between the characteristics of an agricultural product and its geographic origin which might influence these characteristics" (van Leeuwen et al. 2007, 1–5). The primary factor influencing the high quality of a vineyard seems to be related to "a mild water deficit, either because the climate is dry or because the soil water holding capacity is low" (van Leeuwen and Seguin 2006, 8). Both Bordeaux and Burgundy have *terroirs* with great viticultural soils which include a significant amount of rock (Graves, Médoc, Côte de Blaye, Côte d'Or), contain large quantities of clay (Fronsac, Saint-Emilion), or are shallow and rocky (Graves).

The *terroir* of the Bordeaux AOC is divided into six large regions based on its geology and further subdivided into 57 sub-regions defining its 57 appellations. The six regions are Bordeaux and Bordeaux Supérieur, Médoc and Graves, Entre-Deux-Mers with its dry white wine, Libournais (Saint-Emilion/Pomerol/Fronsac), Sauternais with its sweet white, and the Red Côtes (Côtes de Bourg, Côtes de Blayes, Côtes de Francs, Côtes de Castillon, and Premières Côtes de Bordeaux). The largest region is occupied by Bordeaux and Bordeaux Supérieur representing 47% of the area. It produced more than 2.5 million hl of red wine in 2007 on more than 51,000 ha (about 126,000 acres).

The region of Médoc, to the west of the Gironde River, is a strip of land ranging from 3 to 10 km wide and extending 80 km from north to south. It is divided into Médoc in the north, Haut Médoc in the south, and from north to south, the six communes of Saint-Estèphe, Pauillac, Saint-Julien, Listrac-Médoc, Moulis, and Margaux. Médoc means "middle land" in the local dialect, and it is derived from its location between the Gironde estuary and the Atlantic Ocean (Robinson 2006, 89 and 435–436) (Fig. 4.5).

Médoc is famous for the gravel soil which was deposited by the Garonne River on Late Cenozoic Era terraces dominating the estuary. The proportion of sand and silt, mixed with gravel, increases at lower elevation as the soil combines with the recent alluvium of the Garonne River.

The Graves region is located south of Médoc. Its name reflects the large quantity of gravel in its soil.

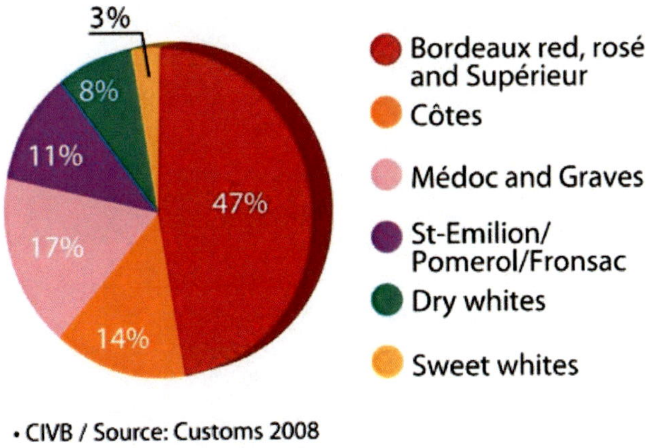

Fig. 4.5 Percentage of the six regions in the 118,900 ha of the Bordeaux AOC in 2008

The gravel and pebbles were brought in the Graves region not only by the Garonne River flooding but also by morainic deposits from the Pleistocene glaciation. The region of Médoc and Graves encompasses 17% of the Bordeaux vineyards and produced more than 700,000 hl of wine in 2007 (Superficies V.Q.P.R.D. 2008) (Fig. 4.6).

The region of Entre-Deux-Mers (meaning "between two seas") is located between the Garonne River to the west and the Dordogne River to the east. It is a vast triangular limestone plateau covered in places by sand, gravel, and clay. This area is an ancient fluvial terrace dating back to the Quaternary Era. The region produces dry white wine (green region on Fig. 4.2), as do two others, Bordeaux and Graves. Combined, these regions produced more than 450,000 hl of dry white wine in 2007 (Johnson and Robinson 2008, 98–99).

Libournais is the region surrounding the town of Libourne. It lies on the right bank of the Dordogne River and is famous for its three sub-regions of Saint-Emilion, Pomerol, and Fronsac. The soils often have a large proportion of clay associated with gravel, limestone, or sand. The Libournais encompasses 11% of the Bordeaux AOC area and produced over 600,000 hl of wine in 2007. Figure 4.7 illustrates the type of soil in the famous Château Cheval Blanc of Saint-Emilion. It is ranked with Château Ausone, within the highest category of Premiers Grands Crus classés A (Johnson and Robinson 2008, 98–99) (Fig. 4.8).

The sweet whites of the Sauternais are divided into semi-sweet and sweet wines. They are produced in a variety of soil types that all have great permeability in an area in the southern and eastern part of the Bordeaux AOC (Johnson and Robinson 2008, 98–99). Representing a mere 3% of its area, it produced 20,000 hl of semi-sweet wine and over 72,000 hl of sweet white wine in 2007 (Conseil Interprofessionnel du Vin de Bordeaux 2009, 1).

The *terroir* of the Red Côtes (Côtes de Bourg, Côtes de Blayes, Côtes de Francs, Côtes de Castillon, and Premières Côtes de Bordeaux) is made of soils rich in clay, with gravel and limestone. There wines are produced on southeast to south-facing slopes in the vicinity or on the banks of the Garonne and Dordogne Rivers. They represent 14% of the area of the Bordeaux AOC and produced about 850,000 hl of red wine in 2007 (Conseil Interprofessionnel du Vin de Bordeaux 2009, 3) (Fig. 4.9).

The *terroir* of Burgundy has diversity of soils but notable homogeneity of climate and geologic setting. Figure 4.10 represents a schematic cross section of the Côte d'Or escarpment. The sedimentary materials present in the escarpment are limestone, marl, marlstone, gravel, and clay deposited during the Jurassic period between 175 million years ago (Mid-Jurassic) and 150 million years ago (Upper Jurassic). Beds of hard limestone alternate with layers of softer marlstone made from a combination of gravel, sand, clay, and marl. Surficial soils have developed from the weathering of the Jurassic rocks. The thinnest soils are found near the summit. Soil thickness increases toward the Bressan Trough which is filled with younger sediments dating back to the Pliocene and the Quaternary Period. At a smaller scale, the mesoclimate of Burgundy, as exemplified by the climate of a small vineyard, may vary according to slope, orientation of the slope, color of the soil, amount of rock in the soil, protection against the wind, and exposure to fog, combined with the movement of these soils on the slope. The various soils and mesoclimates are the basis upon which the Burgundy AOC has been divided into five major regions and further subdivided into 100 sub-regions (Johnson and Robinson 2008, 54–55).

Premiers and grands crus are mostly found at mid- and high slope, at elevations comprised mainly between 250 and 300 m (about 820–980 ft) above sea level. *Appellations communales* occur on the lower third of the escarpment, and regional appellations are found at the western edge of the Bressan Trough.

Côte d'Or is divided into two parts: Côte de Nuits in the north and Côte de Beaune in the south

Fig. 4.6 Vineyard of Château de Portets, located in Portets, in the Graves AOC (Photo by Dana Litwornia, Rowan University)

Fig. 4.7 Soil of Château Cheval Blanc, Saint-Emilion (Photo by Dana Litwornia, Rowan University)

(Johnson and Robinson 2008, 56). Topographically, the regions of Côtes de Beaune and the Côtes de Nuits are similar, but the soil varies in such subtle ways that the vineyard has been subdivided historically into 635 small plots of land called "*climats*." Each *climat* produces a wine of well-defined identity. The smallest appellation is *Romanée* in the Côtes de Nuits. Its *climat* only covers 0.85 ha, making it the smallest in the world. It produced 4,030 bottles of wine in 2007 (Pitiot and Servant 2008, 62). Another famous *climat* is *Clos de Vougeot* in the Côtes de Nuits. Founded in 1110 by Cistercian Monks from the Abbey of Cîteaux, the vineyard passed from Church to private ownership during the

Fig. 4.8 High proportion of clay in the soil of Château La Dauphine, Fronsac (Photo by Dana Litwornia, Rowan University)

Fig. 4.9 Soil of Château Lagarde, Red Côtes de Blayes (Photo by Dana Litwornia, Rowan University)

Revolution of 1789 and is currently the property of a group of 80 winegrowers who produce only Pinot Noir (more than 207,000 bottles in 2007) (Pitiot and Servant 2008, 30). Completely surrounded by a wall, its 50.59-ha plot has three distinct soils. At the top of the vineyard, the soil is about 40 cm thick with gravel and pieces of limestone that weathered from the bedrock of the Jurassic period (175 million years ago) of Bajocian formation. As the vineyard slopes down to the valley below, soils grow deeper, reaching 90 cm. As thickness increases, the proportion of clay, marl, and silt increases.

Figure 4.11 shows the soil of the *Clos de Vougeot Appellation Grand Cru* at mid-slope and its 450-year-old Renaissance castle, the home of the Confrérie des Chevaliers du Tastevin. This group of over 12,000 members was created in 1934 and is comprised of celebrities, scholars, and the owners of the vineyard.

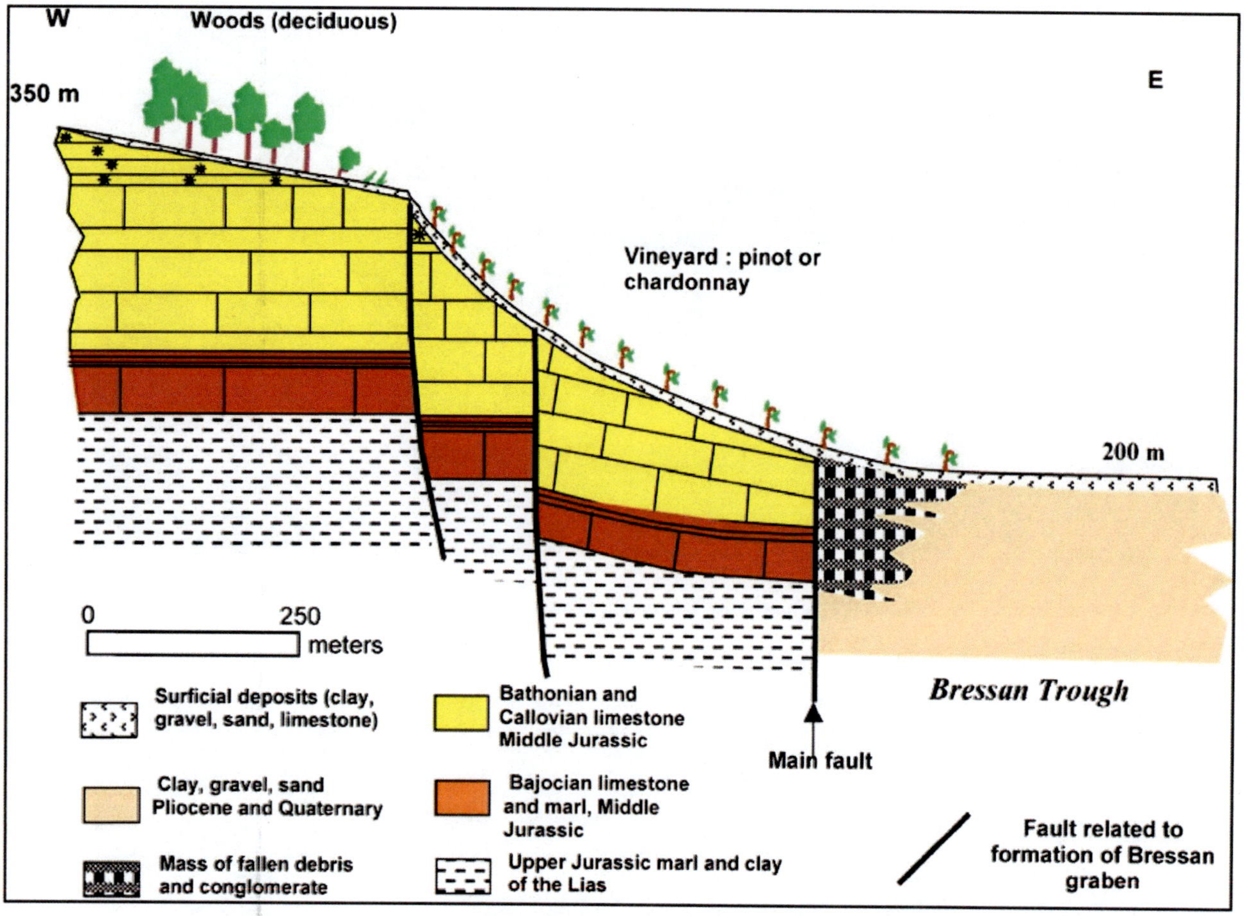

Fig. 4.10 Cross section of the Côte-d'Or (With permission from A. Gallien, SVT Dijon)

Fig. 4.11 Mid-slope soil (brown soil with clay, marl, and limestone gravel) of Clos de Vougeot and its Renaissance castle (Photo by Dana Litwornia, Rowan University)

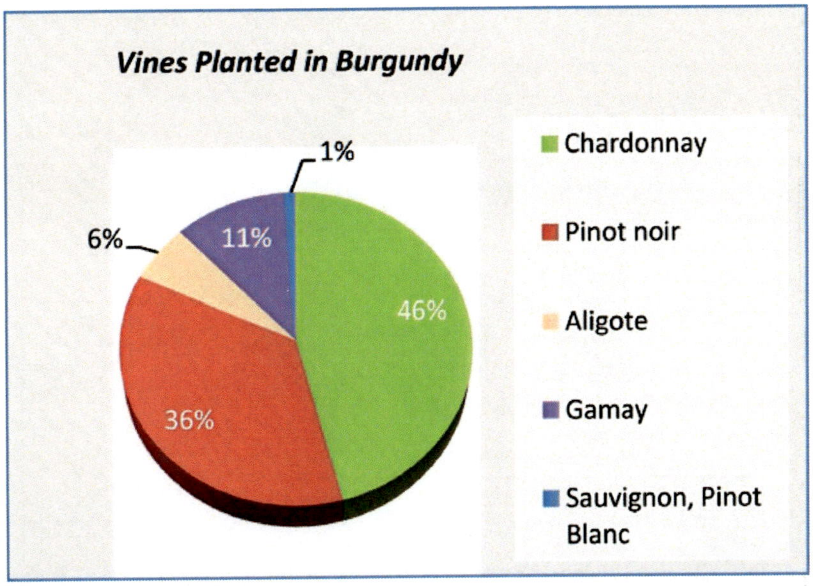

Fig. 4.12 Proportion of vines planted in Burgundy (2008)

They conduct monthly meetings to celebrate the great wines that Burgundy produces.

Bordeaux grows mostly Cabernet Sauvignon, Cabernet Franc, and Merlot, with a little Malbec, Petit Verdot, and Carmenère for the red wines. Bordeaux Blanc is made from Semillion, Sauvignon, and Muscadelle. To these three main cultivars, winegrowers add Colombard, Mauzac, Merlot Blanc, and Ugni Blanc, the proportion of which is not to exceed 30% (Galet 2000, XII–XXXIII).

In Burgundy, 36% of the vines planted are Pinot Noir. It is used for the grands and premiers crus in red wines. Gamay, 11%, is also used for the more regional wines. For white wine of high quality, Chardonnay is the most frequently planted vine (46%). Aligoté, 6%, is grown to make a low-cost white wine which is combined with black currant liquor to produce *Kir*, a local drink made popular by Felix Kir, a catholic priest, who became the mayor of Dijon in 1940 and remained in this position until his death at the age of 92. One percent of the vines are Sauvignon and Pinot Blanc (see Fig. 4.12).

The complexity of the human component of *terroir* – the selection of the cultivars, vineyard management, tradition, and winemaking processes – in Bordeaux and Burgundy can be illustrated through a consideration of the operation of successful vineyards in both regions. The owners of the vineyards described here represent a wide cross section of the wine-producing community in France today. Their vineyards are mapped on Fig. 4.13.

Paul-Henry Nerbusson of Château Lagarde and Jean Halley of Château La Dauphine, two of the vineyard owners, view the industry as a financial investment that can also fulfill their passion for wine (Johnson and Robinson 2008, 104).

Nine owners are the current generation of families that have a long history of winemaking, an unbroken developmental tradition that they plan to continue (Jean Fillastre of Domaine Jaugaret of Saint-Julien in Médoc; Michel Chapard of Château Peymelon in the Red Côtes de Blaye; Alain and Joel Appollot of Clos Trimoulet in Saint-Emilion; Jean-François Meynard in Saint-Magne-de Castillon; Gilles Renversade in the Bordeaux Supérieur AOC; Comte Senard in Aloxe-Corton in Côte de Beaune; Françoise Beaumont in Morey-Saint-Denis, Côte de Nuits; Jean-Marie Ponsot in Morey-Saint-Denis, Côte de Nuits; and Alexandre Moreau and Domaine Bernard Moreau in Chassagne-Montrachet).

Four vineyard owners are conservative innovators who have taken over the family vineyard and intend to return it to preindustrial organic viniculture (Marie-Hélène Théron of Château Portets, Sandrine Haur of Château Haur du Chay, Christelle Gauthier of Château Platon, and Stefan Villate of Château Villate).

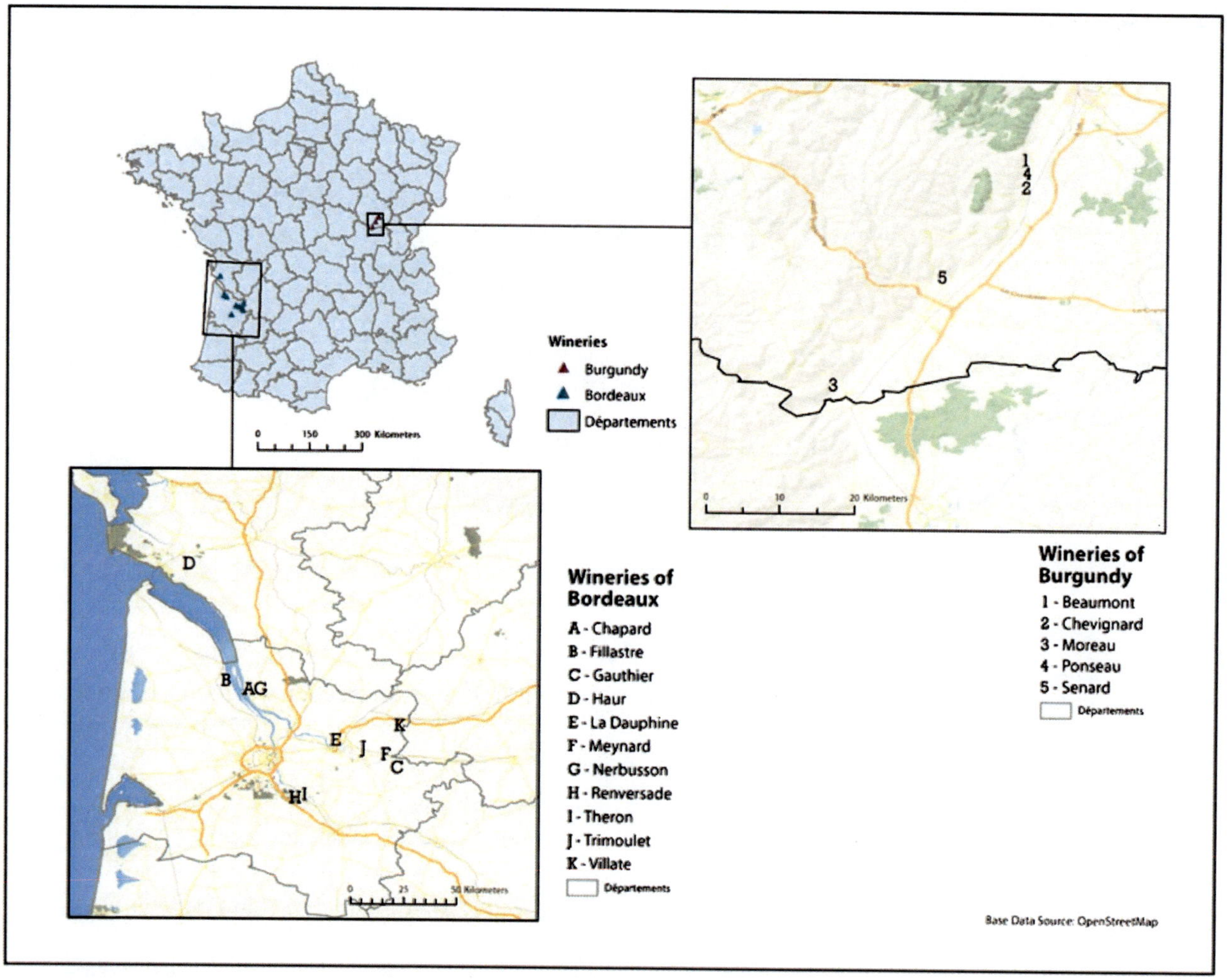

Fig. 4.13 Geographic location of the vineyard owners surveyed in our study (Cartography by Katrina Shand and John Reiser, Rowan University GIS Laboratory)

Entrepreneurial Winemaking

Jean Halley, Château La Dauphine

In the 1970s, the Moueix family, headed by Jean-Pierre Moueix, purchased a number of properties in Fronsac, including Château La Dauphine and Canon de Brem. To many people's surprise, these properties were put up for sale in 2000, the Moueix family having decided to concentrate their efforts on their Saint-Emilion and Pomerol properties. Jean Halley, whose family owns the giant food chain Carrefour, decided to purchase these properties not only as a financial investment but also to satisfy his passion for wine. Installing his son Guillaume as director of the estate with the task of insuring its economic viability, Halley set out to acquire the services of the most competent experts to transform the operations at the property into a model vineyard (conversation with Bernard Lamaud on August 26, 2009, at Château la Dauphine).

Halley's estate manager, Bernard Lamaud, a wine consultant with over 20 years of experience, was given *carte blanche* to eliminate the traditional winemaking building and replace it by the best possible modern facility. Investments of more than 10 million Euros were necessary to buy the equipment, restore the castle (the mid-eighteenth century home of Marie Josèphe de Saxe, wife of one of the sons of Louis XV) (Château de La Dauphine 2009), renovate the *chai* (where the barrels are stored), restructure the vineyard, and improve the physical *terroir*. In 2009, Lamaud altered the drainage pattern of one plot by removing the original soil and building drainage troughs in the subsoil. With the original soil returned and new plantings

Fig. 4.14 *Left*: Drainage channels are built. *Right*: The original soil of hard clayey limestone is piled prior to being returned to the plot (Photo by Dana Litwornia, Rowan University)

established, the water available to the vines will be reduced and the quality of the grapes enhanced. The vines cultivated in the entire vineyard are 80% Merlot and 20% Cabernet Franc. The 20 ha (50 acres) of clayey limestone in Fronsac produce 80,000 bottles yearly of their first wine labeled Château la Dauphine (Fig. 4.14).

The vinification process at the estate is accomplished in a modern facility designed by Lamaud that uses gravity for the efficient movement of materials through the various stages of the process. The grapes, which are hand-picked, plot by plot, are sorted by six skilled workers both before and after de-stemming. They are then delivered by a conveyor belt to a series of individual holding tanks where they are stored, segregated by plot of origin, until they are pressed. Once pressed, the new wine is kept in 34 stainless steel and concrete tanks. The entire process is temperature-controlled and monitored from a single room (Fig. 4.15).

After the initial fermentation process, the wine is aged for 12 months in barrels of French oak, one third of which are replaced every year. The aging of the wine is overseen by Denis Dubourdieu, a research scientist and œnologist at the University of Bordeaux as well as owner of vineyards in Premieres Côtes de Bordeaux and in the Sauternais (Fig. 4.16).

Paul-Henry Nerbusson, Château Lagarde

Paul-Henry Nerbusson invested in the purchase of a 12-ha (30 acres) wine estate, Château Lagarde, in 1990. Nerbusson comes from a family of grain farmers near La Rochelle. He still owns two farms, but like Jean Halley, he has a passion for wine. Ninety percent of the estate's product is red wine (Merlot, Cabernet Franc, Cabernet Sauvignon, and Malbec), with whites (Sauvignon Blanc, Semillon, Ugni Blanc, and Colombard) comprising the rest. Located in the Premiére Côtes de Blaye AOC, the estate already had a reputation for quality when Nerbusson purchased it. Thus, total renovation was unnecessary, and Nerbusson could focus his initial efforts on improving the *chai* (see Fig. 4.17) and renovating the large, nineteenth century "Château." The costs of purchase and renovation

Fig. 4.15 *Left*: View of the upper floor of the building where the fruit is gravity-fed into the stainless tanks. *Right*: Concrete tanks painted in yellow and the stainless steel fermentation tanks (Photo by Dana Litwornia, Rowan University)

Fig. 4.16 *Left*: Temperature controlled cellar. *Right*: New wine press (Photo by Dana Litwornia, Rowan University)

have been covered by other sources of funds, including technology consultation that he performs for the King of Morocco. Because of his wider range of financial resources, Nerbusson can wait patiently for the estate to become profitable without sacrificing quality to the demands of greater commercial success (conversation with Jean Henry Nerbusson, Château Lagarde on August 25, 2009).

Family Vineyards

The interviews of the second group of nine vineyard owners, five in Bordeaux and four in Burgundy, all show a consistent pattern. The owners are proud of the wine they make, aim at perfection, and follow the traditions of production that have been respected in their family for generations.

Fig. 4.17 *Left*: Château Lagarde cellar. *Right*: Redesigned wine label (Photo by Dana Litwornia, Rowan University)

Five Traditional Vineyards in Bordeaux

The smallest estate in this group is Domaine Jaugaret, a 1.5-ha (3.7 acres) plot of Cabernet Sauvignon in the prestigious village of Saint-Julien in Médoc. Jean Fillastre and his family have owned this vineyard since 1654. Production is divided into 80% Cabernet Sauvignon and 20% of Petit Verdot and Malbec. The average age of Fillastre's vines is 60 years, almost as old as he is; some are older than 100 years, having been planted by his grandfather at the turn of the twentieth century. Harvesting by hand, Fillastre continues to employ the wooden fermentation vats and wine press inherited from his ancestors. Rejecting the use of chemical fertilizers, he refuses all the shortcuts of modernism, preferring to adhere to the most classical Bordeaux traditions. His wine spends between 30 and 36 months in barrels of French oak, housed in a cellar in the family home in the middle of the village. Though he produces only 6,000 bottles a year, his wines have found a following in the United States and have been praised by wine critics in *The New York Times* (Asimov 2008) (Fig. 4.18).

Michel Chapard owns 12.5 ha (31 acres) at Château Peymelon, in the Blaye area. A traditional winemaker, Chapard is still interested in improving traditional techniques and experimenting with the density of the cover grass growing between the vines to find the perfect balance between available water, controlled erosion, and the encouragement of early ripening. Close to retirement, this vineyard owner, insurance broker, and treasurer of the local agricultural federation has energetically worked to expand his market by combining with nine of his friends to create a company to export their Cru Bourgeois to China.

Gilles Renversade is a traditional wine grower of Porchères in the Bordeaux and Bordeaux Supérieur Appellations who opened a bed-and-breakfast to complement his income from his 6.5 ha vineyard. Renversade has chosen to sell most of his wine to wine traders instead of having to market it. He only keeps the product of 2 ha for himself, organizing wine tastings for his bed-and-breakfast guests to whom he sells his wine. In this way, he maintains not only the standards of viticulture passed down through his family but also a personal relationship with his customers which has been usually lost in the modern commercialization of the industry.

The other vineyard owners in Bordeaux, Alain Appollot in Saint-Emilion and Jean Meynard in Saint-Magne de Castillon, have similar experiences in the practice of their trade. They follow the same traditions of quality and only differ by the provider of their oak barrels, the number of months the wine spends in them (12–24 months in general), and their reliance on mechanized or manual harvest. Notably, Jean Meynard invented a machine to sort the grapes harvested. His machine was shown on French television and will be

Fig. 4.18 *Left*: Jean Fillastre, accomplished glass blower as well as viticulturist, fills a sample thief of his own manufacture. *Right*: Château Peymelon vineyard (Photo by Dana Litwornia, Rowan University)

used in 2009 for the first time. They share a passion for their work, and they are steadily producing better and better wines.

Four Traditional Vineyards in Burgundy

In Burgundy, four vineyards surveyed for this chapter have each been owned by one family for at least four generations. The vines planted are predominantly Pinot Noir for the red wines and Chardonnay for the white wines. Two additional vines for white wine, Aligoté and Pinot Blanc, and one additional vine for the red, Gamay, are grown in the Bressan Trough.

Only one of these estates, Domaine Bernard Moreau in Chassagne-Montrachet, has 70% of its estate planted in Chardonnay. The *terroir* of this small AOC in the Côte de Beaune has such an incredible quality that it has 19 premiers crus and 3 grands crus "*climats*," of which Domaine Bernard Moreau has 6. Alex Moreau's estate is further distinguished by a monopoly; it includes the entire area of one premier cru, making him the exclusive producer of La Cardeuse red wine. While most of the wines produced at Domaine Bernard Moreau are aged in French oak barrels for 18 months, La Cardeuse is aged for 20. Alex Moreau's goal is to produce "the purest expression of the Pinot Noir from our vineyards," and no doubt he has succeeded (Domaine Bernard Moreau Chassagne-Montrachet 2008).

The three other vineyards surveyed are Domaine Comte Senard, Domaine Beaumont-Palisse, and Domaine Ponsot, all of which mostly produce Pinot Noir grapes. Domaine Comte Senard, which dates back to 1857, is located in the small village of Aloxe-Corton (Domaine Comte Senard 2009). The original 2.5-ha plot (6 acres) was expanded in the course of time to its present size of 9 ha (22 acres). The role of each of the seven generations which contributed to its present success has been to carefully select each additional plot in the most prestigious "*climat*." They produce five grands crus in red, one grand cru monopole in white in Aloxe-Corton, and a premier cru in Savigny-les-Beaunes. The family tradition of exercising extreme care in the vinification process is demonstrated by the use of a cellar, built by Benedictine monks, dating back to the thirteenth century. Their success led them to build an additional cellar in Beaune where a large proportion of the local wine-related businesses are located. Daniel Senard continues to manage the vineyard in Aloxe-Corton, while his son Philippe is in charge of the domaine located near Beaune. His daughter Lorraine, who

Fig. 4.19 *Left*: Jean-Marie Ponsot points to the Middle Jurassic limestone exposed on one wall of his cellar, where vine roots are reaching a depth of 10 m inside the limestone bedrock. *Right*: The Ponsot vineyard surrounding a small castle (Photo by Dana Litwornia, Rowan University)

entered the family business in 2000, oversees the unique vinification process implemented in Domaine Senard, in which the grapes are macerated for 10–12 days at a temperature of 10°C (50°F) to extract the aromas. This technique was suggested by the three œnologists that Daniel Senard consults regularly. The white wine of the Corton Blanc Grand Cru monopole is aged for a year in French oak barrels, 50% of which are replaced each year, while the red of the five grands crus is aged for 18 months. They have decided to stop using the traditional corks and replace them with modern plastic corks that guarantee the long term conservation of this exceptional quality wine.

The other two vineyard owners, Françoise Beaumont-Palisse and Jean-Marie Ponsot, live in Morey-Saint-Denis (Pitiot and Servant 2008, 52). This very small village is famous for the quality of its wine. Domaine Ponsot was created in 1872 when the first member of the family, William Ponsot, purchased a 0.5-ha (1.25 acres) plot near the top of the escarpment. As a result of careful expansion of the vineyard into the best areas of the Morey-Saint-Denis *terroir*, the domaine now comprises plots in seven grands crus, including one, Clos Saint Denis, with vines that are more than 100 years old. Although these extremely old vines produce only 7.5 hl of a dark and fruity wine each year, their product has been glowingly praised whenever it is discussed in wine magazines. The 11-ha domaine is broken into 14 plots located in Morey and the neighboring villages of Clos de Vougeot, Chambertin, Gevrey-Chambertin, and Chambolle-Musigny, producing superlative wines in 7 grands crus and 3 premiers crus. Jean-Marie Ponsot, the patriarch of the family, has maintained the excellent quality of their wines because of his encyclopedic knowledge of wines and commitment to precise standards of production. Fervently adept at preservation of the natural environment, Ponsot and his children (Laurent and Rose-Marie) use chemical treatments only when indispensable. This supervised control of fungi and insects, combined with organic farming, has allowed the Domaine to blossom and to be respected by their friends and competitors (Le Domaine Ponsot 2009) (Fig. 4.19).

Françoise Beaumont-Palisse represents another trend among vineyard owners in Burgundy. At the death of her husband 10 years ago, she allowed her nephew to farm her vineyard under *métayage* – a kind of share cropping (Robinson 2006, 439). In payment for his labor in her vineyard, her nephew receives a portion of the grapes harvested. After vinification and aging, her wine is bottled under her personal label. Like Gilles Renversade, she owns a successful bed-and-breakfast where her guests are offered a free wine tasting on their arrival day and have the opportunity to purchase her award-winning wine. Jean-Marie Ponsot also has four additional parcels under *métayage,* a very common practice in wine production.

Conservative Innovators

Some young wine producers in Bordeaux have taken over the family business with the intention of reinventing it as an organic enterprise, combining an interest in preindustrial technologies and an application of the most modern science of natural agriculture. Marie-Hélène Théron of Château Portets, Sandrine Haur of Château Haur du Chay, and Christelle Gauthier of Château Platon all have taken over family vineyards upon their parents' retirement. Engaging in what traditionally has been a male occupation, all three women have prepared themselves for their work by returning to school to gain an education in modern agronomy, and all three have made a commitment to avoid chemical fertilizers and pesticides, producing their wines through a combination of traditional techniques and modern organic agriculture.

At his father's death in 1982, Stefan Villatte inherited a 30-ha estate in Puynormand, in the Bordeaux AOC. While maintaining the quality of wines produced there, he immediately began to implement more natural procedures, abandoning chemical pesticides and fertilizer. Willing to put in the extra effort needed to maintain an organic regimen, he spends hours daily walking his vineyard to detect early signs of infestation of the vines by fungi or insects. Villatte has also become interested in growing vintage varieties of wheat, from which he produces natural whole wheat bread. Complementing his wine, in 2008, he has expanded into the production of all natural grape juice, a product that has been so successful as to encourage him to dedicate a larger proportion of his harvest to it in the future. Turning some of his woodlands into a wildlife sanctuary, Villatte has converted his father's estate into a productive expression of modern organic agriculture.

Viticultural Pedology

Cornelis van Leeuwen, a researcher and professor at E.N.I.T.A., born in the Netherlands, attended the *Centre de Formation et de Recherche en.Œnologie* at the University of Reims in Champagne before transferring to the University of Bordeaux, where he obtained his degree in 1984. His postgraduate work resulted in a degree in the Sciences of the Vine from the University of Burgundy in Dijon, followed by a Master's Degree in Biology, and a Ph.D. in viticultural pedology in 1986. His dissertation research focused on the study of soils in Saint-Emilion; not surprisingly, in 1992, he was hired as the vineyard manager at the celebrated Château Cheval Blanc in Saint-Emilion (ENITA de Bordeaux 2009). After becoming the estate's technical director, he received a teaching position at the University of Bordeaux, continuing to work for Château Cheval Blanc on a part-time basis. He is known for his extensive research and publications concerning vines and *terroir* and for the professional training he provides to a varied audience of engineers, œnologists, wine estate owners, and their technical directors, not only in France but also internationally. His publications, which emphasize the pragmatic application of his research, are widely read and respected among the wine producers of France; three of the growers interviewed for this chapter (Lamaud, Villatte, and Ponsot) have read his work and applied his ideas in their vineyards.

Grand Connétable of the Confrérie des Tastevins de Clos de Vougeot

The most famous of the people surveyed, in the eyes of the wine connoisseurs, is by far Louis-Marc Chevignard, *Grand Connétable* of the Confrérie des Tastevins de Clos de Vougeot (portrayed in Fig. 4.20). This organization, created in 1934 in the midst of an intense economic crisis and at a time when wine from Burgundy had trouble finding buyers, aims at generating worldwide recognition of the quality of Burgundy wines. Similar confréries exist in the Bordeaux region: the Commanderie du Bontemps in Médoc and the Jurade in Saint-Emilion. Annually, the *Confrérie des Tastevins de Clos de Vougeot* holds 16 meetings with feasting and wine tasting, during which new members are installed. The organization boasts more than 12,000 members on 5 continents. Chapters exist in Russia, the United States, Japan, Singapore, Hong Kong, and, recently, China. Six thousand members live outside of France, 2,300 of these in the United States (Société Bourguignonne de Propagande et d'Edition 2009). Chevignard manages a budget of more than four million Euros and supervises 20 employees in charge of the starred restaurant, maintenance, guides, and activities organized by the Confrérie. As the most recognizable representative of the Burgundy wine industry, Chevignard's greatest contribution to his community is, as was that of the men who preceded him in his position, his untiring promotion of the history and the excellence of the wines of Burgundy.

Fig. 4.20 Louis-Marc Chevignard, *Grand Connétable* of the Confrérie des Tastevins de Clos de Vougeot (Photo by Dana Litwornia, Rowan University)

Marketing the Wines of Burgundy and Bordeaux

Although 26% of the 9,100 wine producers in Bordeaux market some of their wines directly to the public, in both Burgundy and Bordeaux, the number is relatively small (Feredj 2009, 3). Fifty-eight percent of the wines produced annually by Burgundy's 3,800 growers are marketed through 250 wine-trading firms, with another 16% sold through 23 cooperative producers. In Bordeaux, 70% of the wine is sold through 300 trading firms. There are almost twice as many wine cooperatives in Bordeaux (42), representing 43% of the wine producers, employing 22% of the wine-growing area of the region, and generating 23% of the harvest (Feredj 2009, 3–4). This pattern of depending on separate marketing institutions is consistent with the financial needs of agricultural establishments and the requirements of marketing to a world market. It explains as well the value of organizations like the Confrérie des Tastevins de Clos de Vougeot, the Commanderie du Bontemps in Médoc, and the Jurade in Saint-Emilion, which function as virtual advertising agencies to bring international orders to the trading firms that organize and simplify access to the myriad of individual growers. Bordeaux and Burgundy have been able, through the well-established centralization of management of their marketing structures, to maintain the essential characteristics of a cottage industry dependent on the expertise and creativity of individual craftsmen in an age of mass production and mass marketing.

References

Annequin, J. M. 2009. L'agriculture en 2008 en France et en Europe, Le prix des intrants fait baisser les revenus. http://www.insee.fr/fr/ffc/ipweb/ip1247/ip1247.pdf (Last accessed 27 September 2009).

Asimov, Eric. 2008. Domaine du Jaugaret. *The New York Times, Style, Dining and Wine*. 12 October 2009.

Cassagnes, Juliette. 2009. Grêle sur le vignoble bordelais - Et de trois! www.vitinet.com. http://www.viti-net.fr/Outils/Fiches/FichesDetail.asp?idRub=54&id=56259 (Last accessed 10 October 2009).

Château de La Dauphine. 2009. http://www.chateau-dauphine.com/ (last accessed 28 September 2009).

"Chiffres-clés de la Bourgogne Viticole" (Bureau Interprofessionnel des vins de Bourgogne) 2008. http://www.vins-bourgogne.fr/connaitre/ses-reperes/chiffres-cles-de-la-bourgogne/gallery_files/site/321/4635.pdf (Last accessed 20 November 2011).

CIVB. 2009. *Bordeaux Wine: Economic Profile*. Bordeaux: Conseil Interprofessionnel Du Vin De Bordeaux.

Domaine Bernard Moreau Chassagne-Montrachet. 2008. *The Sorting Table*. http://www.thesortingtable.com/V2/brand.php?brandId=6&pageId=6&productId=570&sessionID=Ulmtmtg8sLKLejIR (Last accessed 27 September 2009).

Domaine Comte Senard. 2009. http://www.domainesenard.com/english/premiere.html (last accessed 28 September 2009).

"Eléments clés de la Bourgogne viticole" (Bureau Interprofessionnel des Vins de Bourgogne) 2009. http://www.vins-bourgogne.fr/connaitre/ses-reperes/chiffres-cles-de-la-bourgogne/gallery_files/site/321/12398.pdf (Last accessed 20 November 2011).

ENITA de Bordeaux. 2009. http://www.-ecole.enitab.fr/kees.vanleeuwen/site_kees_page_CV_anglais.html (Last accessed on 10 October 2009).

Feredj, R. 2009. Bordeaux Wine: Economic Profile June 2009, Conseil Interprofessionnel du Vin de Bordeaux.

Galet, Pierre. 2000. *Dictionaire Encyclopédique des Cépages*. Paris: Hachette Livre.

Grêle sur le Vignoble. 2009. *Frédéric Curis, Pouilly-Fuissé, Saint-Véran... Du producteur au consommateur ! Satisfait ou remboursé* ! http://www.domainecuris.com/blog/_divers/grele-sur-le-vignoble.html (Last accessed 10 October 2009).

Guide du demandeur d'une appellation d'origine (A.O.C./A.O.P.). 2009. Institut National de L'Origine et de la Qualité. http://www.inao.gouv.fr/repository/editeur/pdf/GUIDES_et_NT/Guide_dudemandeau_AOC-AOP.pdf (last accessed 10 October 2009).

Johnson, H., and J. Robinson. 2008. *The World Atlas of Wine, Sixth Edition*. New York: Octopus Books.

Le Domaine Ponsot. 2009. http://www.domaine-ponsot.com/ (last accessed 28 September 2009).

Nouveaux dégâts dans les vignes du Bordelais. 2009. *Le Figaro.fr.* http://www.lefigaro.fr/vins/2009/05/ 13/05008-20090513 ARTFIG00482-nouveaux-degats-dans-les-vignes-du-bordelais-.php (Last accessed 10 October 2009).

Pitiot, S., and J. C. Servant. 2008. *A Guide to Burgundy's Appellations, Everything There is to Know About Burgundy Wines*. Beaune Vignole: Promotional Board for Burgundy Wines.

Production du vignoble, – Production de vins, jus et mouts. 2009. *Ministère de l'alimentation, de l'agriculture et de la pêche.* http://www.agreste.agriculture.gouv.fr/IMG/_pdf_saa2009T8a-2.pdf (Last accessed 27 September 2009).

Robinson, J., Ed. 2006. *The Oxford Companion to Wine, Third Edition.* Oxford: Oxford University Press.

Société Bourguignonne de Propagande et d'Edition. 2009. http://www.tastevin-bourgogne.com/ Anindex2.htm (Last accessed 10 October 2009).

Superficies V.Q.P.R.D. 2008. *Etablissement National des produits de l'agriculture et de la mer*. http://www.onivins.fr/pdfs/160.pdf (Last accessed 10 October 2009).

van Leeuwen, C. and Seguin, G.L. 2006. The Concept of Terroir in Viticulture. *Journal of Wine Research* 17 (1): 1–10.

van Leeuwen, C., B. Bois, P. Pieri, J. P. Gaudillére. 2007. Climate as Terroir Component. *Congress on climate and viticulture, Zaragoza 10–14 April 2007* 1–12.

The Okanagan Wine Region of British Columbia, Canada

5

Donna M. Senese, Wayne Wilson, and Bernard Momer

Abstract

This discussion of the Okanagan wine region takes a historical perspective of the political economy of wine in one nascent New World wine region. The Okanagan Valley of British Columbia, Canada, has enjoyed recent and rapid growth as a wine-producing region since grapes were first produced here over 100 years ago. A primarily historical geography of the winescape examines the influence of French Oblate missionaries who introduced wine production, as well as the secular influences of European agricultural settlement and transportation improvements that opened the region to wider markets. The contemporary political history of the Okanagan wine region reveals the strong influence of government protections since the 1960s which have allowed a massive planting of *Vitis vinifera* to satisfy changing domestic market demands for premium wines. Subsequent to the post free trade era, the Okanagan has experienced a rapid acceleration of the industry, considered here as a response to market-driven demands for local products in a globalized industry. Diverse landform and climatic conditions combine with the political economy of *terroir* in the Okanagan where, like other New World wine regions, the existence of place-based authenticity is branded through estate patrimonialization and linked closely to the tourism industry.

Introduction

Wine is a manifestation of culture, and it is a cultural trait that is expressed in unique landscapes that vary from place to place. The Okanagan Valley wine region in British Columbia, Canada, is both unique and representative in this regard. As a unique culture region, the Okanagan is part of a global wine industry transformation that has resulted in the emergence of competitive New World wine regions. In the process of this transformation, the Okanagan has developed an iconic wine landscape that reflects the culture responsible for its production and consumption, as well as the natural attributes of its locality. This Okanagan wine landscape represents an evolution of the wine industry experienced by many New World wine regions where notions of *terroir* require the nuances of time, trial, and error to emerge. In contrast to the long history of viticulture in continental Europe or even parts of Napa, the

D.M. Senese (✉) • B. Momer
Community Culture and Global Studies, IK Barber School of Arts and Sciences, University of British Columbia Okanagan, Kelowna, BC V1V 1V7, Canada
e-mail: donna.senese@ubc.ca; bernard.momer@ubc.ca

W. Wilson
Kelowna Museums, 470 Queensway, Kelowna, BC V1Y 6S7, Canada
e-mail: wwilson@kelownamuseums.ca

P.H. Dougherty (ed.), *The Geography of Wine: Regions, Terroir and Techniques*, DOI 10.1007/978-94-007-0464-0_5, © Springer Science+Business Media B.V. 2012 vineyards and wineries of the Okanagan are largely a cultural manifestation of political and economic evolutions of the past 30 years. The remarkable growth of the Okanagan wine industry coincides with an equally remarkable shift in the consumption of the region's natural and cultural landscape. For decades, the regional economies of the Okanagan have depended on the production value of resources associated with its natural environment. More recently, the warm arid climate, agricultural landscapes, and recreational amenities associated with a natural environment that includes large lakes and scenic mountains have produced a valued tourism economy. Today, much of the tourism culture and economy of the Okanagan centers on the localized wine industry and the associated rural aesthetic that both supports and depends on it. This discussion of the Okanagan wine region traces the historical roots of the wine industry, its context in the natural environment, and the production of a contemporary cultural landscape in the face of local and global viticultural change.

Location and Historical Background

The Okanagan Valley is one of dozens of north–south trending valleys between the Georgia Strait and the British Columbia–Alberta border (Fig. 5.1). There are approximately 450 wineries in Canada today, close to one third of them are located in this unique Okanagan ecumene that capitalizes on a warm dry climate and landscape conditions that produce quality wines and attract increasing numbers of wine tourists and amenity migrants. Like many New World wine regions across the globe, the Okanagan has experienced recent and rapid growth; however, the history of winemaking and grape cultivation reveals a heritage and tradition that continue to evolve. Expansion into previously uncultivated land suitable for viticulture and an accelerating conversion of orchard to vineyard have accompanied the rapid growth and spatial expansion of grape production in the Okanagan Valley and into the Similkameen Valley, flanking the region to the south and west. Today, the Okanagan and Similkameen Valleys boast over 7,000 acres of land cultivated in *vinifera* varieties, and the region is home to more than 130 wineries, many established since 1990 when the synergy of Vintner's Quality Assurance (VQA) standards, free trade agreements, and a global taste for locally produced goods dramatically changed the Okanagan winescape. Like many other New World wine regions, the developmental milestones in the wine region's cultural history are associated with a much longer history of European colonization, regional economic integration, transportation technologies, and widespread culture change.

European settlement began in the region in 1859 and initially grew up around a Catholic Oblate Mission in the central part of the valley. Indeed, popular history has Father Charles Pandosy planting the first grapes in the area. For the following 40 years, the area's economy turned around the rhythms and patterns of extensive agriculture – cattle ranching and grain growing – and the open parkland vegetation was an ideal hearth within which to imagine these activities. In practical terms, there was just enough timber to build houses, barns, and fences, and the remainder of the land was open and available for rapidly expanding herds of cattle. Even more conducive to this agricultural economy was the Okanagan's centrality among the province's burgeoning gold mining settlements in the Kootenay, Similkameen, Fraser, and Cariboo regions.

A new and intensive regional economy began to emerge around the turn of the nineteenth century, and it focused on the tree fruit industry – an industry built largely to meet emerging Canadian and Commonwealth markets. The industry grew because of the new availability of a rail line that favored the shipment of a high-bulk, low-value, and perishable commodity in a cost-effective manner. With completion of the transcontinental Canadian Pacific Railway in the mid 1880s and the construction of a rail spur from this line south to the north end of Okanagan Lake in 1892, the regional economy began to change. Later that same year (1892), this new transport connection was extended to Kelowna and Penticton with the launch of the CPR's steamboat *Aberdeen* on Okanagan Lake. Steamboats quickly became a central force in the landscape as a lifeline to emerging settlements. Steamboat transportation was also a determining factor in the location of many town sites along the lake as people and merchants settled near boat landings. Between 1904 and 1914, the shift to intensive agriculture took on more momentum as tens of thousands of acres of grazing land, hay flats, and grain fields became 5–40-acre orchard lots, supplied with gravity-fed irrigation water and planted to orchards. In half a generation, the area's landscape turned from

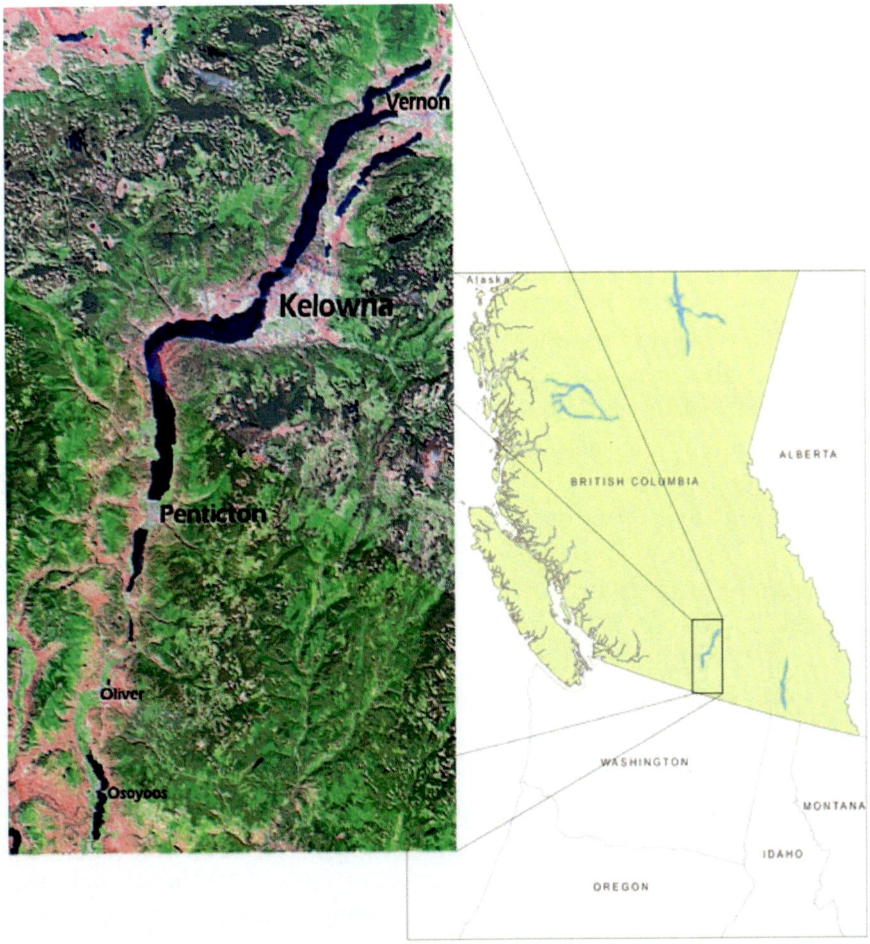

Fig. 5.1 The Okanagan wine region (Source: Department of Natural Resources Canada)

brown to green and took on the look and feel of a well-manicured and bucolic agricultural landscape.[1]

Cultivation of grapes was certainly part of the trial and error process so evident in these types of mid-latitude settlement experiences, but it was not until the latter half of the 1920s that commercial-scale grape growing began. In the late 1920s, J. W. Hughes planted roughly 125 acres of grapes in the Kelowna area and in 1939 added another 75 acres (Agriculture Canada 1984: 2). His market for this early production capacity was both the fresh fruit market and an emerging wine enterprise that began with Growers Wines in Victoria in the 1920s and with Domestic Wine and By-products Ltd (later Calona Wines) in 1932 in the Okanagan. Any relationship between the early grape plantings in the Okanagan in the late 1920s and United States Prohibition events is probably giving too much credit to the few acres of grapes planted in the region, a portion of which were destined for the fresh fruit market.

From this period forward, a succession of provincial governments used an array of protectionist measures to insulate, support, and favor the province's grape growers and wineries. For its purposes, the government's increasing control of and support for the production and distribution of alcohol helped fill government coffers while nurturing a nascent grape industry. Notwithstanding these measures, it would be decades before a viable and successful premium wine industry could emerge and survive (McGuire 1993).

Soon after Hughes planted his first acreage in the Kelowna area, he was able to secure an agreement that guaranteed him $100 per ton (Nichol 1983). Despite the apparent security of this production, grape acreage remained relatively low until the 1960s when further government regulation and the subsequent establishment of the Grape Marketing Board in 1970 successfully pushed a requirement that British Columbia

[1] For a review of the settlement process at work in this region, see Dendy (1976) and Wilson (1989).

wineries use as much as 81% of the grapes grown in the province (McGuire 1993). The shift to increased grape acreage was also driven somewhat by consumer trends that saw domestic wine sales in British Columbia triple between 1964 and 1974 (Nichol 1983).

The expanding grape growing sector also began to change in other significant ways in the 1960s and 1970s, possibly because of the resurrection of American viticulture predominantly in California. Following a global trend, Canadian consumer demand for wine was growing in sheer volume; it also exhibited a more sophisticated palate that demanded better quality wines. For the Okanagan region, this meant a move toward the use of *Vitis vinifera* grape varieties that are suited to the area's climate and soils. Over the next three decades, expansion of the industry continued southward toward Oliver and Osoyoos. Two severe freezes, one in 1968 and another in 1978, wiped out much of the early vinifera varieties cultivated in the valley, including 4,000 vines of Chardonnay, Gewürztraminer, Cabernet Sauvignon, and Merlot imported from California and Washington in 1974. More successful vinifera experiments began in the period 1977–1985, as improving crop science indicated that over-cropping and inappropriate trellising had made the vinifera varieties more susceptible to winter damage. In this setting, grape acreage continued to climb slowly but steadily until the Free Trade Agreement between Canada and the United States in the 1980s saw acreage plummet from more than 3,000 acres in 1988 to approximately 1,000 acres the following year (McGuire 1993). At the same time, the provincial government instituted the Premium Wine Policy Industry Strategy as a mechanism to spur investment in its new estate winery and farm winery models for the production of premium wines. Subsequent expansion in grape acreage has averaged more than 10% per year since then and, at the time of writing, is roughly 7,000 acres (Fig. 5.2).

Wineries in British Columbia have also enjoyed a long history of government protection. For the most part, this took the form of preferential listing opportunities in the provincially controlled liquor distribution and sales system. In addition, markups on wines produced outside the province favored British Columbia wine sales. Finally, government subsidies through agencies such as the Department of Regional Economic Expansion (DREE) supported improved efficiencies in production (McGuire 1993). By the advent of the FTA in the late 1980s, the industry and the market were both poised to benefit from a new focus on premium wines. In 1989, the newly legislated British Columbia Wine Institute was ready to roll out the province's first appellation program of wine standards under the VQA branding. Supported by government legislation and policy that removed barriers to the production of premium wines based exclusively on locally grown grapes, the number of wineries began to expand significantly.

Landform and Climatic Conditions

The landform and climatic conditions of the Okanagan are certainly unique in Canada, but they are also unique in terms of wine-producing regions of North America. The Okanagan Valley is in the southern interior of British Columbia in the Dry Interior Plateau of the province. The valley stretches in a north–south direction for 250 km from a southern border with Washington State and contains a series of long, narrow lakes, which stretch the entire length of the valley. Physiographically, the area is a rolling upland, typical of the province's Interior Plateau. Extensive kame and outwash terraces and an assortment of other glacial and alluvial features that hold generally well-drained soils, an ideal setting for grape production, dominate the lower elevations of the valley. The valley bottom itself is completely dominated by a chain of large oblong lakes that influence warming and cooling trends during the growing season. The upland, particularly in the northern parts of the valley, holds a scattering of small lakes that have been dammed for irrigation purposes.

The moderating effect of the lakes, together with the protective shield of the Monashee Mountain Range to the east and the Coast Mountain Range to the west, provides a protective rain shadow effect setting the stage for a very mild, dry, continental climate (Belliveau et al. 2006). As the northern point of the great arid zone that extends southward into the United States to the Sonoran Desert, precipitation varies somewhat south to north and upslope. The regional values range generally between 300 and 400 mm per year on the agriculturally viable lands, with the higher amounts in the north as the valley trends east toward the Monashee Mountains. Variations in temperature also range over this space, with the highest values in the south of the region. Belliveau et al. (2006) conclude that Okanagan peak summer temperatures match those of other wine

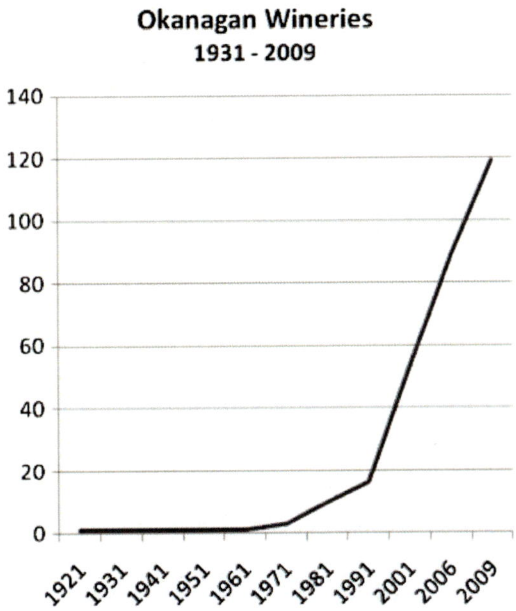

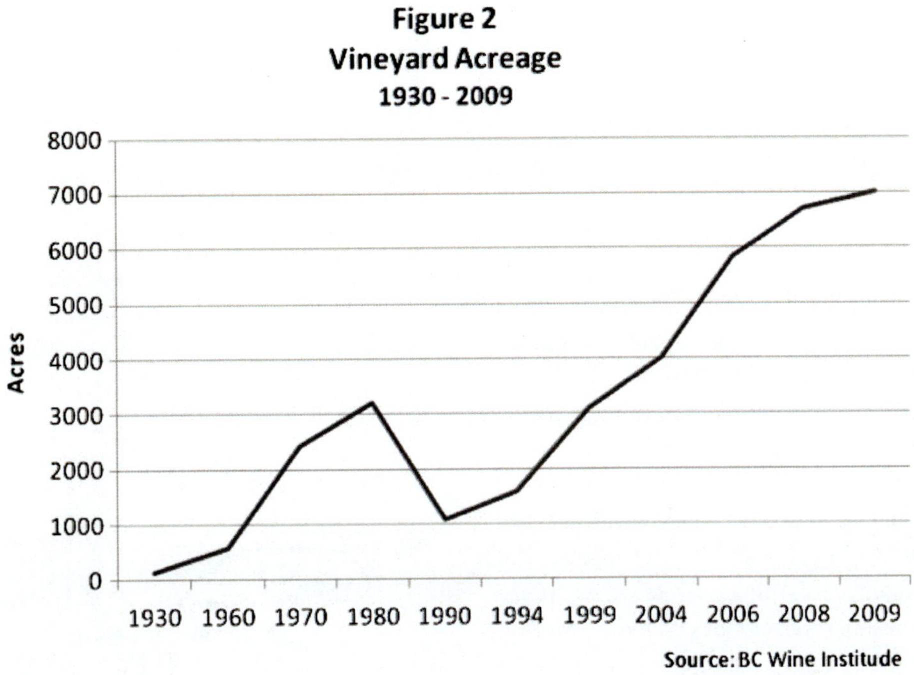

Fig. 5.2 Grape acreage 1930–2009

regions in western France, northwestern California, and Australia; however, the region pushes the geographic limits of where premium grapes can be grown because of fewer degree days in the spring and fall and an overall shorter growing season.

In its 1984 *Atlas of Suitable Grape Growing Locations*, Agriculture Canada combined climatic and soil variables in a first attempt to map the biogeoclimatic opportunities for grape production in the Okanagan and adjacent Similkameen Valleys (Agriculture Canada 1984). The Atlas illustrates several subregions within the valley considered Class 1 or very suitable for premium grape production throughout the valley. However, there are many microclimates and micropedalogic conditions that influence grape production. Microclimates reflect air funneling in mountain passes, differences in elevation, slope angle and direction, and proximity to lakes. Soil types also vary in the valley, with well-drained, sandy soils found in the south and darker, richer, clay, and loam soils to the north and on benchlands (Wilson 1989). The eastern side of the valley is dominated by granitic gneiss of the Precambrian Monashee complex. Rocks on the western side of the valley include younger metamorphic and sedimentary rocks such as sandstone and limestone; however, these are often intruded by Jurassic granites and overlain by tertiary volcanic rocks and sediments. Vineyards of the south Okanagan to Skaha Lake are situated in very-well-drained alluvial deposits; vineyards of the central and north Okanagan are located on mixtures of glacial silt, clay, and till, with good drainage provided by gentle slopes. Climate has created variations in the soil types. Hot and dry conditions in the south Okanagan have produced brown chernozem soils; whereas cooler moister conditions near Kelowna have yielded dark brown chernozems and, in forested areas, eluviated brunisols (Agriculture Canada 1984; Taylor et al. 2002; Roed and Greenough 2004).

Based on the complexity of soil and climatic conditions, Agriculture and Agri-Food Canada further divide the Okanagan into six viticultural regions (Fig. 5.3) based on landforms, climate, and varieties cultivated (Bowen et al. 2005). Kelowna, the northernmost region, is the oldest wine-producing region in the valley and the coolest. Planted in a variety of soils, most vineyards are located on west, facing slopes toward Okanagan Lake, which allows lake-effect moderation and airflow. Cool climate Germanic and French varieties of white wine, Riesling, Chardonnay, Gewürztraminer, and Ehrenfelser, dominate. Further to the south in Penticton and Summerland, temperatures are slightly warmer and sandy eolian soils overlay clay and silty clay loams in most vineyards. Again, cool-climate white wines dominate; however, Bordeaux style varieties including Merlot and Pinot Noir have recently experienced increasing success. Southward into Vaseaux-Oliver, soils become sandier, gneiss and granite bedrock become more common, and temperature degree days climb. Grape varieties are similar to those of the northern regions, with the addition of more red varieties, including Gamay and Cabernet Sauvignon. To the west of Oliver in the benchlands known as the Golden Mile, fluvial fans overlie Paleozoic rocks as soils become darker and give way to loamy gravels.

The very warm climate conditions of the south Okanagan provide suitable habitat for Bordeaux style reds such as Cabernet Sauvignon, Cabernet Franc, and Merlot, while whites typical of the other Okanagan wine regions are present in addition to Pinot Blanc and Chenin Blanc. In the most southerly region of the Okanagan is the Black Sage bench of Osoyoos where over half of the Okanagan's wine grapes are cultivated

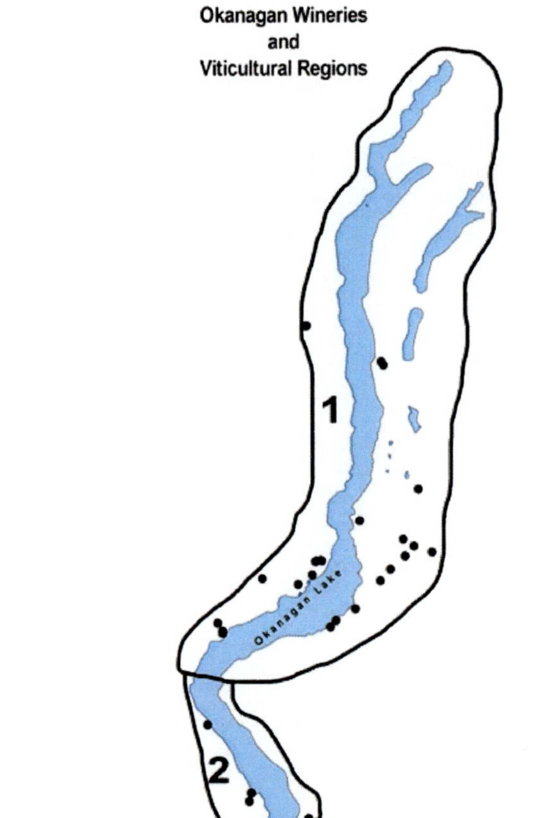

Fig. 5.3 Viticultural regions of the Okanagan

on sandy benchlands over granite bedrock. Here, on the eastern side of Lake Osoyoos, a large variety of reds, including Cabernet Sauvignon, Cabernet Franc, Merlot, Syrah, and Pinot Noir, are grown. The Similkameen Valley to the west is the youngest viticultural region in the area. It boasts a predominance of organic vineyards producing mainly Bordeaux varieties on sloping gravels and loamy fluvial fans (Bowen et al. 2005).

Wine, Landscape, and *Terroir* in the Okanagan

The distinct nature of viniculture and viticulture in the Okanagan is the result of natural conditions and cultural preferences that limit its extent and define its diversity. The tangible imprint of wine production and consumption on the earth's surface is the wine landscape. In the Okanagan, the regional identity of the wine landscape borrows heavily from the localism associated with *terroir*. *Terroir* is a unique Gallicism that has not translated easily outside of France, yet discussion of its existence, character, and role in the resistance to globalization has been widely debated, especially among geographers in New World wine regions (see for example Moran 1993; Morris 2000; Gade 2004). Further complicating the understanding *terroir* is the fact that there is no satisfactory translation in any language (Vaudour 2002). The complications of translation add to the mystery, attraction, and associations of quality that make the *terroir* brand valuable. The traditions of place-based authenticity associated with *terroir* (Gade 2004) seem intrinsic to wine-producing landscapes of Old World wine regions. Application of the term in New World wine regions like the Okanagan is more recent and, some might argue, problematic. Given the relative short history of quality wine production and the variation of natural landscape and climatic conditions in the valley, any notion of an Okanagan *terroir* should be considered organic as it evolves at a variety of scales.

There is a strong relationship between the wine and tourism industries in the Okanagan, which relies on the branding power of *terroir*. Tourism is an unusual industry in that it is consumed at the point of production (Britton 1991). Wine tourism facilitates the consumption–production overlap through tours, tasting rooms, and cellar door sales. Local consumption at increasingly smaller scales has become central to both quality wine production and quality tourism production. Wine tourism regions with appellation designations and defined *terroir* have emerged as the logical receptors of increasing adoration by aging baby boomers and urbanites (see for example: Bell and Valentine 1997; Bessiere 1998) in search of the class attachment and health benefits associated with epicurean consumption. In many respects, the wine tourism experience also fulfills the desires of an increasing number of Western consumers to decrease their carbon footprint while resisting the homogenizing effects of globalization by consuming locally.

Evolution of the Okanagan wine and wine tourism landscape is a unique outcome of a contemporary culture that demands localized experiences, discovery, and consumption of cultural products along with the lifestyles associated with them. The synthesis of wine and tourism demand has pushed expansion of the wine industry outside of its traditional European home base (Senese 2008). It is the agglomeration appeal of *terroir*, regional branding (Hall et al. 2000), and "VQA" and appellation that has created an upward spiral of wineries and their ancillary attractions, culinary schools, and bed and breakfasts in many New World winescapes like the Okanagan.

The changing Okanagan wine landscape reflects the New World revolution of the industry in many parts of the world where branding strategies have become essential in the competitive global market. The wave of wineries that emerged in the post Canada–USA free trade period between 1988 to approximately 1995 relied on the Old World appeal of rural romanticism found in the place names, images, and architecture of the Germanic wine regions where many of the winery proprietors and wine varieties produced in the valley originated (Fig. 5.4). While lacking the authenticity associated with Old World wine regions, this is a classic form of cultural reproduction so often witnessed in the North American landscape where international migration has taken place. It was also a branding strategy used by early producers in the valley who, like Old World producers, maintained a claim to quality based on tradition that stresses regions and brand names suggestive of the prestige of the producer (Heslop et al. 2009).

The winescape of this era also reflects patrimonialization of the wine industry as discussed by Gade (2004) where traditional place-based productions of wine from

Fig. 5.4 Cultural reproduction in the Okanagan winescape

the Old World is reproduced in this nascent though modern wine region. Here, the images of estate-based, family-run viticultural tradition give prominence in the branding of the wine industry (Fig. 5.5). The class attachment and association with wealth and family-based viticultural heritage elicit images of tradition in a wine region where, in reality, relatively little existed. The first attempts to apply geographical indications to the BC wine industry also reflect the imprint of patrimonialization by government designation of wineries as farmgate, estate, or commercial, based on the sourcing of its grapes:

> Commercial wineries do not necessarily have vineyards attached to them and are free to buy from other vineyards and to import grapes and juice from other countries. Estate wineries must cultivate and harvest a vineyard of no less than 20 acres, and half of all grapes used in wine production must come from the winery's own property. Three quarters of all grapes used in farmgate wine production must come from the winery's own vineyards and the majority of their wines must be sold on-site. Not unexpectedly, because of these regulations, most estate and farmgate wineries are small, family run operations. (Wine Country 1993: 3)

Wine producers in Australia, California, and New Zealand rejected the designation of region as the basis for classifying wines, opting instead to differentiate by grape variety. Further, not until the post 1990 VQA standardization of wine quality in British Columbia had been firmly established and quality wines from the Okanagan began to win medals in international competitions that a turn in the wine landscape of the Okanagan began to emerge. Like other New World wine regions in Australia, New Zealand, and Argentina, wineries of the Okanagan were also keen to appeal to new, younger, less-sophisticated wine buyers with considerable disposable incomes (Heslop et al. 2009). Two brand strategies began to emerge in the landscape, both with characteristics of an indigenous New World landscape that is neither Old World traditional or highly sophisticated. Some new wineries (for instance: Silkscarf, 2006; Oliver Twist, 2007; Forbidden Fruit, 2005; Therapy Vineyard, 2005; Zero Balance, 2008; Hijas Bonitas, 2008; Laughing Stock, 2008; Dirty Laundry, 2005; and Blasted Church, 2005) chose the fresh approach and cheekier attitude of some Australian and Californian producers who adopt less pretentious, fun names, often associated with "critters" (e.g., kangaroos, birds, penguins, lizards, etc.) (Heslop et al. 2009). Other new producers replaced the cultural reproduction of Germanic winery architecture, place names, and images with those sourced in the local area.

Fig. 5.5 Estate patrimonialization in the Okanagan winescape

These new wineries increasingly relied on the images of the natural landscape and local heritage of the region to design wineries, labels, and brand images (Fig. 5.6). For instance, Burrowing Owl, Crowsnest, Silver Sage, Blue Mountain, and Mount Boucherie borrowed place names and landscape icons of flora and fauna to re-imagine the wine industry. Locally sourced rock and wood enhance the architectural design of wineries that capitalize on fantastic views of the natural countryside and benefit from winery locations on the sloping vineyards above Okanagan, Skaha, and Osoyoos Lakes. In 2002, Nk'Mip, the first aboriginal winery in North America opened in Osoyoos (Fig. 5.7). The winery, developed by the Osoyoos Indian Band is now collaborating with Vincor, one of Canada's largest wine producers. Despite the change in proprietorship, branding of the winery still impresses a sense of place-based production that capitalizes on both the natural landscape and cultural heritage of the region:

> The Band shares its history and traditions with those who visit the winery. Tour experiences at the winery are enhanced through the architecture, designed and constructed to display native art and artifacts, and enhance the beauty and unique nature of the wild sage desert surroundings. From the early years of ranching, trading and small farms, the people of the Osoyoos Indian Band have continued to change with the times. The Band has a long history of growing grapes and owns 230 acres at the Inkameep Vineyard that were first planted in 1968. Under the expertise of trained viticulturalist Sam Baptiste, new varieties are being planted each year to meet current market demands. Pinot Blanc, Chardonnay, Riesling, Pinot Noir, Merlot and Cabernet Franc are all sourced from Inkameep Vineyard. In addition to the original site, huge vineyards have since been planted and now sprawl across the Band's 32,000 acres. (http://www.nkmipcellars.com/)

Early European settlers imagined the greening of the Okanagan Valley and transformed a naturally arid landscape into a rich agricultural landscape that, with the help of transportation ingenuity, fed a growing population of settlers. In a similar light, the Okanagan winescape is testament to the determination of early viticulturalists of the region who succeeded in establishing a wine industry in a unique landscape where few believed sustained cultivation of vinifera varietals was possible. The contemporary Okanagan wine region also reflects the global wine revolutions of the past 50 years that have resulted in a new wave of wine traditions in many New World wine regions. In the Okanagan, as in Australia, New Zealand, and California, the symbiosis of agricultural tradition with

Fig. 5.6 Indigenous images in the Okanagan winescape

Fig. 5.7 Nk'Mip Winery, Osoyoos British Columbia

the consumptive value of the winescape aesthetic have resulted in a unique and competitive wine industry that has appealed to a new age of wine consumer and changed the face of the cultural landscape.

References

Agriculture Canada 1984. Atlas of suitable grape growing locations in the Okanagan and Similkameen Valleys, British Columbia. Canada: Agriculture Canada.

Bell, D. and G. Valentine 1997. Consuming Geographies – We are where we eat. London: Routledge

Belliveau, S., B. Smit, and B. Bradshaw 2006. Multiple exposures and dynamic vulnerability: evidence from the grape industry in the Okanagan Valley, Canada. *Global Environmental Change* 16(4): 364–378.

Bessiere, J. 1998. Local development and heritage: traditional food and cuisine as tourist attractions in rural areas. *Sociologia Ruralis* 38 (1): 21–35.

Bowen, P., C. Bogdanoff, B. Estergaard, S. Marsh, C. Usher, C. Smith, and G. Frank 2005. Geology and wine 10: Use of Geographic Information System technology to assess viticulture performance in the Okanagan and Similkameen Valleys, British Columbia. *Geoscience Canada* 32(4): 161–176.

Britton, S. 1991. Tourism, capital and place: towards a critical geography of tourism. *Environment and Planning D: Society and Space* 9: 452–478.

Dendy, D. R. B. 1976. One Huge Orchard. B.A. Graduating Essay, University of Victoria, Victoria British Columbia.

Gade, D. W. 2004. Tradition, territory and terroir in French viniculture: Cassis, France and Appellation Controlee. *Annals of the Association of American Geographers*, 94(4): 848–867.

Hall, C.M., B. Sharples, B. Cambourne, and N. Macionis eds. 2000. *Wine tourism around the world: development, management and markets.* Oxford: Elsevier Science

Heslop, L., D. Cray, and A. Armenakyan 2009. Brand And Country-Of-Origin Effects In Wine Decision Making: Is Incongruity A Problem In A Wine World Turned Upside Down? *ASAC 2009 Administrative Sciences Association of Canada,* Niagara Falls, Ontario.

McGuire, D. W. 1993. The Political Economy of the Grape and Wine Industry in British Columbia and the Impact of the Free Trade Agreement. Okanagan University College, Department of Economics, Kelowna British Columbia.

Moran, W. 1993. The wine appellation as territory in France and California. *Annals of the Association of American Geographers,* 83(4): 694–717.

Morris, A. 2000. Globalisation and regional differentiation: the Mendoza Wine Region. *Journal of Wine Research,* 11(2): 145–153.

Nichol, A. E. 1983. Wine and Vines of British Columbia. Bottesini Press, Vancouver. 1983.

Roed, M. A. and J. D. Greenough eds. 2004. Okanagen Geology Kelowna: Okanagan Geology Committee.

Senese, D. 2008. Turismo enogastronomico Canadese, la nuova frontiera del turismo combinata all'ecoturismo e al turismo culturale" *Biteg*: *Congresso Internatzionale Sul Turismo Enogastronomico,* Saint Vincent, Italia.

Taylor, V.F., H. Longerich, and J.D. Greenough 2002. Geology and Wine 5. Provenance of Okanagan Valley Wines, British Columbia, Using trace elements: promise and limitations' *Geoscience Canada* 29(3):110–120.

Vaudour, E. 2002. The quality of grapes and wine in relation to geography: notions of terroir at various scales. *Journal of Wine Research,* 13(2): 117–141.

Wilson, K.W. 1989. Irrigating the Okanagan. MA Thesis. University of British Columbia, Vancouver, BC

Wine Country 1993. The BC Guide to Wine and Wineries of the Okanagan and Beyond. Premier 1993 Edition. Kelowna, British Columbia.

Part III

Physical

6 The Importance of Soil and Geology in Tasting *Terroir* with a Case History from the Willamette Valley, Oregon

Scott Burns

Abstract

Wines differ from each other based on seven different factors: the type of grape, the bedrock geology and resulting soils, the climate, the soil hydrology, the physiography of the site, the winemaker, and the vineyard management techniques. The first five of these factors make up what the French call *terroir*, "the taste of the place." An important process is the bedrock weathering into soils which then liberate chemical nutrients to the grape vines. Twelve of the sixteen essential elements for wine grapes come from the soil. Soil physical characteristics, such as soil depth, pH, drainage, salinity, color, and particle size, can affect the quality of grapes. Control of vigor in the plants is so important and is accomplished through irrigation or the use of poor soils. Certain grapes do better on different soils; therefore, it is apparent that all around the world, the geology and soils make up an important component of the *terroir* of the wine. In the Willamette Valley of Oregon in the United States, the *terroir* is strongly influenced by the bedrock geology and soils. The two dominant groups are the volcanic soils, the Jory Series, which developed on the Columbia River Basalts, and the Willakenzie Series, which developed on uplifted marine sedimentary rocks in the foothills of the Oregon Coast Range. The wines made from the grapes of these two soils are very different – so different that the Willamette Valley AVA has been subdivided into six new American Viticultural Areas based primarily on *terroir*.

Introduction

The soil and the underlying geology are very important factors in the making of a fine wine. A central theme of this paper explores how the physical environment of the vineyard shapes the character of great wines. Are soil and the geology the most important factors? The answer is probably no, but they are important (Gergaud and Ginsburgh 2005). The French have a term for all of the physical parts of the vineyard: *terroir* (Wilson 1998, 2001; Unwin 2012; Haynes 1999). This paper focuses on the geology and soil components of *terroir*.

First, the stage is set by discussing seven important factors that affect the quality of wine. Following is a brief section on the meaning of *terroir* and how it relates to those seven factors, plus a section on the relation of geology and soils to winemaking with a survey of famous wine making areas around the world

S. Burns (✉)
Department of Geology, Portland State University, Portland, OR 97207-0751, USA
e-mail: burnss@pdx.edu

P.H. Dougherty (ed.), *The Geography of Wine: Regions, Terroir and Techniques*,
DOI 10.1007/978-94-007-0464-0_6, © Springer Science+Business Media B.V. 2012

highlighting some of the relationships between the geology and soils and the overall *terroir* of the sites. Finally, the *terroir* of the northern Willamette Valley is explored and related back to its different geological parent materials and the resulting soils.

Important Factors in the Making of a Wine

There are seven factors that influence the character of each wine. Each has its importance in contributing to a particular wine's unique taste. The first factor is the *grape* itself. A Cabernet Sauvignon tastes different than a Pinot Noir. A Pinot Gris tastes different than a Merlot. Even finer distinctions can be made based on the clones of each grape.

The second factor is *bedrock geology,* which through weathering will produce a *soil* with a particular character. Limestone soils will produce completely different wines than vineyards on granite bedrock. Sometimes, the soil will differ from the bedrock because it is a sedimentary deposit left there by a glacier, landslide, stream, or wind. These are surficial deposits. White (2009) states that the five important soil factors are: depth, structure, strength, chemistry, and organisms.

Third is *climate* (Jones 2006). A climate that is dry at the end of the grape-growing season is desirable. A significant difference between daytime high and nighttime low temperatures helps to build complexity in the grape juice. Most of the world's wine-producing regions are located between 30° and 50° latitude in both the northern and southern hemispheres (Neirynck 2009). Wine grapes do not do well in extremely cold or extremely warm and wet regions.

Soil hydrology is the fourth factor that influences wine production. The texture and structure of the soil affects its drainage characteristics. Clay-rich soils are poorly drained. Grape plants prefer well-drained soils. In the Napa Valley, soil drainage is very important in determining the best sites (Swinchatt and Howell 2004; Swinchatt et al. 2006).

Fifth is the *physiography* of the site. What is the orientation of the vineyard? In the northern hemisphere, a south-facing slope can maximize solar radiation. What is the elevation? There is a top elevation above which the grapes do not ripen at the end of the summer.

The sixth factor is the *winemaker,* who has great influence. The vintner chooses what types of yeast and what types of oak barrels to use and if there is to be malolactic fermentation or not, whether the bottle will be corked or closed with a screw top, and how long the wine should be kept in the barrel.

Last are *vineyard management techniques*. Do the rows go north–south or east–west? What type of trellising is used? How many times are the vines trimmed? Should the vines be irrigated? All of these factors are important! Each has a different importance depending upon where the vineyard is located.

Terroir

An excellent paper by Tim Unwin (2012) discusses the meaning of *terroir*, which he emphasizes has been debated for years. Everyone has a different interpretation of its meaning, and this author agrees that the *terroir* is more than just the soil. It is a complex interaction of all of the physical aspects of the vineyard: the geology, soils, climate, geomorphology, and grapes. Wilson (1998) agrees, noting that, to him, *terroir* includes the vine, the subsoil, siting, drainage, and microclimate. To me, it is the grape, the geology/soil, climate, physiography, and soil hydrology.

The French belief in the importance of *terroir* is evident. On their wine labels, they do not put the name of the grape (except in the Alsace region). It is the location of the vineyard, the *terroir*, which is important. If the wine is red, and it is from Burgundy, then it should be known that the grapes used are Pinot Noir. If it is white wine from Burgundy, then the grapes have to be Chardonnay. But, a red from Beaujolais can only be the Gamay grape.

In Burgundy, *terroir* has been studied for over 400 years, first by monks and then by winemakers. Consistently, the best wines were produced from grapes grown at the top of slopes in the Corton area. They were given the classification of "Grand Cru" in the mid-1800s when France decided to brand the best wines. The next region down the slope was the next best wine zone year after year and was given the classification of "Premier Cru." Produced at the bottom of the slope were the common wines.

Geologists found that the boundaries of the three classification zones were geological (Wilson 1998). The *terroir* is controlled by the underlying rock units (Fig. 6.1). The Grand Cru zone is underlain by marl limestone (Argovian facies of the Pernand Marl), beneath the Premier Cru is Dalle Nacree flagstone limestone, and

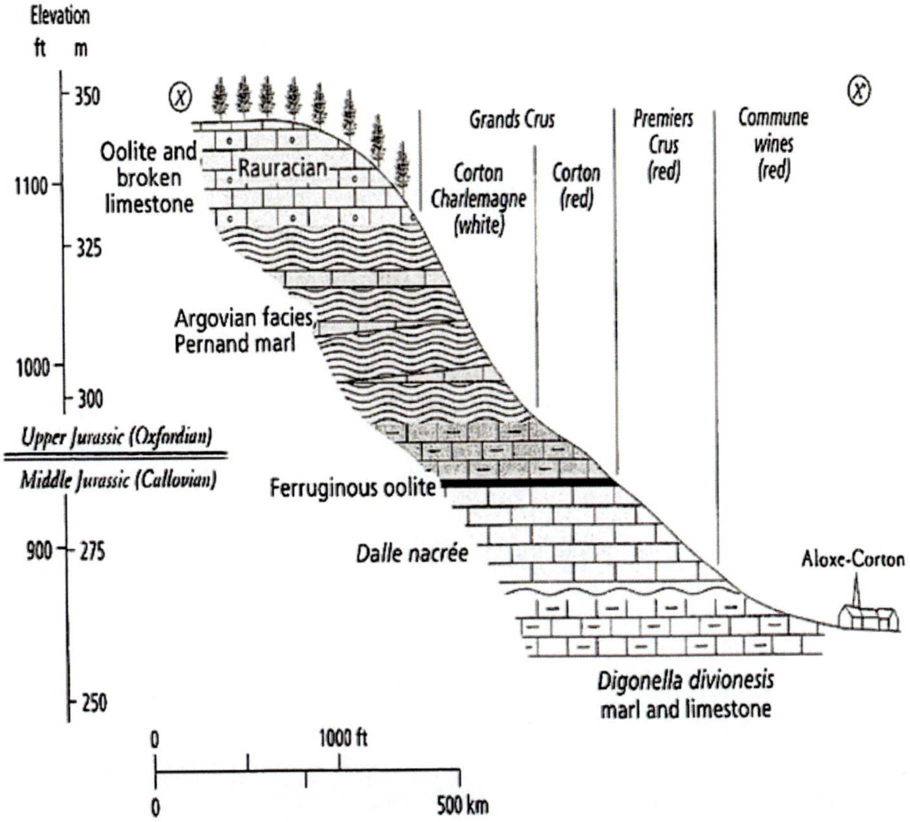

Fig. 6.1 Relationship between the wine quality classification and the underlying bedrock (Wilson 1998)

under the common wine zone is a mixture of limestone and marl (Digonella divionesis) (Fig. 6.1).

In Oregon, one of the four "founding fathers of modern winemaking," Dick Erath, has said many times, "Eighty percent of the quality of the wine comes from the vineyard and twenty percent from the winery." Clearly, he is a believer in *terroir*.

Geology and Soils in Viticulture

After centuries of wine making, vintners have found that certain physical soil characteristics are important for quality vineyards. Seguin (1986) and Pomerol (1989) argue that these physical characteristics are more important than chemical characteristics of the soil in producing distinctive *terroirs*. Soil depth affects the rooting depth and possible waterlogging of the plants (White 2009). Grape root systems have been known to penetrate to 5–6 m depths in gravelly and sandy soils such as Medoc, France, and Sunraysia, Australia (White 2009). Grape vines can perform well in shallow soils as evidenced by the shallow soils over limestone of the midslopes of Burgundy (White 2009). Some grape varieties like different soil depths, e.g., Chardonnay does better on deep soils, whereas Syrah (Shiraz) does best on shallow soils (White 2009).

Soil pH is important for it affects the availability of several nutrients and extremes in pH inhibit root growth (White 2009). Values below 5.5 or above 10 are deemed too acidic or too basic to sustain grape-growing activity (Neirynck 2009). High pH values reduce the availability of iron, copper, zinc, and manganese, whereas low pH values reduce phosphorus uptake (White 2009). Common salt (excess sodium and chloride) is not good for soils because it reduces the availability of potassium to the vines which is an essential nutrient (Neirynck 2009).

Soil texture (amount of gravel, sand, silt, and clay) influences the soil's water-holding capacity, drainage, and ease of cultivation (White 2009). Soil structure and consistence relate to how the soil particles are held together and also affects the strength of the soil, the drainage, and the ease of root penetration (White 2009). Soils need to be well drained because excess water causes grapes to swell. Soils with too high a clay content tend to be compact which prevents vine roots from penetrating to depths with essential nutrients and water (Neirynck 2009). The color of the soil affects the temperature of the soil; the darker the soil, the warmer the soil temperatures and the faster the maturation of the grapes (Neirynck 2009). Red colors indicate good drainage (White 2009). Rocky soils also hold warmth and promote maturation. Stony soils force grapes to search deeply for water and nutrients and are particularly suitable for grapes such as Syrah (Neirynck 2009). Presence of limestone indicates a neutral to alkaline pH, good soil structure, and possible restriction in rooting depth (White 2009). Presence of hardpans such as caliche horizons can also restrict rooting depths and cause the roots to extend laterally which affects the grapes quality.

White (2009) discusses the soil and geological characteristics of low and high potential sites for growing grapes. Low potential sites have clastic sedimentary or metamorphic rocks with little weatherable mineral remains, and exhibit shallow soils, sandy soils with little water holding capacity, low or high pH, little organic matter with low nitrogen content, weak soil structure, and poorly drained subsoil. High potential sites are often on igneous rocks, especially mafic (dark colored) ones, or unweathered shales, or limestones, plus metamorphic rocks and alluvium from the above rocks. The soils are generally deep with some fine particles for water-holding capacities, pH from 5.5 to 7.5, no exchangeable aluminum, no salts, high organic matter in the A horizon, well-aggregated soils, and well drained. Low potential sites need closer vine spacing to increase production per hectare; high potential sites need wider spacing (White 2009).

Vigor in a grape plant is important to the production of great wines. Excess vigor reflects a lack of balance between the vegetative and reproductive parts of the vine where too much vegetative growth happens. It results in poor fruit set and shading that slows the fruit ripening and lowers the quality of the grape (White 2009). Excess vigor in red grapes creates excess shading of the grapes during ripening which retards the development of color and phenolic compounds (White 2009). Excess vigor can result from too many nutrients coming from the soil or from too much water through irrigation. In moist climates, a perfect soil has low nutrient content, and in the dry climates, one restricts the water available no matter what the soil nutrient status.

On the other hand, some white grape varieties, such as Sauvignon Blanc, need a good supply of nitrogen and therefore vigor to develop the full aromatic potential of the fruit in these wines that are meant to be consumed without aging (White 2009). For white wines like Riesling and Semillon that benefit from aging, the use of fruit from less vigorous soils is preferred. Stress reduces vigor, and Seguin (1986) believes that the stress that contributes to the *terroir* of the region may be induced by the rate at which the vine can withdraw water from the soil which goes back to the physical characteristics of the soil.

Grapevines depend upon the soil for two things: water and nutrients (White 2009). Of all of the elements known to humans, there are 16 elements that act as nutrients that grapevines need to grow normally, flower, and produce fruit (White 2009). The ten macronutrients that are required in relatively large concentrations are: carbon, hydrogen, nitrogen, oxygen, phosphorus, sulfur, calcium, magnesium, potassium, and chlorine. The six micronutrients which are required in smaller concentrations are iron, manganese, zinc, copper, boron, and molybdenum. The carbon and the oxygen come mainly from the atmosphere through photosynthesis, and oxygen and hydrogen from water through the root system, so the rest of the nutrients come from the soil. The soil gets its nutrient supply from the weathering of the bedrock below the site, from sediments of the surficial deposit of the site, or from windblown dust. Other trace elements (found in low concentrations) are not essential to the plants but do give different flavors to the grapes. Examples of trace elements are chromium, selenium, iodine, cobalt, arsenic, mercury, cadmium, lead, and nickel (White 2009). Soil mycorrhizas are important parts of the soil biota because they enhance the vine's uptake of the immobile elements to the plants.

Tables 6.1 and 6.2 illustrate the soils upon which different grapes do well (Neirynck 2009). White wines do best on calcareous soils (chalks, limestone, and marl). Rieslings also do well on slates, and Sauvignon Blanc and Semillon do well on gravel soils. White wine grapes do not do well on granitic soils (which tend to be sandy) or on clay-rich soils. Red wines seem to do better on calcareous soils, but also mature well on granitic soils, especially Syrah and Grenache. Red wines do not, however, generally do well on slate soils (Neirynck 2009). Gladstones (1992) argues that rocky and gravelly soils produce the best wines across the different climates of the New World.

Certain parent materials produce flavors and aromas in the wines (Neirynck 2009). Calcareous soils produce lemon and citrus flavored wines with a long acid finish. Marl soils and silty calcareous soils tend to generate peppery flavors. Sandstone soils tend to bring a "nervy"

Table 6.1 Growing white wine grapes in various soil compositions (Source: Neirynck 2009)

Soil (light to heavy)	Riesling	Chardonnay	Sauvignon Blanc	Semillon	Trebbiano	Muscat
Granite						
Chalk			X	X		
Limestone	X	X			X	X
Marl		X				X
Loam						X
Gravel			X	X		
Clay						
Sandy						
Slate	X					

Table 6.2 Growing red wine grapes in various soil compositions (Source: Neirynck 2009)

Soil (light to heavy)	Cabernet Sauvignon	Syrah	Pinot Noir	Merlot	Grenache	Cabernet Franc
Granite		X			X	
Chalk			X			
Limestone	X	X	X			
Marl			X			
Loam			X		X	
Gravel	X			X	X	X
Clay				X		
Sandy						
Slate						

character to the wines, and the fine-grained clay soils emphasize tannic characteristics. Schist-rich soils stress the austere nature of the wine, and volcanic soils produce full-bodied wine with smoke-based aromas. John Livingston (1998) states that volcanic parent material in the soil gives a distinctive character to the Cabernet Sauvignon wines of the Stag's Leap area in the Napa Valley of California.

W.H. Terry Wright (2002) investigated the impact of differences of particle size in the soils at the Iron Horse Vineyards in Sebastopol in the Sonoma Valley of California on Pinot Noir wines. Vineyard A had 42% clay in its soil, and Vineyard B had 20% clay. Vineyard A grapes yielded wines with a deep-cherry color and deep berry and cherry-cola flavors. Vineyard B yielded grapes that produced wine with a light cherry color, light berry taste, softer tannins, a bigger nose (stronger aromatic components), and generally more elegance.

Geology and Soil Differences of *Terroir* in Vineyards of the World

There are many grape-growing regions around the world, each with a particular climate that matches well with a particular grape variety (Jones 2006). Climate is very important, but the geology and soil add a special taste to each of the wines (Macqueen and Meinert 2006). A survey of selected regions shows the impact of geology and soil on wines produced in those regions.

Europe

France contains a number of famous wine-growing areas. The *terroir* of the *Burgundy* region is dominated by marl limestone and alluvial deposits (Wilson 1998). The best wines come from vineyards on the marl, the next best are over pure limestone, and the wines of lesser quality are produced from the alluvial soils at the bottoms of the slopes. To the south, one finds the *Beaujolais* region where the Gamay grape is dominant, where the bedrock geology is mainly granites and schists (White 2009). In the *Bordeaux* region, particularly the *Medoc* region, the best wines come from vineyards along the river containing mainly cobbles, gravels, and sand. In the largest wine-producing region in the world, the *Languedoc-Roussilion* region of southwest France, most of the soils have developed on sediments that have collected in this basin in the last 100 Ma (Suthren et al. 2006). There, most vineyards are separated into "slopes" and "plains." Soils tend to be shallow, very stony, and excessively drained, and so are in need of irrigation (White 2009). In the *Rhone* region, most of the soils are developed on limestone bedrock and have abundant glacial sediments from the Alps (White 2009).

Italy is a very diverse wine-producing region containing many *terroirs* (Cita et al. 2004). In the north, the Alps and their valleys are dominated by sedimentary rocks, of the uplifted Tethys Sea, and metamorphosed sediments with granitic intrusions. The middle of the country is dominated by the Apennine Mountains and the sedimentary and volcanic bedrocks associated with them. In the south and on Sicily, one finds abundant volcanic bedrock associated with sedimentary rocks. Soils vary considerably. Cita and Fiore (2006) have contrasted the *terroirs* of the Adige River Valley in the Southern Alps and their coarse-grained calcareous alluvial fan soils, with the vineyards in the southern Apennine's fertile, potassium-rich volcaniclastic terrain. Both areas produce red wines, but the northern area has low fertility in the soils compared to the volcanic sediments of the southern site where high fertility demands more vineyard management.

Spain is a country with more land under vineyards than any other country in the world (Johnson and Robinson 2009). Most of the grape-growing regions, called bodegas, are on limestone bedrock with red, clay-rich soils which are common on calcareous soils. The Toro, Ribera del Duero, Rioja, Navarra, and Catalunya wine regions are typical of the limestone bedrock. An interesting *terroir* region with very distinctive wines is the Priorat region southwest of Barcelona where the soil is called "llicorella" which is dark brown in color because it comes from slate bedrock. Slate is also found in the soils of the Axarquia Mountain bodega in the south, and chalk soils dominate the vineyards of the Andalucia region where Spain's famous sherry comes from. In the northwest of the country in the Galician region, one finds granite bedrock and associated alluvial soils. The Somontano region northwest of Barcelona has very sandy soils that are low in potassium and very gravelly soils in the lower reach of the River Vero.

Portugal is probably most famous for its fortified wine called port. The geology affecting the soils and the *terroir* is quite variable (Johnson and Robinson 2009). Granite, slate, and schist dominate the bedrock in the north and the central inland areas of the country. Limestone bedrock and clay and sandy soils are more characteristic of the central coastal area and the southern part of the country.

Southwestern *Germany* is a cool climate wine-producing area where white wines dominate, especially Riesling (Johnson and Robinson 2009). Slate bedrock dominates the Middle Mosel Valley, the Saar and Ruwer tributary valleys of the Mosel, the Nahe region, and the central Rheingau area. In the Urzig region of the Middle Mosel, the slate is mainly red instead of gray, and the wines are very different. The Upper Mosel Valley is mainly limestone, so instead of Riesling and Muller-Thurgau grapes being grown there like the rest of the Mosel Valley, Pinot Gris and Pinot Blanc grapes are grown. Quartzite is the main bedrock in the Lower Mosel Valley and parts of the Rheingau. In the Franken region, the soil is more limestone and clay-rich which are preferred by the Silvaner grape. These grapes produce wines that seize the minerals, producing a wine that resembles Chablis. In the Bad Wurttenberg region in the south, the soils are mainly loess on which many red varieties are produced, while the Pinot Gris grapes grow best on the volcanic soils.

United States

In the *Napa Valley of California,* the substrates of the vineyards are divided into three groups: residual, alluvial, and fluvial (Swinchatt et al. 2006). The residual soils are the colluvial soils of the slopes of the Vaca and Mayacamas mountains on each side of the valley. Bedrock is mainly the Sonoma Volcanics, but in some places, the Franciscan Formation of uplifted marine rocks can also be found. Soils tend to be thin, less than a meter in depth. The dominant mapping unit is the alluvial substrate of the alluvial fans found on the eastern and western sides of the valley. Soils are all less than 15,000 years old. Differences in the fans depend on the particle size of the sediments. The third division of substrates is fluvial deposits of the Napa River – mostly fine-grained soils that are deep and easily penetrated by the roots. Certain bedrock is not good for grape production, and an example is found at the northern end of the valley where serpentinite bedrock which is high in nickel and magnesium adversely affects the growth of the plants (White 2009).

Washington is the second largest wine-producing state in the United States. Over 95% of the vineyards are on sediments associated with the great Missoula

Floods or loess from the floods (Allen et al. 2009). The Walla Walla Valley and the Yakima Valley are the major regions of eastern Washington wine production. Meinert and Busacca (2000) list over 75 different soil series found on these glacial rhythmites, dunes, gravel bars, and loess in the Walla Walla Valley alone – a region that produces superb Syrah, Merlot, and Cabernet Sauvignon wines. Volcanic ash is also found in some of the soils where the sediments overlay Columbia River Basalt.

One special region of *eastern Washington* that consistently receives scores from wine rating services in the 90s is the *Red Mountain Appellation* (Meinert and Busacca 2002). The steady westerly winds, the cool nights and hot, dry days, along with the sandy to gravelly glacial slack-water sediments from the Missoula Floods, produce superb wines. Caliche horizons and volcanic ashes contribute to the *terroir* of this appellation. Red Mountain is one of the warmest and driest viticultural sites in the state of Washington, with total sunshine similar to the Napa Valley in California.

The *Umpqua Valley AVA in Oregon* lies at the intersection of the Coast Range, made of uplifted marine sediments about 50 Ma old, and the Klamath Mountains, accreted terrane sediments over 100 Ma old composed of complex folded igneous and metamorphic rocks. Vineyards are mainly on terraces of the valley and upland surfaces on the bedrock (Jones et al. 2004). Soils are more acidic in the northern part of the valley.

The majority of the vineyards in *Idaho* are found in the Western Snake River Plain at the western side of the state (Gillerman et al. 2006). Some of the vineyards are developed on basalt from the Snake River Floodplain. Sediments were deposited in ancestral Lake Idaho and after the lake drained. The ancestral Boise River incised into the sediments producing stream terraces where many wineries are located. The oldest is the Tenmile Gravel Terrace, and the youngest is the Deer Flat Terrace which contains loess deposits. The wineries are located on other catastrophic flood deposits from the Bonneville Flood which occurred 14,500 years ago. Soils developed on the lowest terraces and the youngest deposits are mainly Inceptisols, but the highest terraces show well-developed soils with caliche and duripan soils.

Canada

The *Niagara Peninsula* is the main wine-producing area in *Canada*, accounting for up to 80% of Canadian production (Haynes 2000). The bedrock of the region is mainly Paleozoic carbonates, with lesser amounts of shales, sandstones, and dolomites. Resting on top of the bedrock are mainly glacial deposits of till, but also some glaciolacustrine and glaciofluvial deposits. Two major soil *terroirs* are noted there. The Lake Iroquois Plain, the area between Lake Ontario and the Lake Iroquois shoreline bluff, is characterized by heavy clay soils and is mainly flat. The Escarpment, the terraces of the Lake Iroquois Bench and the slopes above the Niagara Escarpment north of the Vinemount Moraine, is favored because of the excellent slope and groundwater drainage.

The second main wine-producing area in *Canada* is the *Okanagan Valley of British Columbia* (Taylor et al. 2002). The underlying bedrock is mainly composed of Paleozoic and Mesozoic terranes accreted onto North America consisting of mafic metavolcanics, granites, and gneisses. A Pleistocene basalt flow underlies one vineyard. Pleistocene glacial deposits up to 100 m thick cover most of the bedrock with glaciofluvial deposits along the sides of the valley. Soils tend to be sandier in the south and more clay-rich in the north and a combination of mainly Inceptisols with some Alfisols with a few carbonate horizons in them. Because of the glacial mixing of the sediments in the soils, there seems to be very little chemical variation between vineyard soils. Bowen et al. (2005) found that vineyards sited on coarse-textured, loamy soils appear to be the best suited for producing high-quality wines.

Southern Hemisphere

Wines of the *Coastal Region of South Africa*, primarily the Stellenbosch (Fig. 6.2) and Paarl districts, are the most famous in Africa (Bargmann 2003). Bedrock falls into five groups: clay-rich sediments of the Malmsberry Group, granitic intrusives of the Cape Granite Suite, conglomerates and mudstones of the Klipheuwel Group, sandstones of the Table Mountain Group, and alluvium on the bottom of the valleys. Soils formed in these vary from very well-drained soils to poorly drained ones on the clay-rich bedrocks.

Chile and Argentina are the main wine-producing areas in South America. Many of the vineyards in *Chile* are also located on alluvium from rivers (Fig. 6.3) and alluvial fans from the surrounding mountains (White 2009). In neighboring *Argentina*, the story of

Fig. 6.2 Vineyards in the Stellenbosch area of South Africa

Fig. 6.3 Vineyard on alluvial soils north of Santiago, Chile

Fig. 6.4 Vineyard on the alluvial fans of the Andes in Mendoza, Argentina

terroir is also in alluvial fans and terraces (Fig. 6.4) incised into them at the edge of the Andes in places like Mendoza (White 2009). The upper terraces have good caliche soils.

Many of the vineyards in *New Zealand* are dominated by alluvial soils made of sediments on stream terraces and in floodplains. Soils are deep, well-drained, and not very fertile, e.g., those found in the regions of Hawke's Bay, Martinborough, Gisborne on the North Island, and Marlborough (Wairau and Awatere valleys), Nelson, and Canterbury regions on the South Island (White 2009). An area called "Gimblett Gravels" in Hawke's Bay claims to be the first New World wine region defined by *terroir*. The very stony and gravelly soils have low fertility but act as a heat store that tempers the cool sea breeze that Hawke's Bay experiences. The alluvial soils are dominated by volcanic sediments on the North Island and greywacke sediments on the South Island. The Kawarau Valley of *Central Otago* on the South Island of New Zealand has an interesting *terroir* – mainly thin soils developed on schist bedrock (White 2009). Waiheke Island off the coast of the North Island (a short distance from Auckland) is mainly greywacke soils. Another interesting *terroir* on the South Island is found in the Omihi Hills where the bedrock is a part of the Torlesse Group of limestone deposits. The alluvial soils of the nearby Waipara River Valley have high calcium carbonate values because of these calcareous deposits.

Three of the main wine regions in *Australia* have distinctive geologies and soils associated with them (White 2003). In the *Coonawara* region of South Australia, the clay loam soils of the upland limestone bedrock forms a wonderful Terra Rossa, a very red soil that produces excellent wines. In the *Barossa Valley* of South Australia, probably the country's most famous region, soils are old and are developed on bedrock of marble, gneiss and granites, and alluvium from those three bedrocks. The soils are dark red with caliche subhorizons and are famous for producing robust red wines. The *Heathcote* and *Nagambie Regions* of Victoria are located on a volcanic intrusion of dolerite and diabase with conglomerates and cherts. Many of the vineyard soils are alluvium derived from these bedrocks.

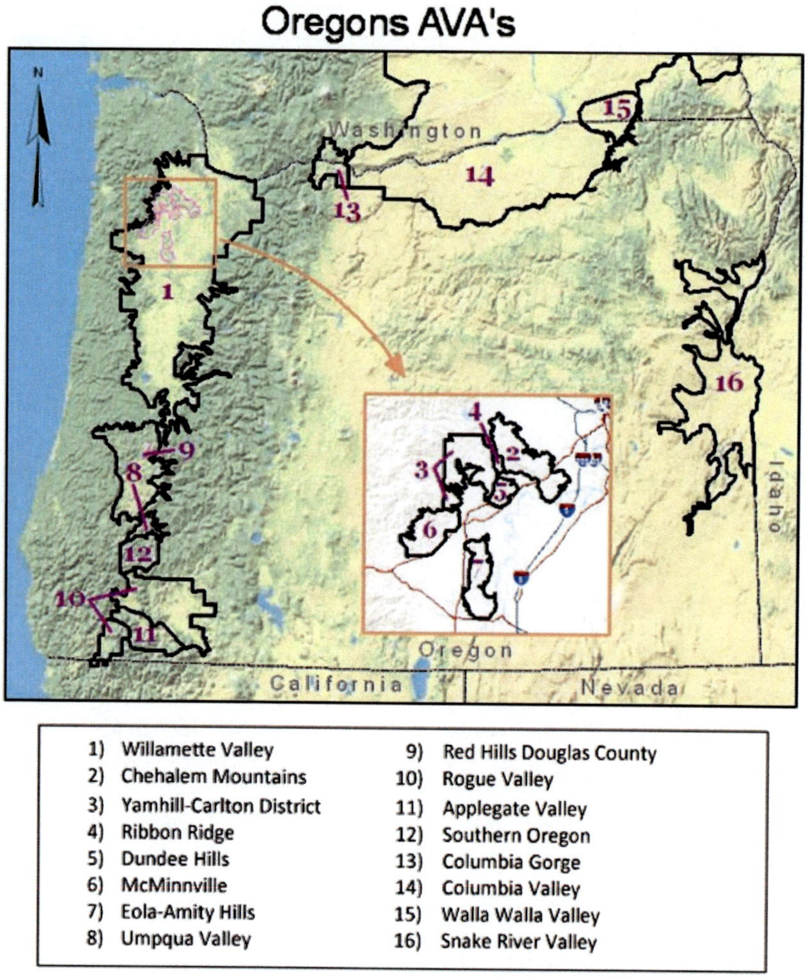

Fig. 6.5 Oregon AVAs with the newly subdivided Willamette Valley AVAs, based on different terroirs

The Northern Willamette Valley, Oregon: Best Place in the World to Taste Differences in *Terroir*

The northern Willamette Valley is the heart of wine country in Oregon (Fig. 6.5). It is where David Lett started modern winemaking in Oregon in 1961. It has a cool climate region similar to Burgundy in France and is a perfect place to grow Pinot Noir (Fig. 6.6), Pinot Gris, Chardonnay, and German-style wines like Riesling, Gewürztraminer, and Muller-Thurgau. It stretches from Portland in the north to Salem in the south, and occupies the wide valley from the Cascade Mountains to the Coast Range.

The wine industry in Oregon has matured significantly since the early days of the 1960s. Today, there are over 400 wineries, placing it third in the nation with over 800 vineyards (Oregon Wine Board 2011). Over 40 grape varieties are grown producing annual sales of over \$300 million. Annual production ranks fourth in the nation at 1.6 million cases. The impact on the economy is over \$1.4 billion annually. The ratio of red to white wine is 60–40%. In total, 17,400 acres are planted in grape vines with 13,800 acres harvested every year.

In the late 1990s, Dionne Starr-Peace (2002) and Burns (2002) surveyed the vineyards in the northern Willamette Valley to see which soils were being used to grow wine grapes. The summary of the findings is found in Table 6.3. A total of 23 soil series were found. They are classified to the order level of Soil Taxonomy (Soil Survey Staff 2009) with determinations of the bedrock under each vineyard and a calculation of the total acreage of that soil. It is a summary of the data from over 5,250 acres of soils on over 200 vineyards.

The primary soil, the Jory series, with 1,504 acres formed on Columbia River Basalt, is an Ultisol (Fig. 6.7). This Xeric Palehumult is the state soil of

Fig. 6.6 Pinot Noir grapes which are the leading grape produced in the Willamette Valley

Table 6.3 Distribution of soils in vineyards in the northern Willamette Valley

Soil series	Parent material	Order	Acres
Jory	Basalt	Ultisol	1,504
Willakenzie	Marine sediment	Alfisol	1,245
Laurelwood	Basalt	Alfisol	825
Yamhill	Basalt	Mollisol	333
Woodburn	MISSOULA FLOODS	Alfisol	298
Nekia	Basalt	Ultisol	195
Cornelius	Old loess	Alfisol	141
Saum	Basalt	Inceptisol	126
Bellpine	Marine sediment	Ultisol	118
Willamette	MISSOULA FLOODS	Mollisol	110
Dupee	Marine sediment	Alfisol	58
Peavine	Marine sediment	Ultisol	50
Helvetia	Mixed	Mollisol	45
Cascade	Young loess	Inceptisol	31
Wellsdale	Marine sediment	Alfisol	31
Steiwer	Marine sediment	Mollisol	28
Carlton	MISSOULA FLOODS	Mollisol	25
Amity	MISSOULA FLOODS	Mollisol	24
Melbourne	Marine sediment	Alfisol	18
Aloha	MISSOULA FLOODS	Inceptisol	15
Hazelair	MISSOULA FLOODS	Mollisol	12
Chehalem	Mixed	Mollisol	10
Dayton	MISSOULA FLOODS	Alfisol	6

23 different soils found–being remapped

Fig. 6.7 Jory Soil, the state soil of Oregon

Oregon and has cousins like Nekia, a shallow Jory, Saum, and Yamhill series. The second most cultivated soil is an Alfisol of the Willakenzie Series with 1,245 acres formed on uplifted marine sediments of the

Fig. 6.8 Cross-section view of Willakenzie soil taken at the Elk Cove Winery with a wine bottle for scale

Oregon Coast Range and composed of mainly sandstones and shales (Fig. 6.8). This Ultic Haploxeralf has similar cousin soils like Bellpine, Dupee, Peavine, Wellsdale, Steiwer, and Melbourne. Third is the Laurelwood Series with 825 acres. This Ultic Haploxeralf soil developed on the basalt bedrock, but has mid-Pleistocene loess mixed into the soil providing more nutrients and therefore a classification of an Alfisol. It is a hybrid between a Jory and Willakenzie soil. Other loess soils are the Cornelius, with old loess mixed in and thus a classification of a Mollic Fragixeralf, and the Cascade Series, with young loess and a classification of Typic Fragixerept, making up most of the soil.

The last group of soils developed on the valley bottoms is not favorable for wine production. These are in the nutrient-rich Missoula Flood deposits (Allen et al. 2009) and are primarily the Woodburn Series (Aquic Xerept), but also include the Willamette, Carlton, Aloha, Amity, Dayton, and Hazelair Series. The primary bedrock of wineries in the northern Willamette Valley is basalt (Table 6.4).

In the last 15 years, a "great debate" has developed in the northern Willamette Valley as to which soil series produces the best Pinot Noir wine (Burns 2002). Some say it is the Jory soil and its cousins developed on the basalt bedrock, while others say that it is the Willakenzie soil and its cousins developed on marine sediment soils. There is a big difference in the wines produced on these two soils. If all of the factors are kept constant except the soil series, i.e., the same grape, the same year, the same climate, the same vineyard management, and the same winemaker, two completely different Pinot Noir wines result. Jory soils produce a wine that is light red in color, has a strong bouquet, and has fruit flavors of red cherries, raspberries, red plums, and red currents, whereas Willakenzie soils produce wine that is dark red in color, has a strong finish, and has fruit flavors of dark cherries, blackberries, and black plums.

Table 6.4 Distribution of parent material in vineyards of the northern Willamette Valley

Parent Material	Acres	Percent
Basalt	2,158	41
Basalt and old loess	825	15
Marine sediment	1,562	30
MISSOULA FLOODS	475	9
Old loess	141	3
Young loess	31	1
Mixed bedrock	55	1

Ken Wright, a leading winemaker in Oregon, says that the Jory soils produce fruit-driven flavored wines and Willakenzie soils produce floral and spice flavors like lavender, cola, tobacco, cedar, and anise. Opinions of flavor differ, but there is no question that there is a

Fig. 6.9 Red Jory soils of the Dundee Hills AVA in the Willamette Valley

major difference in the two wines and that it comes from differences in soil. Since Jory and Willakenzie soils are located so near to one another, it is easy to isolate the effect of the soil factor and to compare wines from both soils in the same day. The close proximity of different bedrocks and the big differences between the two wines make the northern Willamette Valley one of the best places in the world to taste *terroir* differences based on the soil variation.

Why are there differences in the wines produced on these soils? At this point, there are no good explanations. A study has just started at Portland State University to explore the differences using a mass spectrometer to characterize the elements, especially the heavy metals and trace elements that are found in the soils and the juices as well as the wines produced in selected blocks of grapes where the soils were sampled.

Until a couple of years ago, the whole Willamette Valley was one AVA (American Viticultural Area). In 2008, it was subdivided into six sub-AVAs based on the different local *terroirs*. First, the vineyards of the Dundee Hills declared that their Jory soils developed on basalt and needed to have their own label (Fig. 6.9). Then, the rolling hills of the area to the west of the Dundee Hills declared that they wanted Yamhill-Carlton as their AVA because their geology is marine sediments with Willakenzie soils. Then, the Eola-Amity Hills in the middle of the valley claimed that their Jory/Nekia soils were special, and they wanted their own AVA. Ribbon Ridge, at the north end, has mainly Willakenzie soils and marine sediments, and they applied for their own AVA based on the soil, geology, and the physiography of the ridge. Chehalem Mountain followed suit with their declaration of having mainly Laurelwood soils and basalt bedrock being special. Finally, McMinnville asked for its own AVA.

Conclusions

Seven main factors have been shown to affect the taste of a wine. They are the grape variety, geology and resulting soils, climate, soil hydrology, the physiography of the site, the winemaker, and vineyard management techniques. The first five of these factors comprise what the French call "*terroir*," "the taste of the place." *Terroir* differs from wine region to wine region around the world as has been shown in this work. The examples discussed in this chapter definitely show that geology and soils are important in affecting *terroir*.

In reviewing the wine-producing areas of the world, one finds that many of the well-known wines are grown on alluvial deposits, often with a limestone component. The importance of the physical characteristics of the soil and the chemical components of the bedrock and parent material of the soils greatly influence the local terroirs (Burns 2009).

In Oregon's Willamette Valley in the United States, there are two main *terroirs* based mainly on the geology and resulting soils. The Jory Series of soils developed on the Columbia River Basalts in western Oregon produce Pinot Noir wines that are light red in color and have red fruit flavors. The Willakenzie Series of soils developed on the uplifted marine sediments of the nearby Coast Range produce Pinot Noir wines that are dark red in color and have dark red fruit flavors. Because of these differences in wines, the Willamette Valley AVA has been subdivided into six new AVAs (American Viticultural Areas) based on the geology/soil characteristics of the Willamette Valley *terroir*.

References

Allen, J.E., M. Burns, and S. F. Burns. 2009. Cataclysms on the Columbia: The great Missoula floods. Portland, OR: Ooligan Press.

Bargmann, C. J. 2003. Geology and wine 7. Geology and wine production in the Coastal Region, Western Cape Province, South Africa. *Geoscience Canada* 30: 161–182.

Bowen, P. A., C. P. Bogdanoff, B. P. Estergaard, S. G. Marsh, K. B. Usher, C. A.S. Smith, and G. Frank. 2005. Geology and wine 10. Use of Geographic Information system technology to access viticultural performance, Okanagan and Simalkameen valleys, British Columbia. *Geoscience Canada* 32: 161–176.

Burns, S. F. 2002. The great debate – is the best Pinot Noir in Oregon produced on soils from sedimentary or volcanic soils? *GSA Abstracts with Programs* 34 (5): A21

Burns, S. F. 2009. Northern Willamette Valley – the best place in the world to taste the difference in *terroir*. *Geological Society of America Annual Meeting, Abstracts with Programs* 31 (7): 696.

Cita, M. B., S. Chiesa, R. Colacicchi, G. M. Crisci, P. Massiotta, and M. Parotto. 2004. *Italian Wines and Geology*. Milano, Italy: Paesaggi Geologici.

Cita, M. B., and A. Fiore. 2006. Grape growing and wine making in two unique Italian *terroirs*: Teroldego Rotaliano and Aglianico del Vulture. *Geoscience Canada Reprint Series* 9: 165–181.

Gergaud, O. and V. Ginsburgh. 2005. La qualite des vins de Bordeaux se resume-t-elle uniquement a celle de son *terroir*? *La Journee Vinicole* 29 April 2005:13–15

Gillerman, V. S., D. Wilkins, K. Shellie, and R. Bitner. 2006. *Terroir* of the western Snake River Plain, Idaho, USA. *Geoscience Canada* 33: 37–48.

Gladstones, J. S. 1992. *Viticulture and the environment. A study of the effects of environment on grape growing and wine quality with emphasis on the present and future areas for growing wine grapes in Australia*. Adelaide, Australia: Winetitles.

Haynes, S. J. 1999. Geology and wine 1. Concept of *terroir* and the role of geology. *Geoscience Canada* 26: 190–194.

Haynes, S. J. 2000. Geology and wine 2. A geological foundation for *terroirs* and potential sub-appellations of Niagara Peninsula wines, Ontario, Canada. *Geoscience Canada* 27: 67–87.

Johnson, H. and J. Robinson. 2009. World Atlas of Wine. Seventh Edition. London: Mitchell Beazley.

Jones, G. V., N. Snead, and P. Nelson. 2004. Geology and wine 8. Modeling viticultural landscapes: A GIS analysis of the *terroir* potential in the Umpqua Va1ley of Oregon. *Geoscience Canada* 31: 167–178.

Jones, G. V. 2006. Climate and *terroir*: Impacts of climate variability and change on wine. *Geoscience Canada Reprint Series* 9: 203–217.

Livingston, J. 1998. The geology of fine wine. *California Wild*. California Academy of Sciences Quarterly

Macqueen, R. W. and L. D. Meinert. 2006. *Fine wine and terroir: The geoscience perspective*. St. Johns, Newfoundland: Geological Association of Canada, Geoscience Canada Reprint Series 9.

Meinert, L. D., and A. J. Busacca. 2000. Geology and wine 3. *Terroirs* of the Walla Walla Valley Appellation, southeastern Washington State, USA. *Geoscience Canada* 27: 149–171.

Meinert, L. D., and A. J. Busacca. 2002. Geology and wine 6. *Terroir* of the Red Mountain Appellation, central Washington State, USA. *Geoscience Canada* 29: 149–168.

Neirynck, Baudouin. 2009. *The Grapes of wine, The fine art of growing grapes and making wine*. New York: Square One Publishers, Garden City Publishers.

Oregon Wine Board. 2011. http://industry.oregonwine.org/wp-content/uploads/2011/08/Or-EconReport-2010-Final.pdf

Pomerol, C. 1989. *The wines and winelands of France. Geological journeys*. London: Robertson McCarta.

Seguin, G. 1986, "*Terroirs*" and pedology of wine growing. *Experientia* 42: 861–873.

Soil Survey Staff. 2009. *Keys to soil taxonomy*. 11th Edition. Washington, D.C: USDA.

Starr-Peace, D. 2002. *Distribution of soils in the northern Willamette Valley in vineyards*. Unpublished M.A. Thesis, Department of Geography, Portland State University.

Suthren, R. J, M. B. Fowler, and P. D. Guion. 2006. *Terroir*, wine and geoscience fieldwork education in Languedoc-Rouissillon, southern France. *Geoscience Canada Reprint Series* 9: 233–246.

Swinchatt, J. and D. G. Howell. 2004. *The winemaker's dance: Exploring terroir in the Napa Valley*. Berkeley, California: University of California Press.

Swinchatt, J., D. G. Howell, and M. Silacci. 2006. A review of Napa Valley, California, *terroir*. *Geoscience Canada Reprint Series* 9: 217–231.

Taylor, V. F., H. P. Longerich, and J. D. Greenough. 2002. Geology and wine 5. Provenance of Okanagan Valley wines, British Columbia, using trace elements: Promise and limitations. *Geoscience Canada* 29: 110–120.

Unwin, T. 2012. *Terroir*: At the heart of geography. In P. Dougherty, ed. *The Geography of Wine: Studies in Viticulture and Wine*. Springer Verlag Scientific.

White, R. E. 2003. *Soils for fine wines*. New York: Oxford University Press.

White, R. E. 2009. *Understanding vineyard soils*. New York: Oxford University Press.

Wilson, J. E. 1998. *Terroir*: The role of geology, climate and culture in the making of French wines. London, UK: Mitchell Beazley.

Wilson, J. E. 2001. Geology and wine 4. The origin and odyssey of *terroir*. *Geoscience Canada* 28: 139–141.

Wright, W. H. Terry, 2002. Diverse geology/soils impact wine quality. Practical Winery and Vineyard. September/October. p. 40–49.

Climate, Grapes, and Wine: Structure and Suitability in a Variable and Changing Climate

7

Gregory V. Jones, Ryan Reid, and Aleksander Vilks

Abstract

The geography of wine production worldwide results largely from the dependence on climates that provide the conditions needed to produce a quality product. Overall, the climate structure in a given wine-producing region largely determines the suitability of the region to different winegrape cultivars, the overall wine style produced there, and the economic viability of the industry. While the baseline climate determines the suitability, climate variability has a dominant effect on vintage-to-vintage variations in both wine quality and production. Additional challenges come from changes in climate that have altered the suitability of some regions for the better and challenged others.

Introduction

Climate is a pervasive factor in the success of all agricultural systems, influencing whether a crop is suitable to a given region, largely controlling crop production and quality, and ultimately driving economic sustainability. Climate's influence on agribusiness is never more evident than with viticulture and wine production, where climate is arguably the most critical aspect in ripening fruit to optimum characteristics to produce a given wine style. Any assessment of climate for wine production must examine a multitude of factors that operate over many temporal and spatial scales. Namely, climate's influences must be considered at the macroscale (synoptic climate), the mesoscale (regional climate), the toposcale (site climate), and the microscale (vine row and canopy climate). In addition, climate's influences come from both broad structural conditions and singular weather events manifested through many temperature, precipitation, and moisture parameters. To understand climate's role in growing winegrapes and wine production, one must consider: (1) the weather and climate structure necessary for optimum quality and production characteristics; (2) the climate's suitability to different winegrape cultivars; (3) the climate's variability in wine-producing regions; and (4) the influence of climate change on the structure, suitability, and variability of climate.

Weather and Climate Structure for Wine Quality and Production

As a component of terroir, climate arguably exerts the most profound effect on the ability of a region or site to produce quality grapes. Worldwide, the average climatic conditions of wine regions determine, to a large degree,

G.V. Jones (✉) • R. Reid • A. Vilks
Departments of Geography and Environmental Studies, Southern Oregon University, Ashland, OR, USA
e-mail: GJones@sou.edu; RReid@sou.edu; AVilks@sou.edu

P.H. Dougherty (ed.), *The Geography of Wine: Regions, Terroir and Techniques*, DOI 10.1007/978-94-007-0464-0_7, © Springer Science+Business Media B.V. 2012 the grape cultivars that can be grown there, while wine production and quality are chiefly influenced by site-specific factors, husbandry decisions, and short-term climate variability (Jones and Hellman 2003). At the broadest scale, wine region climates have been historically compared using the Köppen climate classification system (Köppen 1931; de Blij 1983). The Köppen climate classification system is broadly based on the expression of climate in the native vegetation and combines average annual, seasonal, and monthly temperatures and precipitation to define homogeneous regions. A modified version of the system divides all climates in seven broad groups: (a) tropical climates, (b) dry climates, (c) mild mid-latitude climates, (d) severe mid-latitude climates, (e) polar climates, and (f) highland climates that are further divided according to various temperature and precipitation relationships.

Due to the geography of the old world wine regions of Europe and the use of the Köppen system, wine production has become synonymous with Mediterranean climates (Csa and Csb). However, it is clear that today's wine regions span a wide range of Köppen climate types. Reid and Jones (2009) analyzed Köppen climate types of the majority of the world's wine regions using the updated Köppen–Geiger gridded data (Peel et al. 2007) and found that the mid-latitude C climates make up roughly 55% of the world's defined wine-growing regions. However, Mediterranean climate types (Csa, Csb) only make up roughly 15% of the surface area globally. Other important C climate types include humid subtropical (Cfa, Cwa 24%), maritime temperate (Cfb, Cwb 12%), and maritime subarctic (Cfc 4%). Examples of wine regions within the mild mid-latitude climate types include Bordeaux (Cfb), Napa Valley (Csb), Coonawarra (Csa), and Barolo (Cfa). Other prominent Köppen climate types that major wine regions fall within include mid-latitude dry climate (BSk 16%) and subtropical dry climate (BSh 3%) (Priorat, Spain and Mendoza, Argentina) and humid continental (Dfa, Dfb 13%) severe mid-latitude climates (Finger Lakes Region, New York and Niagara, Canada).

However, while the Köppen climate classification system provides a broad regional to global generalization of climates, it lacks the specific detail in weather/climate parameters and their use in viticulture and wine production suitability. Individual weather/climate factors affecting grape growth, production, and wine quality include solar radiation, average temperatures, and temperature extremes – including winter freezes and spring and fall frosts and summer heat stress – heat accumulation, wind, precipitation, humidity, and soil water balance characteristics. General descriptions of these factors are given below.

Solar Radiation

Incoming solar radiation (insolation) provides the energy necessary for grape growth and maturation (Mullins et al. 1992). Throughout the growth stages of the grapevine (Fig. 7.1), the amount of insolation is critical in maintaining the optimum levels of photosynthesis. The most critical stages come during the development of the berries starting at bloom (*floraison*) and continuing through the harvest. During bloom, high amounts of insolation result in effective plant tissue differentiation into flowers (Crespin et al. 1987). Low absolute insolation during this stage can influence "coulure" or the failure to fully flower and set berries. The relationship between low amounts of insolation and coulure is neither linear nor predictable but is more tied to varietal characteristics. During the ripening of the berries, insolation mainly acts to control the amount of sugar in the grapes and, therefore, the wine's potential alcohol content (Fig. 7.1).

Controls on the amount of insolation include (1) those that are inherent with Earth/Sun relationships, such as overall amount received by any point on the surface of the Earth, seasonal variations in the angle of incidence of the sun's rays, and the day length and (2) those that are controlled by variations at or near the Earth's surface, such as cloud cover, the reflective nature of the surface of the soil, and the role topographic variations (slope, aspect, and obstructions) have on the relative amount of insolation received.

Average Temperatures

Growing season length and temperatures are a critical aspect because of their major influence on grape ripening and fruit quality and, therefore, cultivar adaptation to specific regions or sites (Gladstones 1992). It is in their ideal climates that a given cultivar can achieve optimum ripening profiles of sugar and acid that can be naturally timed with flavor component development to maximize a given style of wine and the vintage quality. The growing season necessary for the cultivation of winegrapes varies from region to region but averages

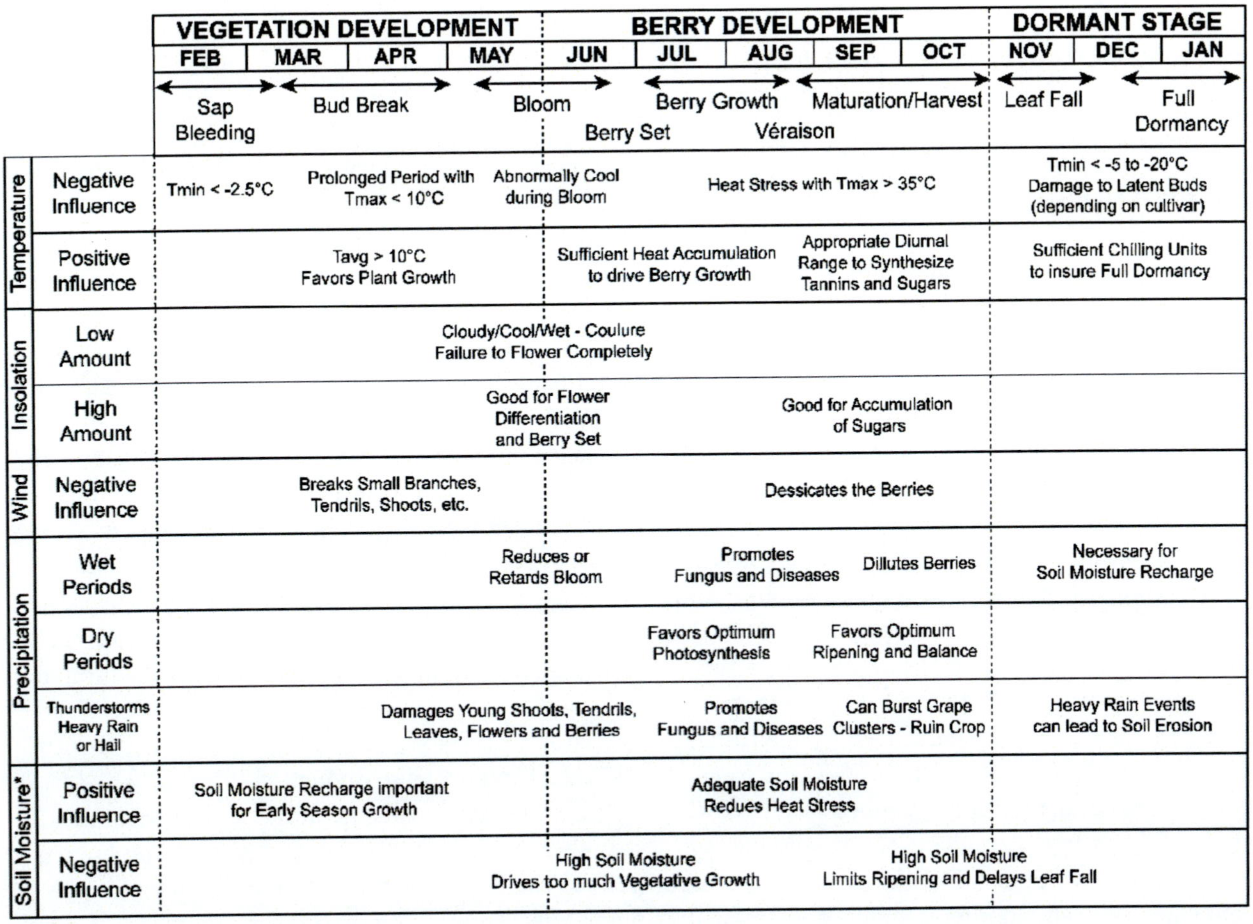

Fig. 7.1 Weather and climate influences on grapevine development and phenological growth stages (Crespin et al. 1987; Jones 1997)

approximately 170–190 days (Mullins et al. 1992). Prescott (1965) notes that an area is suitable for grape production if the mean temperature of the warmest month is more than 18.9°C and that of the coldest month exceeds −1.1°C. The general thermal environment for grapevines has numerous influences which can be positive or detrimental depending on the timing with plant growth. Negative influences typically come from extremes (see below) but can also come from prolonged periods with average temperatures below normal during growth events such as bloom (Fig. 7.1). Positive influences include how temperatures above 10°C can initiate plant growth in the spring and how temperatures influence heat accumulation, which, in turn, drive ripening potential (see below).

In addition, the less variability in the temperature leading up to harvest, both on a diurnal and day-to-day basis, the better the wine quality (Gladstones 1992). This is evident in that the majority of the most renowned and established vineyards in the world are in regions with the most equable day-to-day and diurnal climates. However, differences do occur where some cultivars ripen better in higher diurnal temperature ranges (moderate days, cool nights for cool climate cultivars) while others do better in lower diurnal temperature ranges (moderate to warm days, warm nights for warmer climate cultivars). Overall, the relative amount of insolation, the composition and color of the soil, the local topography and slope aspect, and drainage capabilities can all be major factors in the temperature structure of a vineyard, especially at night. In addition to the air temperature, both within and outside the canopy, the temperature of the soil can have a strong influence on vine growth and fruitfulness. This is especially important during the spring where warmer soils initiate root growth sooner and, when combined with warm air temperatures, hasten bud break. During later growth stages, warm soil surfaces, enhanced by heat retention from rocky material, aid in ripening by warming the vine canopy during the day and into the night. Furthermore, during the dormant stage (from after leaf fall through bud break the next year), an average temperature minimum or chilling (hours below a certain temperature) is generally needed to effectively set the latent primary buds for the following year.

Temperature Extremes

In contrast to average temperature influences on potential vine growth and wine style, some of the most important individual temperature aspects include the potential of midwinter, low-temperature injury, late spring frosts, and the influence of excessive summer heat on grape quality (Fig. 7.1). In regions that have a more continental climate, low-temperature injury to grapevines during the winter is often a limiting factor to production viability. Research has also shown that there is a minimum winter temperature that the grapevines can withstand. This minimum ranges from −5°C to −20°C, with some cultivars and hybrids more cold hardy than others, and is chiefly influenced by microscale climate variations controlled by location and topography (Amerine et al. 1980; Winkler et al. 1974). Temperatures below these thresholds will damage plant tissue by the rupturing of cells, the denaturing of enzymes by dehydration, and the disruption of membrane function (Mullins et al. 1992).

In spring, prolonged temperatures above 10°C initiate vegetative growth (Amerine and Winkler 1944). However, during this stage, temperatures below −2.5°C can adversely affect the growth of the vegetative parts of the plant, and hard freezes can reduce the yield significantly. Nearing maturation, early frost or freezes can lead to the rupture of the grapes along with significant loss of volume. Frost and/or freeze occurrence during the spring and fall generally comes in two forms: (1) advection frosts and (2) radiation frosts. An advection frost occurs as cold air masses are brought into a region with the passage of a cold front. Frosts and freezes associated with cold air masses occur sporadically during the spring and fall and can cause problems over the majority of a region.

Radiation frosts, on the other hand, occur throughout most of the fall, winter, and spring and are a much more common problem in wine regions. Radiation or ground frosts occur as the ground and the air in the lower layers of the atmosphere within and just above a grapevine canopy give off heat, warming the air in successive layers upward. If the dew point temperature is low enough, the result is that the air near the ground is cooled to the frost point. As the ground and lower layers of the atmosphere cool down, the heat energy lost is conveyed upward to form what is called a radiation inversion – a situation in which temperature increases with height from the surface. On nights when inversions form, a warmer thermal zone or belt develops upslope that provides a measure of protection from the coolest valley bottom sites (Wolf 1997). The thermal zone varies from region to region but is generally found from 50 to 250 m off the valley floor. In narrow valleys, vineyards need to be situated higher up in elevation than in broad valleys. Inversions are common in many winegrape-growing regions globally and occur most frequently on long, calm, cloud-free nights.

At the other end of the spectrum of temperature influences, extreme heat – temperatures greater than 35°C – in either the growing season or ripening season, negatively impacts winegrape production through inhibition of photosynthesis (Gladstones 1992) and reduction of color development (Kliewer and Torres 1972) and anthocyanin production (Mori et al. 2005). While a few days of temperatures greater than 30°C can be beneficial to ripening potential, prolonged periods can induce heat stress in the plant and lead to premature "véraison" (color change and start of ripening), a possible abscising of the berries, and partial or total failure of flavor ripening (McIntyre et al. 1982).

Heat Accumulation or Bioclimatic Indices

The characterization of mesoclimates in viticultural areas is commonly done as integrated mathematical expressions of temperature that permit the calculation of bioclimatic indices (Fregoni 2003). These indices are typically summed over a period of time important to the vine's growth and production (usually the 6 or 7 months of the vine's growth and development cycle; see Fig. 7.1). The relationships between heat accumulation, vine growth, and maturation potential was first postulated by A. P. de Candolle in the nineteenth century where he observed that vine growth started when the mean daily temperature reached 10°C. As the various indices were created, they were each typically related to the typicity of the wine that can be produced with classes associated with cool climate cultivar wines to warm climate cultivar wines to fortified wines and table grapes. However, each of these indices has been criticized in one way or another, i.e., too simplistic or appropriate for one region only, and no single, universally validated index has been developed (Gladstones 1992; Jones et al. 2010).

Various forms of these bioclimatic indices have been created and include the degree-day formulation of the Winkler Index (Amerine and Winkler 1944), different forms of a heliothermal index (Branas 1974; Huglin 1978), the quality index of Fregoni (2003), a latitude–temperature index (Jackson and Cherry 1988; Kenny and Shao 1992), and a growing season average temperature index (Jones 2006) with each helping to define the suitability of a region to the planting of certain winegrape cultivars. For arguably a more holistic multivariate index, Smart and Dry (1980) developed a simple classification of viticultural climates that uses five dimensions of mean temperatures, continentality (defined as the difference between the average mean temperature of the warmest and coolest months), sunlight hours, aridity (based upon the difference between rainfall and evaporation), and relative humidity. Gladstones (1992) developed a classification similar to Amerine and Winkler (1944) but refines it by imposing an upper limit on mean temperatures, a correction factor for latitude, and a correction for each month's temperature range.

More recent research (Tonietto 1999; Tonietto and Carbonneau 2004) has combined a reduced number of bioclimatic indices that account for heat accumulation ripening temperatures and drought variability and provide a classification of viticulture climates that are arguably more universally valid. The multicriteria climatic classification system (Geoviticulture MCC System) results in 36 different climatic types from a summation of 3 indices; the Huglin index (HI), a cool night index (CI), and a dryness index (DI). The classification has been successfully tested to differentiate the climate of 97 viticulture zones worldwide (Tonietto and Carbonneau 2004) and for specific wine regions such as Galicia (Blanco-Ward et al. 2007). However, these more holistic climate measures are more complex to calculate and compare, require parameters not readily available everywhere, and are less usable by growers (Jones et al. 2010). In addition, more work needs to be done with the best resolution gridded climate data products (Jones et al. 2009; Daly et al. 2008) to better characterize the spatial characteristics of wine region climates at the region to global scales.

Wind

The role that the wind plays in the growth of the grapevine and the production of fruit is mainly through the effects on vine health and yield, but can also play a role on the heat budget of a vineyard. This is manifested in both a physical nature, through direct contact with the vines and through physiological effects of photosynthesis disruption (stomata closure), and reduced disease infestations (Winkler et al. 1974). During the early stages of vegetative growth, high winds can break off the new shoots, delaying and even reducing the amount of flowering (Fig. 7.1). As the berries proceed through véraison and into the maturation stage, high winds can be very effective at desiccating the fruit and can result in lower volume and quality. However, drying winds that occur at night and early morning can help reduce the occurrence of fungus-borne diseases through limiting the formation of dew on the leaves and berries. Nighttime winds can also be beneficial in that they can help limit the occurrence of radiation frosts.

Local winds, generated from a region's topography, are very common in viticulture areas worldwide. The most common local winds are the general land–sea breeze which affects coastal regions or those near large bodies of water and the mountain–valley breeze which affects inland areas with substantial topographical relief and provides the dry-summer viticulture regions with some relief through late afternoon advection from the coast or down the mountains. While the overall occurrence of winds can have both positive and negative effects on the growth and maturation of grapevines, they can be mitigated somewhat by location, topography, and the use of natural and man-made windbreaks (Hellman and Jones 2003). However, windbreaks in a region also might serve as an obstruction for cold air drainage and could worsen frost or freeze conditions.

Precipitation, Humidity, and Water-Balance Characteristics

Given its importance to vine balance, fruit quality and yield, and disease pressure, understanding water relationships in any wine region is very important. As such, these factors should be assessed from many viewpoints: (1) ambient atmospheric moisture, (2) local rainfall frequency and timing, and (3) soil water holding capacity. In addition, each of these aspects of water availability can be evaluated in terms of a water balance or budget.

Atmospheric humidity is very important in regulating the evaporative demands put on the grapevines and the occurrence of fungal diseases. During the phenological stages of the grapevine, some of the climatic conditions that can most severely afflict the vines and berries are associated with moisture. Atmospheric moisture is commonly measured as relative humidity and as such displays a distinct diurnal and seasonal cycle. Relative humidity is normally highest early in the morning, when temperature is lowest and at a minimum during the maximum heating of the day. The contrasts between morning and afternoon relative humidity are the least along coastal regions and greatest inland. Since lower temperatures give higher relative humidity, and since cold air pools in valleys and in obstructed areas on slopes, humidity levels and the associated fungal problems can be enhanced in these zones. Over the course of the season, relative humidity is lowest during the summer and highest during the winter with the contrast between the seasons being the greatest for the afternoon values. However, in extreme cases, water stress resulting from low relative humidity can manifest itself in leaf loss, severe reductions in vine metabolism, and fruit damage or loss (Mullins et al. 1992). Even moderate periods of moisture stress can substantially reduce the relative level of photosynthesis, resulting in lower fruit yields and quality. However, controlled moisture stress through irrigation applied at the right times during growth and ripening can enhance quality and control yield.

As high levels of atmospheric moisture allow certain fungal problems to develop, the occurrence of rain during critical growth stages can lead to devastating effects (Fig. 7.1). While ample precipitation during the early vegetative stage is beneficial (Jones and Davis 2000a, b), during bloom, it can reduce or retard flowering; during berry growth, it can enhance the likelihood of fungal diseases; and during maturation, it can further fungus maladies, yellow, and dilute the berries and thus reduce the sugar and flavor levels, and severely limit the yield and quality (Mullins et al. 1992). Examination of the world's viticulture regions suggests that there is no upper limit on the amount of precipitation needed for optimum grapevine growth and production (Gladstones 1992). On the other hand, grapevine viability seems to be limited in some hot climates by rainfall amounts less than 500 mm, although this can be overcome by regular irrigation, if allowed. Extreme meteorological events, such as thunderstorms and hail, while generally rare in most viticultural regions, are extremely detrimental to the crop. Both events can severely damage the leaves, tendrils, and berries during growth and, if they occur during maturation, can split the grapes, causing oxidation, premature fermentation, and a severe reduction in volume and quality of the yield (Fig. 7.1).

As an integration of many climate parameters, a soil water balance takes into account seasonal variations in temperature, precipitation, and available soil moisture to give an estimation of water requirements (either natural or via irrigation). A water balance essentially defines the "water need" by plants and the atmosphere in any region. In most grape-growing regions, there is a period of soil water surplus from late fall through late spring, followed by a period of drawdown of soil moisture through evaporation by the atmosphere and transpiration by plants during the summer through the early fall when precipitation begins replenishing the soil. Adequate soil moisture recharge during spring can drive vine growth and result in more effective bloom and berry set (Williams 2000). While some soil moisture during the summer growth period can reduce heat stress, too high soil moisture can drive too much vegetative growth and lead to inadequate ripening (Matthews and Anderson 1988) along with delayed leaf fall, putting the vines at risk of late fall frost/freeze events (Fig. 7.1).

Climate Suitability for Winegrape Cultivars

Winegrape cultivar suitability to a given region is controlled by the baseline climate. Historically, there have been numerous temperature-based metrics, e.g., degree-days, mean temperature of the warmest month, average growing season temperatures, etc., that have been used for establishing optimum climates for the range of winegrape cultivars (Gladstones 1992). This is not to say that precipitation or any other weather/climate factor is not important (see the previous section) but that temperature is the most influential factor in overall growth and productivity of winegrapes. At the global scale, the general bounds on climate suitability for viticulture are found between 12°C and

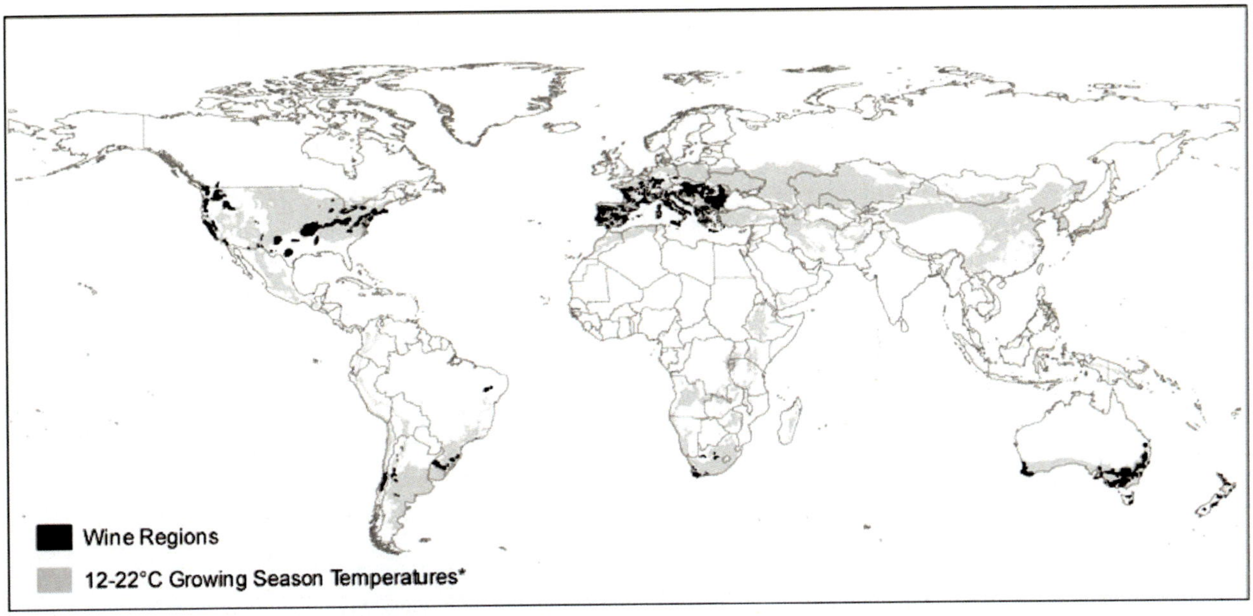

Fig. 7.2 Global wine regions and general climate zones. The wine regions are derived from governmentally defined boundaries (e.g., American Viticultural Areas in the United States, Geographic Indicators in Australia and Brazil, and Wine of Origin Wards in South Africa) or areas under winegrape cultivation identified with remote sensing (e.g., Corine Land Cover for Europe) or aerial imagery (e.g., Canada, Chile, Argentina, and New Zealand). The general climate zones are given by the 12–22°C growing season (Apr–Oct in the Northern Hemisphere and Oct–Apr in the Southern Hemisphere) average temperatures derived from the WorldClim database (Hijmans et al. 2005)

22°C for the growing season in each hemisphere (Gladstones 2004; Jones 2007; Fig. 7.2). As seen in Fig. 7.2, the 12–22°C climate bounds depict a largely mid-latitude suitability for winegrape production.

However, many subtropical to tropical areas at higher elevations also fall within these climate zones (similar to the Köppen climate classification scheme). Furthermore, any general depiction of average temperatures will also include large areas that have not been typically associated with winegrape production. This is evident in Fig. 7.2 where large areas of eastern Europe, western Asia, China, the midwestern and eastern USA, southeastern Argentina, southeastern South Africa, and southern Australia fall within the 12–22°C thresholds. While many of these regions may have the growing season temperatures conducive to growing winegrapes, other limiting factors such as winter minimum temperatures, spring and fall frosts, short growing seasons, and water availability would limit much of the areas mapped to the average conditions.

Furthermore, while the vast majority of the world's wine regions are found within these average growing season climate zones, there are some exceptions. For example, there are defined winegrape growing areas in the USA (Texas, Oklahoma, and the Mississippi delta region), in Brazil (São Francisco Valley), and in South Africa (Lower Orange River in the Northern Cape) that are warmer than 22°C during their respective growing seasons. However, these regions have different climate risks, have developed viticultural practices to deal with the warmer climates, e.g., two crops per year, irrigation, etc., or produce table grapes or raisons, and do not necessarily represent the average wine region.

Further refining the climate suitability for many of the world's most recognizable cultivars, Jones (2006) shows that high-quality wine production is limited to 13–21°C average growing season temperatures (Fig. 7.3). The climate-maturity zoning in Fig. 7.3 was developed based upon both climate and plant growth for many cultivars grown in cool to hot regions throughout the world's benchmark areas for those winegrapes. While many of these cultivars are grown and produce wines outside of their individual bounds as depicted in Fig. 7.3, these are more bulk wine – or high yielding – for the lower end market and do not typically attain the typicity or quality for those same cultivars in their ideal climate. Furthermore, growing season average temperatures below 13°C are typically limited to hybrids or very early ripening cultivars that do not necessarily

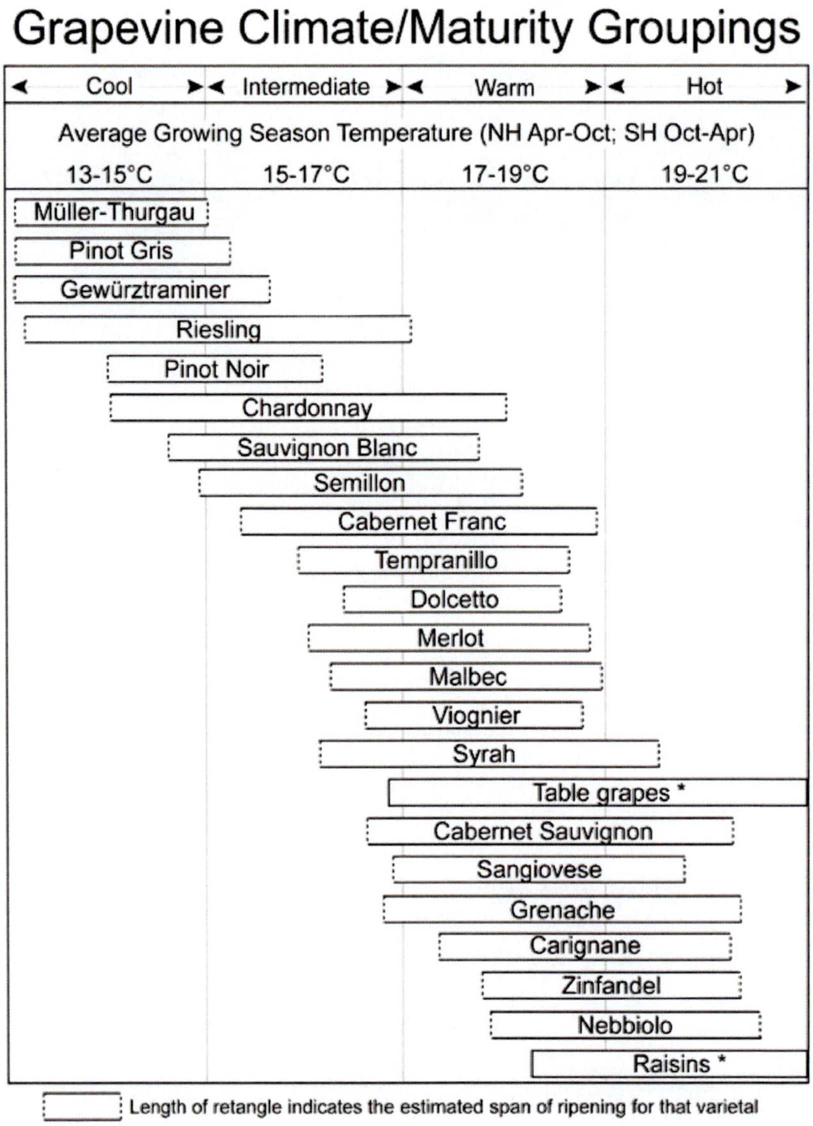

Fig. 7.3 Climate-maturity groupings based on relationships between phenological requirements and growing season average temperatures for high- to premium-quality wine production in the world's benchmark regions for many of the world's most common cultivars. The *dashed line* at the end of the bars indicates that some adjustments may occur as more data become available, but changes of more than ±0.2–0.5°C are highly unlikely (Jones 2006)

have large-scale commercial appeal. At the upper limits of climate, some production can also be found with growing season average temperatures greater than 21°C, although it is mostly limited to fortified wines, table grapes, and raisons. Hall and Jones (2010) and Jones et al. (2010) have mapped these climate limits for Australia and the western USA, detailing the within region climate suitability across the cool, intermediate, warm, and hot climate types. This work helps depict the true wine region spatial climate structure, instead of the common practice of using stations which clearly do not properly characterize the climates experienced in the actual vineyard areas.

An example of cool climate suitability is found with the widely recognized Pinot Noir cultivar which is typically grown in regions that span from cool to lower intermediate climates with growing seasons that range from roughly 14.0°C to 16.0°C (e.g., Burgundy or Northern Oregon). The coolest of these is the Tamar Valley of Tasmania, while the warmest is the Russian River Valley of California. Across this 2°C climate niche, Pinot Noir produces the broad style for which it is known with the cooler zones producing lighter, elegant wines and the warmer zones producing more full-bodied, fruit-driven wines (Table 7.1). While Pinot Noir can be grown outside the 14.0–16.0°C growing season average temperature bounds, it is typically unripe or overripe and readily loses its typicity.

As examples of intermediate to warmer climate cultivars, the noble winegrapes Cabernet Franc and Cabernet Sauvignon have been widely planted worldwide and are clearly two of the most recognized in

Table 7.1 Climate's general influence on wine characteristics

Wine characteristic	Cool climate	Intermediate to warm climate	Warm to hot climate
Fruit style	Lean, Tart	Ripe, Juicy	Overripe, Lush
White flavors	Apple, Pear	Peach, Melon	Mango, Pineapple
Red flavors	Cranberry, Cherry	Berry, Plum	Fig, Prune
Body	Light	Medium	Full
Acidity	Crisp, Tangy	Integrated	Soft, Smooth
Alcohol	Low to moderate	Moderate to high	High to very high
Overall style	Subtle, Elegant	Medium intensity	Bold

the world. The spread of these two cultivars worldwide has produced an assortment of wine styles from quite diverse regions. Figure 7.3 shows this wide diversity with both Cabernet Franc and Cabernet Sauvignon having roughly 3.5°C climate ranges, nearly double that of Pinot Noir. Cabernet Franc can be grown in intermediate to warm climates (15.4–19.8°C) as evidenced by its quality production in the Loire Valley of France which has a 16.2°C growing season average temperature. Cabernet Sauvignon, on the other hand, is grown in regions that span from intermediate to hot climates with growing seasons that range from roughly 16.8°C to 20.2°C (e.g., Bordeaux or Napa). The lower climate limit for Cabernet Sauvignon suitability is found in Hawke's Bay, New Zealand, while the upper climate limit is found in Robertson, South Africa.

Another example of both broad climate and wine styles diversification can be found with Chardonnay. As Chardonnay production expanded worldwide, it was found to be one of the most flexible and forgiving cultivars, producing a range of wine styles. Chardonnay can be grown in relatively cool climates (~14–16°C), creating an elegant, crisp style that is flavored more by apple, pear, and fig, while in warmer climates (~16–18°C), it produces a bolder style with more peach to honey notes that could be enhanced via the use of oak barrel aging. Cultivar diversification within regions is also evident in Fig. 7.3 with two notable examples. The Piedmont region of Italy has become known for its red wines which are made from either Dolcetto or Nebbiolo. While both cultivars are grown over a relatively narrow climate range, Dolcetto does better in a slightly cooler climate (16.4–18.4°C) while Nebbiolo does better in a warmer climate (17.8–20.4°C) with a long growing season. The Piedmont region has, therefore, diversified its production and quality potential in that during cool years Dolcetto does best, during warm years Nebbiolo does best, and during average years they both do relatively well. Another example is in Bordeaux where the average growing season temperatures of 16.8–17.8°C makes it ideal to Cabernet Franc, Cabernet Sauvignon, and Merlot, albeit with warmer years favoring higher Cabernet Sauvignon quality.

Another characteristic of climate that helps define where winegrape cultivars do best is the ripening period day–night temperature pattern (Gladstones 1992; Tonietto and Carbonneau 2004; Jones et al. 2009). The ripening period is broadly defined as the growth stage that starts with véraison or the initiation of sugar accumulation and the change in berry color to harvest and averages 40–60 days for most cultivars and wine regions (Jones et al. 2005b). However, the majority of research points to the importance of the last 30 days when the most of the polyphenols, color, flavor, and aroma compounds develop (Gladstones 1992). Measured either as the diurnal temperature range (Gladstones 1992; Jones et al. 2009) or simply the average nighttime temperature during the last month of the vintage (Tonietto and Carbonneau 2004), the effect is on how the diurnal conditions influence the quality potential of the fruit and resulting wine. For example, the majority of the cooler climate cultivars achieve greater typicity in ripening periods with higher diurnal temperature ranges whereby daytime temperatures are high enough to facilitate ripening while nighttime temperatures slow respiration and metabolism. These conditions bring about the more delicate aromas and wine styles associated with cool to intermediate climate cultivars.

Many warmer climate cultivars, on the other hand, typically require warmer nighttime temperatures to more fully ripen and to metabolize compounds that are considered detrimental to quality. For example, Cabernet Sauvignon contains a high concentration of methoxypryazines which are responsible for the characteristic "green, herbaceous, or vegetative" aromas from under ripe fruit. While many factors such as vine training, sunlight exposure, and canopy management can influence methoxypryazine content, regions with lower diurnal temperatures or higher nighttime temperatures tend to allow for greater metabolism of the compounds producing less "green" aromas (Roujou de Boubee et al. 2000).

Interestingly, another compound has been tied to the climate structure in regions that produce Syrah or Shiraz. The compound that distinguishes the peppery aroma and flavors of Syrah has been identified as *rotundone* by Siebert et al. (2008). Syrah with greater peppery characteristics comes from cooler climates and higher diurnal temperature ranges during ripening while those with very little peppery notes and more fruit-driven flavors come from warmer climates with lower diurnal temperature ranges during ripening.

Climate Variability in Wine Regions

While the average climate structure in a region determines the broad suitability of winegrape cultivars, climate variability influences issues of production and quality risk associated with how equitable the climate is year in and year out. Climate variability in wine regions influences grape and wine production through cold temperature extremes during the winter in some regions, frost frequency and severity during the spring and fall, high temperature events during the summer, extreme rain or hail events, and broad spatial and temporal drought conditions. Climate variability mechanisms that influence wine regions are tied to large-scale atmospheric and oceanic interactions that operate at different spatial and temporal scales (Fig. 7.4). The most prominent of these is the large-scale Pacific sector El Niño–Southern Oscillation (ENSO) (Glantz 2001), which has broad influences on wine region climates from North America (Jones and Goodrich 2008), Australia and New Zealand (Gordon 1986; Power et al. 1999), South Africa (Tyson 1986), South America (Garreaud et al. 2009), and Europe (Rodó and Comín 2000). However, the effects of ENSO on wine region climate variability vary tremendously in magnitude and are of opposite sign depending on the location of the wine region and are often coupled with other more influential regional mechanisms (Jones and Goodrich 2008).

For Europe, the dominant climate variability mechanism is the North Atlantic Oscillation (NAO; Fig. 7.4), which has a large climatic influence on the North Atlantic Ocean and surrounding land masses (Hurrell 2003). The NAO is associated with changes in the surface westerlies across the North Atlantic due to a large-scale seesaw in atmospheric mass between the subtropical high and the polar low. The corresponding index that numerically characterizes the NAO varies from year to year, but also exhibits a tendency to remain in one phase for intervals lasting several years. The positive phase of the NAO index exhibits a stronger than usual subtropical high-pressure center (Azores high) and a deeper than normal Icelandic low. The increased pressure difference over the North Atlantic results in more and stronger winter storms crossing the Atlantic Ocean on a more northerly track. For Europe, the positive phase results in warmer and wet winters across northern areas and warmer and drier conditions across the Mediterranean basin. In addition, Northern Canada and Greenland experience cold and dry winters, while the Eastern USA experiences mild and wet winter conditions. The negative phase of the NAO index exhibits a weak subtropical high and a weak Icelandic low, and the reduced pressure gradient results in fewer and weaker winter storms crossing on a more west–east pathway. The result is typically moist air advection and wetter conditions over the Mediterranean and cold, dry conditions in Northern Europe. The eastern US coast experiences more cold air outbreaks and, hence, snowy weather conditions, while Greenland typically has milder winter temperatures.

Relationships between the NAO and winegrape production in Europe are mixed with most showing little to no correlation (Jones 1997). This is likely due to the fact that the NAO is largely a wintertime mechanism and its effects diminish over the growing season. However, there is some evidence that the positive NAO phase brings drier growing seasons to the Iberian Peninsula and the Mediterranean that result in lower quality and yield (Esteves and Manso-Orgaz 2001; Grifoni et al. 2006). For the northern vineyards of Europe, either phase can be problematic, whereby a positive NAO brings wetter conditions and greater disease pressure while the negative phase brings greater frost pressure during spring. In addition, Souriau and Yiou (2001), using historical grape harvest dates from Northeastern France and Switzerland, showed significant correlations between harvest dates and the NAO and suggested using the record "as an interesting proxy" to reconstruct the NAO back in time. Although

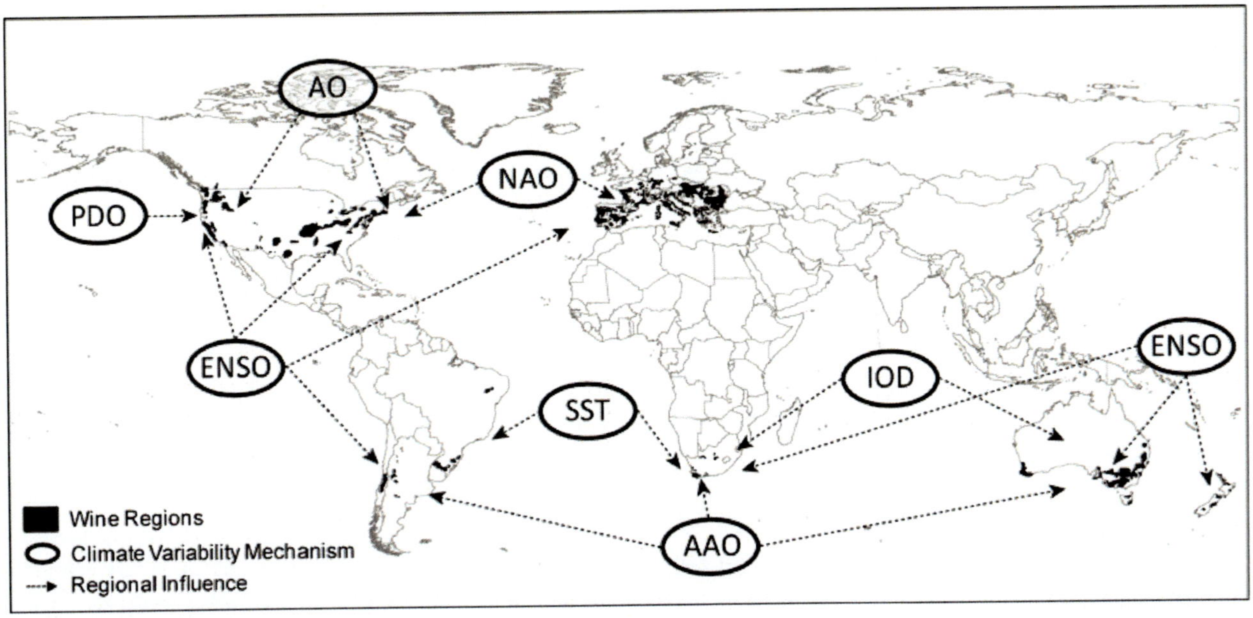

Fig. 7.4 Global wine regions, climate variability mechanisms, and their areas of known influences as described in the text. *ENSO* El Niño–Southern Oscillation, *PDO* Pacific Decadal Oscillation, *NAO* North Atlantic Oscillation, *IOD* Indian Ocean Dipole, *AO* Arctic Oscillation, *AAO* Antarctic Oscillation, *SST* sea surface temperatures. The wine regions are as described in Fig. 7.2

ENSO plays a significant role in determining the inter-annual variability of climate in the lower latitudes, its influence on European climate is weak (Mathieu et al. 2004) or hard to differentiate from the effects of the NAO (Rodó and Comín 2000).

For North America, the effects of climate variability mechanisms are different depending on whether it is the west coast or east coast (Fig. 7.4). Western USA and Canadian wine regions are largely influenced by conditions in the North Pacific and tropical Pacific Oceans which, in turn, affect the circulation of the atmosphere (Mantua and Hare 2002; Jones and Goodrich 2008). The two main mechanisms in the Pacific are sea surface temperatures (SSTs) that drive variations in the Pacific Decadal Oscillation (PDO, North Pacific) and El Niño/La Niña (ENSO, tropical Pacific). The main difference between the two, besides location, is that the PDO is long term (multi-decadal swings) and ENSO is short term (2–5 years). During El Niño conditions (ENSO warm phase), areas from Southern California north to the Oregon border are typically wetter and warmer than normal.

While these conditions vary by event, during extreme El Niño events, e.g., 1982–1983, high precipitation anomalies are observed over the whole region. Points north of the Oregon border and into Canada, on the other hand, typically experience drier conditions during El Niño events. During La Niña (ENSO cold phase), points from Northern California and into Canada typically experience cooler and wetter winters, while Central to Southern California is cooler but significantly drier. For eastern North American wine regions, ENSO effects are minor to moderate and often masked by NAO effects (see above). However, El Niño conditions are typically correlated with warmer than average winters in the upper Midwest Northeast of the USA. La Niña conditions, on the other hand, tend to bring cooler winters with more snow across northeastern wine regions. Another aspect of ENSO effects on eastern North American wine regions is its influence on tropical storm systems. During El Niño years, the tropical Atlantic Ocean experiences lower activity which could allow growers in the eastern USA to make it through to harvest without heavy rainfall impacts from tropical storms.

As the dominant long-term climate variability mechanism in western North American, the PDO is a result of low-frequency variability in sea surface temperatures (SSTs) in the extratropical Pacific Ocean (Frauenfeld et al. 2005). The PDO index represents the main variability structure in mean monthly SSTs in the Pacific Ocean north of 20° N latitude (Mantua and Hare 2002) and is related to the circulation of the atmosphere as described by the North Pacific Index

(Trenberth and Hurrell 1994). Positive or warm phase values of the PDO refer to above normal SSTs along the west coast of North America and along the equator and below normal SSTs in the central and western North Pacific centered around 45°N latitude (Fig. 7.4). Negative values of the PDO refer to the opposite distribution of SSTs in these same areas. During the positive PDO phase, the western North American fall, winter, and spring tend to be warmer and drier than average, while during the negative PDO phase, the region is colder and wetter.

In terms of winegrape production impacts, Nemani et al. (2001) found that a recent warm phase of the PDO (1977–1998) was associated with higher winter and spring temperatures, fewer frost days, and longer growing seasons than the preceding cold phase of the PDO. All of the changes to Napa and Sonoma Valley climate structure were beneficial to winegrape production and wine quality. Jones and Goodrich (2008) examined climate parameters important for wine production in the western USA and found that ENSO variations were not as critical as the PDO; however, when they were important, it was La Niña conditions that were more significant than El Niño conditions. The PDO, on the other hand, had a strong influence on west coast wine region climates where the cold phase of the PDO produces cooler springs with substantially greater spring frost, later frost events, and cooler growing seasons. However, some of the greatest impacts on west coast wine region climates occurred when ENSO was in either a neutral or La Niña phase and the PDO was in a cold phase or where one area of the Pacific Ocean was in phase with the other (both cold or neutral cold). These conditions typically result in lower yields and significantly lower wine quality that comes about from frost impacts and cooler growing conditions. No studies have been done to document the impact of any of the climate variability mechanisms on eastern North American wine regions. However, cold air outbreaks from the Arctic Oscillation (related to the NAO) appears to be important for ice wine production in Eastern Canada and winter freeze and spring frost damage (Gu et al. 2008).

For Australia and New Zealand, climate variability over the region is largely influenced by a similar suite of atmospheric and oceanic mechanisms (Fig. 7.4). The three most prominent include ENSO, the Antarctic Oscillation (AAO, also called the Southern Annular Mode), and the Indian Ocean Dipole (IOD), but as discussed previously regarding the ENSO–PDO modulation, their interactions over the Australia–New Zealand region appear to be most important. ENSO impacts over Australia show that during El Niño years, rainfall is reduced, and both the daily maximum temperature and the diurnal temperature range over much of Australia are increased (Power et al. 1999). La Niña years are often wetter over much of Australia, but as the recent long-term severe drought conditions show (the so-called "big dry"), not all La Niña years bring adequate rainfall to the continent (Ummenhofer et al. 2009). New Zealand is not normally affected as strongly by El Niño conditions as are parts of Australia. However, during El Niño years, New Zealand tends to experience stronger or more frequent winds from the west in summer, typically leading to drought in east coast areas and more rain in the west. During winter, the winds tend to be more southerly, bringing colder conditions to both the land and the surrounding ocean (Gordon 1986). La Niña years in New Zealand bring more northeasterly winds and moist, rainy conditions to the northeast of the North Island and reduced rainfall to the South Island. Temperatures during La Niña years tend to be warmer over much of the country (Gordon 1986).

The Antarctic Oscillation (AAO) is a measure of changes in windiness and storm activity between the middle and high latitudes over the Southern Hemisphere continents, oceans, and Antarctic sea-ice zone (Gillett et al. 2006). AAO influences intra- and inter-seasonal climate variability throughout the Southern Hemisphere (Fig. 7.4). During the positive phase of the AAO, the mid-latitude westerlies are contracted poleward which results in significant cooling over Antarctica and much of Australia along with significant warming over southern South America, Tasmania, and New Zealand (Hendon et al. 2007). Due to the poleward shift in the westerly storm tracks, precipitation during the positive phase of AAO tends to decrease over much of southern South America, New Zealand, southeastern Australia, and Tasmania, while increasing over Western Australia and South Africa (Gillett et al. 2006). During its negative phase, the AAO exhibits a stronger and more mid-latitude position of the westerlies, increasing the storm activity over the southern land regions.

Ummenhofer and England (2007) and Pezza et al. (2008), show that ENSO, AAO, and Antarctic sea ice also interact as a complex physical system whereby the North Island of New Zealand and eastern Australia are more influenced by the tropical ENSO while the

South Island of New Zealand and southern to western Australia are more influenced by the extratropical AAO. However, recent research by Ummenhofer et al. (2009) has revealed the importance of tropical moisture transport to Australian precipitation variability via Indian Ocean sea surface temperatures. Captured in the Indian Ocean Dipole (IOD), the coupled ocean–atmosphere mechanism in the region brings more moisture into Australia during its negative phase (warmer SSTs in the eastern Indian Ocean) and drier conditions during its positive phase. Ummenhofer et al. (2009) show that the recent drought, i.e., the "big dry", is largely the result of the IOD being in its positive or neutral phase since 1992 where it has caused acute water shortages, record agricultural losses, and the drying-out of two of Australia's major river systems. Furthermore, Nicholls (2009) also ties these interactions to a strengthening of the subtropical ridge over the continent which is likely influenced by variations in each mechanism but appears to be most influenced by AAO.

Similar with many of the other wine regions of the world, South American climate variability, especially precipitation anomalies, is largely influenced by ENSO (Paegle and Mo 2002). These effects are even more pronounced due to the proximity to the eastern tropical Pacific basin nexus of ENSO (Fig. 7.4). The spatial diversity of precipitation, temperature, and wind patterns over the wine regions in southern South America is largely due to the long meridional extent of the continent and the disruption of the large-scale circulation caused by the Andes Mountains (Garreaud et al. 2009). The structure of the land mass and its proximity to large oceans create considerable geographical variations in the timing and intensity of interannual and interdecadal climatic fluctuations with precipitation variability highest in Northeastern Brazil, Central Chile, and central to southern Argentina. Similar to other regions, e.g., Australia, South Africa, and Europe, the nature of climate variability over the southern South American is driven by both tropical (ENSO) and extratropical mechanisms (PDO, AAO). Garreaud et al. (2009) provide a comprehensive overview of South American climate showing the intra- and interannual relationships between these mechanisms. In general, ENSO has the most dominant influence with El Niño episodes typically associated with below normal rainfall and warmer than normal conditions in the northern part of South America, as well as anomalously wet conditions in the southeastern portion of the continent and Central Chile (Paegle and Mo 2002; Garreaud et al. 2009).

Grimm et al. (2000) found that Southern Brazil has the strongest signal in El Niño events while Northeastern Brazil is more prone to droughts caused by interactions with South Atlantic SSTs and the African continent. In addition, Grimm and Tedeschi (2009) find that ENSO cycles have a strong effect on the frequency of extreme precipitation events over most of South America with the largest changes occurring in Southern Brazil, Uruguay, and Argentina. Opposite rainfall anomalies are typically observed in northern South America and the southeastern part of the continent during La Niña events. Garreaud et al. (2009) also found that there is some decadal and interdecadal variability evident in many climate records across the continent with the strongest connection to the PDO. Andreoli and Kayano (2005) and Kayano et al. (2009) show that the relationship between ENSO and PDO over South America is one of modulation such that when they are in phase, i.e., warm and warm or cold and cold, their effects are greatest, which is similar to what Jones and Goodrich (2008) found for the Western USA.

Higher latitude influences of the AAO drive some of the largest temperature variations in South America with warm events across most of the southernmost areas (south of 35°) during the positive phase of the AAO. Conversely, the negative phase of the AAO is responsible for large, synoptic-scale cold air outbreaks over southern South America with frost and freeze impacts (Garreaud 2000; Vera and Vigliarolo 2000). AAO-related precipitation variability is significant over the wine regions and the Lake District of southern Chile and along the east coast of the continent into Brazil (Garreaud et al. 2009).

Climate variability for South African wine regions is driven by interactions between ENSO and sea surface temperature anomalies in the Indian and South Atlantic Oceans (Fig. 7.4), which together influence both the tropical and the temperate atmospheric circulation and moisture fluxes over the subcontinent (Tyson 1986). As such, these mechanisms have significant influences on the position and strength of heat-driven continental low pressure systems, the location and amplitudes of the westerly storm troughs, and rainfall variability over South Africa and surrounding countries (Mason and Jury 1997). Aspects of year-to-year climate variability in South Africa tied to ENSO, AAO, and IOD have been well studied due to their impacts on drought and crop failure over Southern Africa (Dilley 2000; Reason and Rouault 2002; Todd et al. 2002; Kandji et al. 2006). Typically, El Niño years are anomalously warm and dry, with drought disasters tending to occur the year following the onset of an El Niño with seasonal forecasting efforts being implemented to better prepare for some of the impacts (Dilley 2000). Wetter than normal years tend to occur during La Niña events and have been tied to extreme locust outbreaks in Southern Africa (Todd et al. 2002). The roles of other regional mechanisms on South African climate variability have been studied less than ENSO impacts (Kandji et al. 2006). However, other Southern Hemisphere studies show that during the positive phase of Antarctic Oscillation (AAO), precipitation tends to increase over South Africa while the negative phase shows no significant relationship (Gillett et al. 2006). The Indian Ocean Dipole (IOD), on the other hand, appears to play a significant role on both temperature and precipitation variability over Southern Africa. During the positive phase of the IOD, warmer SSTs in the western Indian Ocean shift tropical moisture toward Southern Africa and away from Australia (Ummenhofer et al. 2009). Misra (2003) and Valimba et al. (2006) show that the ENSO and IOD effects on South Africa are a result of the modulation of the zonal Walker circulation by the variability of Pacific Ocean SST. The Walker circulation adjusts longitudinally to changes in SSTs, influencing the dominant modes of SST anomalies over the Eastern Indian and South African precipitation variability.

While aspects of year-to-year climate variability in Australia, South America, and South Africa have been studied with respect to ENSO or other regional mechanism, no studies have investigated relationships between temperature and precipitation variability and grape and wine characteristics, e.g., crop growth cycle, fruit composition, yield/production, and wine quality. Research in these regions on other crops has shown strong influences on yield variability. For example, Podesta et al. (2002) found that South American national level yields of corn, soybeans, and sorghum are lower during La Niña and higher during El Niño years. For South Africa, Rosenzweig and Hillel (2007) and Kandji et al. (2006) have detailed the impacts of drought on corn and other crops during El Niño years, indicating that 50–70% reductions in crop yields can be experienced during extreme events. For Australia, significant changes in the gross value of the majority of crops grown on the continent have been shown (Nicholls, 1985), mostly driven by the low grain yields that typically occur during El Niño events (Garnett and Khandekar 1992). However, the recent long-term drought, which has been tied to interactions between ENSO, IOD, and the subtropical ridge, has all but wiped out the Australian rice industry (Ummenhofer et al. 2009).

While the impacts of climate change on the average climate of wine regions worldwide are evident (see below), arguably just as important is how that climate varies. Increases in climate variability in a given region would bring about greater risk associated with climate extremes, which, in turn, would strain the economic viability of wine production in any region. Both observations and models indicate that climates experience changes in both the mean and the variability of temperatures in wine regions and elsewhere (Jones 2007). For example, if the change response of a warming climate was only in the mean, then there would be less cold weather and more hot and record hot weather. On the other hand, increases in the temperature variance alone would result in more cold and hot weather and record conditions. However, evidence points to increases in both the mean and variance which would bring about less change for cold weather events and much more hot weather and record hot weather (IPCC 2007). For example, Schär et al. (2004) demonstrate that the European summer climate structure is expected to experience a pronounced increase in year-to-year variability in response to greenhouse gas forcing. While heat waves are caused by entrenched high pressure and blocking patterns, such an increase in variability might be able to explain the unusual European summer 2003 and would strongly affect the incidence of heat waves and droughts in the future. Evidence of changing climate variability in wine regions was also found by Jones (2005) and Jones et al. (2005a) where the coefficient of variability in the growing season climates throughout the western USA and many other wine regions globally has increased over the last 50 years. Jones et al. (2005a) also found that model projections through 2050 show a continued increase in the coefficient of variability of growing season temperatures in 20 of 27 wine regions globally.

Climate Change, Viticulture, and Wine

From the discussion above on the climate structure, suitability, and variability associated with regional to global wine production, it is clear that viticultural regions are located in relatively narrow geographical and climatic ranges. In addition, winegrapes have relatively large cultivar differences in climate suitability further limiting some winegrapes to even smaller areas that are appropriate for their cultivation. These narrow "*niches*" for optimum quality and production put the cultivation of winegrapes at greater risk from both short-term climate variability and long-term climate changes than other more broad-acre crops (Jones 2007).

In general, the overall wine style that a region produces is a result of the baseline climate, while climate variability determines vintage quality differences. Climatic changes, which influence both variability and average conditions, therefore, have the potential to bring about changes in wine styles (Jones 2007). Our understanding of climate change and the potential impacts on viticulture and wine production has become increasingly important as the earth system undergoes natural cycles and fluctuations and changing levels of greenhouse gases, and alterations in Earth surface characteristics bring about changes in the Earth's radiation budget, atmospheric circulation, and hydrologic cycle (IPCC 2007). Observed warming trends over the last 100 years have been found to be asymmetric with respect to seasonal and diurnal cycles, with greatest warming occurring during winter and spring and at night (Karl et al. 1993; Easterling et al. 2000). The observed trends in temperatures have been related to agricultural production viability by impacting winter hardening potential, frost occurrence, and growing season lengths (Carter et al. 1991; Menzel and Fabian 1999; Easterling et al. 2000; Nemani et al. 2001; Moonen et al. 2002; Jones 2005).

To place viticulture and wine production in the context of climate suitability and the potential impacts from climate change, Fig. 7.3 provides the framework for examining today's climate-maturity ripening potential for premium quality wine cultivars grown in cool, intermediate, warm, and hot climates (Jones 2006). For example, Cabernet Sauvignon is grown in regions that span from intermediate to hot climates with growing seasons that range from roughly 16.8°C to 20.2°C, e.g., Bordeaux or Napa. Cooler climate cultivars such as Pinot Noir are typically grown in regions that span from cool to lower intermediate climates with growing seasons that range from roughly 14.0°C to 16.0°C, e.g., Northern Oregon or Burgundy.

From the general bounds that cool to hot climate suitability places on high quality wine production, it is clear that the impacts of climate change are not likely to be uniform across all cultivars and regions but are more likely to be related to climatic thresholds whereby any continued warming would push a region outside the ability to produce quality wine with existing cultivars. For example, if a region has an average growing season temperature of 15°C and the climate warms by 1°C, then that region is climatically more conducive to ripening some cultivars while potentially less for others. If the magnitude of the warming is 2°C or larger, then a region may potentially shift into another climate-maturity type, e.g., from intermediate to warm. While the range of potential cultivars that a region can ripen will expand in many cases, if a region is a hot climate-maturity type and warms beyond what is considered viable, then grape growing becomes challenging and maybe even impossible. Furthermore, observations and modeling have shown that changes in climate have not and will not likely just be manifested in changes in the mean but also in the variance where there are likely to be more extreme heat occurrences but still swings to extremely cold conditions (IPCC 2007). Therefore, even if the average climate structure gets better in some regions, variability will still be very evident and possibly even more limiting than what is observed today.

History has shown that winegrape-growing regions developed when the climate was most conducive and that shifts in viable wine-producing regions have occurred in the past due to changes in climate, making production more difficult or easier (Le Roy Ladurie 1971; Pfister 1988; Gladstones 1992). In Europe, records of dates of harvest and yield have been kept for nearly a thousand years (Penning-Roswell 1989; Le Roy Ladurie 1971), revealing periods with more beneficial growing season temperatures, greater productivity, and arguably better quality in some regions. Other evidence has shown that vineyards were planted as far north as the coastal zones of the Baltic Sea and Southern England during the medieval "Little Optimum" period (roughly 900–1300 AD) when temperatures were up to 1°C warmer (Gladstones 1992). During the High Middle Ages (twelfth and thirteenth centuries), harvesting occurred in early September as compared to early to late October during much of the twentieth century (Pfister 1988; Gladstones 1992). However, during the fourteenth century, dramatic temperature declines lead to the "Little Ice Age" which extended into the late nineteenth century and resulted in most of the northern vineyards dying out and growing seasons so short that harvesting grapes in much of the rest of Europe was difficult. In addition, research has used contemporary grape harvest dates from Burgundy to reconstruct spring–summer temperatures from 1370 to 2003, and while the results indicate that temperatures as high as those reached in the warm 1990s have occurred several times in the region since 1370, the extremely warm summer of 2003 appears to have been higher than in any other year since 1370 (Chuine et al. 2004).

More recent research of the impacts of climate change on wine quality by Jones et al. (2005a) analyzed growing season temperatures in 27 of arguably the best wine-producing regions in the world and found that average growing season temperatures warmed 1.3°C over the last 50 years. However, the warming was not uniform across all regions with greater magnitudes in the western USA and Europe and less warming in Chile, South Africa, and Australia. The greatest warming was seen in the Iberian Peninsula, Southern France, and parts of Washington and California with warming greater than 2.5°C. For example, Jones et al. (2005a) found that the observed warming during 1950–1999 for the Burgundy, Rhine Valley, Barolo, and Bordeaux regions ranged from 0.7°C to 1.8°C. The study also found that vintage ratings in these same regions (Sotheby's and the Wine Enthusiast: Stevenson 2001; Mazur 2002) have shown trends of increasing overall quality with less vintage-to-vintage variation and that growing season temperatures were important factors in vintage ratings across many regions, albeit not uniform across the regions and not always linear.

Depending on the region and wine type, the marginal effects of the growing season temperatures show that a 1°C warmer vintage can result in 10–22 ratings point increases (Jones et al. 2005a). However, the research also notes that the role of factors other than growing season temperatures such as technology and familiarity has important influences on vintage ratings. Furthermore, the research found that climate thresholds are evident in many regions where, once past a given growing season temperature, quality declines are seen. Therefore, the general rule of thumb "the warmer the better" does not apply for all wine regions where some are near or at the optimum growing season temperatures for some cultivars to achieve the highest quality wine.

More regionally specific and temporally resolved analyses concur with the global observations of wine region temperature trends (Jones and Davis 2000a, b; Jones et al. 2005b; Jones 2007; Webb et al. 2008; Hall and Jones 2008; Ramos et al. 2008). Overall, during the last 30–70 years, many of the world's wine regions have experienced a decline in frost frequency, shifts in the timing of frosts, and warmer growing seasons with greater heat accumulation. In North America, research has shown significant changes in growing season climates, especially in the western USA. For example, during 1948–2002 in the main grape-growing regions of California, Oregon, and Washington, growing seasons warmed by 0.9°C, driven mostly by changes in minimum temperatures, with greater heat accumulation, a decline in frost frequency that is most significant in the dormant period and spring, earlier last spring frosts, later first fall frosts, and longer frost-free periods (Jones 2005). Temporal changes for the Napa Valley since 1930 (Jones and Goodrich 2008) show that heat accumulation is over 350 units higher (degree-days in °C units) and has been the result of significant warming at night where the minimum temperatures have climbed 3.0°C while daytime temperatures have not changed significantly.

Precipitation amounts and timing are highly variable in the western USA, being more tied to larger-scale climate variability mechanisms such as El Niño or the Pacific Decadal Oscillation than structural trends (Jones and Goodrich 2008). A focused study for Napa and Sonoma, California, found that higher yields and quality over the last 50 years were influenced by asymmetric warming (at night and in the spring) where a reduction in frost occurrence, advanced initiation of growth in the spring, and longer growing seasons were the most influential (Nemani et al. 2001). In addition, recent analyses of wintertime extreme freeze events for two important growing regions in North America – Eastern Washington and the Niagara Peninsula of Canada – reveal that although there has been some warming in moderate minimum temperature levels (days with temperatures less than 0°C), extreme low temperatures (–5°C or less) have not changed in frequency over the last 75 years (own data). Furthermore, from the limited data available across the USA, observed changes in grapevine phenology document changes on the order of 2–5 days earlier per decade over the last 25–35 years depending on cultivar and region (Wolfe et al. 2005; Jones 2007) and are strongly correlated to warmer springs and summers.

Recent research for Europe shows similar results as those found in North America detailed above (Jones et al. 2005b). An examination of climate and phenology trends over the last 30–50 years for 11 locations across a range of climate types in Europe (cool to warm) and for 16 cultivars shows that warming occurred across most seasons but is strongest in spring and summer. The growing seasons, over the studied wine regions, warmed by an average of 1.7°C, with most of the observed warming coming at night. Heat accumulation increased as well with degree-days rising by 250–300 units (°C units) while precipitation frequency and amounts did not change significantly. A study in Burgundy, France, by Chabin et al. (2007) found significant changes in temperature with elevation where La Rochepot, 200 m higher than Beaune (at similar elevations to the main vineyards in the region), experienced temperatures during 1988–2006 that were the same as those in Beaune during 1973–1987.

Specifically for Spain, Jones et al. (2005b) found that growing seasons warmed on average by 0.8–1.2°C for the Galicia and Valladolid regions with the warming being much more significant at night (minimum temperatures increasing 1.1–2.1°C) than during the day (not significant). Heat accumulation, either measured by the Huglin Index or Winkler Index (see below), increased inland but did not change significantly in the more coastal region of Galicia. Ramos et al. (2008) also found overall growing season warming in the Penedès, Priorat, and Segrià wine regions of northeast Spain of 1.0–2.2°C.

The work also revealed the potential for added moisture stress where declining precipitation during spring and summer combined with the observed warming results in an increased water demand of 6–14% in an already semiarid region. Along with changes in many temperature parameters in the northeast of Spain, there are concomitant changes in vine and wine parameters, including earlier phenological events, higher wine quality with higher ripening diurnal temperature ranges, and reduced production in the warmest vintages (Ramos et al. 2008). Furthermore, for Europe, in general, grapevine phenological timing showed strong relationships with the observed warming with trends ranging 6–25 days earlier over numerous cultivars and locations (Jones et al. 2005b). Changes are greatest for véraison and harvest dates which typically show a stronger, integrated effect of a warmer growing season. Interval lengths between the main phenological events also declined with bud break to bloom, véraison, or harvest dates shortening by 14, 15, and 17 days, respectively. Averaged over all locations and cultivars, grapevine phenology shows a 3–6-day response per 1°C of warming over the last 30–50 years.

Projections of future climates are produced through models based upon knowledge of how the climate system works and are used to examine how the environment, in this case viticulture and wine production, are likely to respond to these changes. These climate models are complex 3-D, mathematical representations of our Earth/atmosphere system that represent spatial and temporal analyses of the laws of energy, mass, moisture, and momentum transfer in the atmosphere and between the atmosphere and the surface of the Earth. Additionally, climate models are based upon IPCC emissions scenarios (IPCC 2001) which reflect estimates of how humans will emit CO_2 in the future. The many models in use today, combined with the fact that they are modeling a nonlinear system and using different emission scenarios, result in a range of potential changes in temperature and precipitation for the planet (IPCC 2001). Work over the last three decades using model projections shows that the observed warming trends in wine regions worldwide are predicted to continue. From one of the early analyses of the impacts of climate change on viticulture, it was suggested that growing seasons in Europe should lengthen and that wine quality in Champagne and Bordeaux should increase (Lough et al. 1983). These results have largely been proven correct (Jones et al. 2005a).

Furthermore, spatial modeling research has also indicated potential shifts and/or expansions in the geography of viticulture regions with parts of Southern Europe predicted to become too hot to produce high-quality wines and northern regions becoming more stable in terms of consistent ripening climates and/or viable once again (Kenny and Harrison 1992; Butterfield et al. 2000). Examining specific cultivars (Sangiovese and Cabernet Sauvignon), Bindi et al. (1996) found that climate change in Italy should lead to shorter growth intervals but increases in yield variability. Other studies of the impacts of climate change

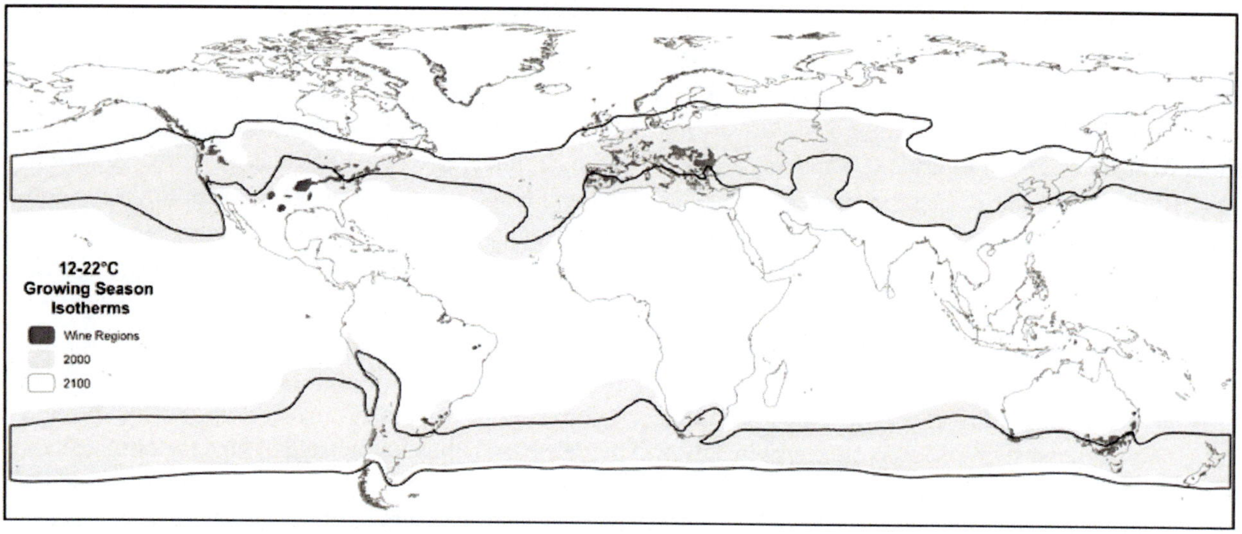

Fig. 7.5 Map of growing season average temperatures (Northern Hemisphere Apr–Oct, Southern Hemisphere Oct–Apr) derived from observations and model runs from the Community Climate System Model (*CCSM*). This figure shows only the shifts in the isotherms for the 2000–2100 period and the A1B emission scenario (moderate future consumption). The wine regions are as described in Fig. 7.2

on grape growing and wine production reveal the importance of changes in the geographical distribution of viable grape-growing areas due to changes in temperature and precipitation, greater pest and disease pressure due to milder winters, changes in sea level potentially altering the coastal zone influences on viticultural climates, and the effect that increases in CO_2 might have on both vine growth and grape quality and even the texture of oak wood used for making wine barrels (Tate 2001; Renner 1989; Schultz 2000; McInnes et al. 2003).

As discussed in the climate suitability section previously, the broadest scale of global suitability for viticulture shows that optimal zones are found between either the mean annual 10–20°C isotherms (de Blij 1983; Johnson 1985) or the growing season 12–22°C isotherms (Gladstones 2004; Jones 2006). To examine these global latitudinal bounds of viticulture suitability due to climate, Jones (2007) used output from the Community Climate System Model (CCSM) on a 1.4°×1.4° latitude/longitude resolution and B1 (moderate), A1B (midrange), and A2 (high) emission scenarios to depict the 12–22°C isotherms shifts for three time periods 2000, 2050, and 2100. Changes from the 2000 base period show both shifts in the amount of area suitable for viticulture and a general latitudinal shift poleward. By 2050, the 12°C and 22°C isotherms shift 150–300 km poleward in both hemispheres depending on the emission scenario (not shown). By 2100, the isotherms shift an additional 125–250 km poleward (see Fig. 7.5 for the midrange A1B scenario). The shifts are marginally greater on the poleward fringe compared to those on the equatorial fringe in both hemispheres. However, the relative area of land mass that falls within the isotherms across the continents expands slightly in the Northern Hemisphere while contracting in the Southern Hemisphere due to land mass differences (Fig. 7.5). Similar shifting is seen by 2100 for all emission scenarios (not shown).

Using Hadley Centre Climate Model (HadCM3) output and an A2 emission scenario (Pope et al. 2000) to 2049 for 27 of the world's top wine-producing regions, Jones et al. (2005a) compared the average climates of two periods, 1950–1999 and 2000–2049. The results suggest that mean growing season temperatures could warm by an average 1.3°C over the wine regions studied with notable wine areas such as Burgundy Rhine Valley, Barolo, and Bordeaux potentially seeing warming ranging from 0.9°C to 1.4°C. Also, the projected changes are greater for the Northern Hemisphere (1.3°C) than the Southern Hemisphere (0.9°C). Examining the rate of change projected for the 2000–2049 period only reveals significant changes in each wine region with trends ranging from 0.2°C to 0.6°C per decade. Overall trends during the 2000–2049 period average 2°C across all regions with the smallest warming in South Africa (0.9°C/50 years) and greatest warming in Portugal

(2.9°C/50 years). For the Burgundy, Rhine Valley, Barolo, and Bordeaux regions, decadal trends are modeled at 0.3–0.5°C while the overall trends are predicted to be 1.5–2.4°C.

In addition, Jones et al. (2005a) showed that many of the wine regions may be at or near their optimum growing season temperature for high-quality wine production and further increases, as predicted by the differences between the means of the 1950–1999 and 2000–2049 periods, will place some regions outside their theoretical optimum growing season climate. The magnitude of these mean growing season changes indicates potential shifts in climate-maturity types for many regions at or near a given threshold of ripening potential for cultivars currently grown in that region. Refer back to Fig. 7.3, Bordeaux's growing season climate of the last 50 years averaged 16.5°C and add to it the overall trend in projected warming in Bordeaux of 2.3°C by 2049. An 18.8°C average growing season would place Bordeaux at the upper end of the optimum ripening climates for many of the red cultivars grown there today and outside the ideal climates for the main white cultivars grown in the region. Still more evidence of these impacts come from Napa, where a 17.5°C historical average is projected to warm by 2.2°C to 19.7°C by 2049. This would place Napa at the upper end of optimal ripening climates for nearly all of the most common cultivars (Fig. 7.3). Finally, the results also show warming during the dormant periods which could influence hardening potential for latent buds, but observations and models indicate continued or increased seasonal variability which could spell problems in freeze or frost prone regions.

For the USA as a whole, White et al. (2006) used a high-resolution (25 km) regional climate model forced by an IPCC A2 greenhouse gas emission scenario and estimated that potential premium winegrape production area in the conterminous USA could decline by up to 81% by the late twenty-first century. The research found that increases in heat accumulation will likely shift wine production to warmer climate cultivars and/or lower-quality wines. Additionally, the models show that while frost constraints will be reduced, increases in the frequency of extreme hot days (>35°C) in the growing season have the potential to severely challenge or completely eliminate winegrape production in many areas of the USA. Furthermore, grape and wine production will likely be restricted to a narrow west coast region and the northwest and northeast areas where excess moisture is already problematic (White et al. 2009).

In another regional analysis for the west coast of the USA, Lobell et al. (2006) examined the impacts of climate change on yields of perennial crops in California. The research combined the output from numerous climate models testing climate uncertainty with multiple statistical crop models testing crop response uncertainty for almonds, walnuts, avocados, winegrapes, and table grapes. The results show a range of warming across climate models of ~1.0–3.0°C for 2050 and 2.0–6.0°C for 2100 and a range of changes in precipitation from –40% to +40% for both 2050 and 2100. Winegrapes showed the smallest yield declines compared to the other crops, but showed substantial spatial shifts in suitability to more coastal and northern counties. The authors also note that yield trends have low attribution to climate trends and are more due to changes in technology (mostly) and an increase in CO_2 (likely).

Examining the structure and suitability for viticulture in the western USA (California, Oregon, Washington, and Idaho), Jones (2007) and Jones et al. (2010) used various temperature metrics such as growing season average temperatures and degree-days to depict the spatial suitability for viticulture across cool to hot climates and the cultivars that grow best in those regions. Using gridded climate data at a 400-m spatial resolution (PRISM; Daly et al. 2008), the work shows the generally known pattern of cooler regions being found higher in elevation, more coastal, and more northerly, e.g., the Willamette Valley, while the warmest regions are mostly confined to the central valley and further south in California, e.g., the San Joaquin Valley. For the 1971–2000 climate normal time period, 51% of the total area in western USA falls into Winkler degree-day regions I–V with 42% of the total area being too cold (<850°C units) and 7% too hot (>2,700°C units). Limiting the data to only those areas in regions I–V finds that region I encompasses 58.8%, region II 11.3%, region III 7.1%, region IV 7.9%, and region V 14.9%. Therefore, the western USA is predominately at the margins of suitability with 73.7% in the coolest and hottest regions (regions I and V, respectively). Using projections for increases in average growing season temperatures from the Community Climate System Model (CCSM) of 1.0–3.0°C for 2049 results in a range of increases in growing degree-days of 15–30% (Jones, 2007). At a +1.0°C warming (roughly a 15% increase in growing degree-days) by 2049, the area of the western USA in regions I–V increases 5% from 51% to 56% and at +3.0°C warming (roughly a 30% increase in growing degree-days), increases by 9–60%. Overall, the changes show a reduction in the areas that are too cold from 42% to 24% while the areas that are too hot increase from 7% to 16% in the greater warming scenario (Jones 2007). Similarly, by individual Winkler regions, there are shifts to predominately more land in region I with smaller changes to regions II–V. Spatially, the shifting of regions occurs toward the coast, especially in California, and upwards in elevation, most notably in the Sierra Nevada mountains. Other regions show large-scale shifting from one Winkler region to another, e.g., Willamette Valley shifting from predominately region I to region II.

Other regional work in Europe (Kenny and Harrison 1992; Butterfield et al. 2000; Stock et al. 2005), Australia (McInnes et al. 2003; Webb et al. 2005; Hall and Jones 2008), and South Africa (Carter 2006) has examined climate change through different modeling approaches but has come up with similar results as discussed above. Kenny and Harrison (1992) did some of the early spatial modeling of future climate change impacts on viticulture in Europe and indicated potential shifts and/or expansions in the geography of viticulture regions with parts of Southern Europe predicted to become too hot to produce high-quality wines and northern regions becoming viable once again. The latter has largely been seen with Southern England once again producing wines that are gaining distinction and quality (Selley 2008).

Examining changes in the Huglin Index of suitability for viticulture in Europe (Huglin 1986), Stock (2005) shows increases of 100–600 units that result in broad latitudinal shifts with new areas on the northern fringes becoming viable, changes in varietal suitability in existing regions, and southern regions becoming so hot that overall suitability is challenged. Specifically in Spain, Rodriguez et al. (2005) examine different emission scenarios to place lower and upper bounds on temperature and precipitation changes and find trends of 0.4–0.7°C per decade with summer warming greater than in the winter. Overall, the changes result in warming by 2100 of between 5–7°C inland and 3–5°C along the coast. Concomitant with these temperature projections, Rodriguez et al. (2005) show much drier springs and summers and lower annual rainfall which is less spatially homogeneous across Spain than is temperature. Furthermore, to examine grapevine responses to climate change, Lebon (2002) used model output to show that the start of Syrah ripening (véraison) in southern France would shift from the second week of August today to the third week of July with a 2°C warming and to the first week of July with a 4°C warming. Additionally, the research found that significant warming during maturation and especially at night would disrupt flavor and color development and ultimately the wine's typicity.

In Australia, Webb et al. (2005) analyzed climate change scenarios for viticulture showing that temperatures by 2070 are projected to warm in Australia by 1.0–6.0°C increasing the number of hot days and decreasing frost risk, while precipitation changes are more variable but result in greater growing season stress on irrigation. The changes projected for Australia have tied future temperature regimes to reduced wine quality with southerly and coastal shifts in production regions being the most likely alternative to maintaining viability. Hall and Jones (2008) in modeling growing season climates for Australia found that 8 of the 61 recognized wine regions in the country would be warmer than the known growing season temperature threshold for suitability by 2030, 12 by 2050, and 21 by 2070 without further adaptive measures. In South Africa, regional projections of rising temperatures and decreased precipitation have been shown to put additional pressure on both the phenological development of the vines and on the necessary water resources for irrigation and production (Carter 2006). The research implies that the practice of winemaking in South Africa is likely to become riskier and more expensive with the most likely effects being shifts in management practices to accommodate an increasingly limited water supply. The author notes that the situation will likely exacerbate other economic issues such as increases in the price of wine, a reduction in the number of wine growers, and need for implementation of expensive and yet unknown adaptive strategies (Carter 2006). Together these studies, and those detailed previously, indicate that the challenges facing the wine industry include more rapid phenological development, changes in suitable locations for some cultivars, a reduction in the optimum harvest window for high-quality wines, and greater management of already scarce water resources.

The observations and modeling efforts described above related to climate change have had and will likely have further impacts and challenges to viticulture and wine production. Namely, the changes strongly suggest that shifts in climate-maturity potential will likely be evidenced mostly through more rapid plant growth and out of balance ripening profiles. For example, if a region currently experiences a maturation period (véraison to harvest) that allows sugars to accumulate to favorable levels, maintains acid structure, and produces the optimum flavor profile for that cultivar, then balanced wines result. In a warmer than ideal environment for a given cultivar, the grapevine will go through its phenological events more rapidly, resulting in earlier and likely higher sugar ripeness, and while the grower or winemaker is waiting for flavors to develop, the acidity is lost through respiration, resulting in unbalanced wines without greater after-harvest inputs or adjustments in the winery. As a result, higher alcohol levels have been observed in many regions. For example, Duchêne and Schneider (2005) found that potential alcohol levels of Riesling at harvest in Alsace have increased by 2.5% (by volume) over the last 30 years and was highly correlated to significantly warmer ripening periods and earlier phenology. Godden and Gishen (2005) summarize trends in composition for Australian wines, and while not attributing any influence to the much warmer conditions experienced in Australia today (McInnes et al. 2003; Webb et al. 2005), they show increases in the alcohol content of 12.3–13.9% for red wines and 12.2–13.2% for white wines from 1984 to 2004. For Napa, average alcohol levels have risen from 12.5% to 14.8% from 1971 to 2001 while acid levels fell and the pH climbed (Vierra 2004).

While higher alcohol can be viewed by some as a good thing, alcohol makes wine "hotter" and less food friendly. Furthermore, while Vierra (2004) argues that this trend is due to the tendency for bigger, bolder wines driven by wine critics and the economics of vintage rating systems, Jones (2006) and Jones (2007) find that climate variability and change are likely responsible for over 50% of the trend in alcohol levels. It appears that the result is a combined effect of growers/producers responding to wine writers/raters trending toward bigger wines, but only able to do so in a warmer climate that allows the style to be produced. Besides changes in wine styles, one of the more germane issues related to higher alcohol levels is that wines typically will not age as well or as long as wines with lower alcohol levels. Finally, harvests that occur earlier in summer, in a warmer part of the growing season, e.g., August or September instead of October in the Northern Hemisphere, will result in hotter harvested fruit and potentially desiccated fruit without greater irrigation inputs.

Conclusions

The geography of wine is intimately tied to climate. The climate structure of the various wine-growing regions worldwide is the critical determinant of the suitability of these regions to given cultivars, their potential wine style, and the region's overall productivity. While our understanding of the role climate structure and suitability plays in viticulture and wine production is generally well developed, the roles that climate variability and climate change have on wine quality and productivity are now being more studied due to the economic impacts that they have and/or potentially could have on the industry. Regional mechanisms for climate variability are better understood and defined today, providing insights into vintage to vintage and region to region influences. However, more applied research at the regional level is needed to better understand the range of variability that a region, or even an individual cultivar, can reliably produce quality wine in.

While it is clear from recorded history and proxy records that the climates of the Earth have varied and changed on both long and short timescales (Le Roy Ladurie 1971; Pfister 1988), the current rate of change and future predicted changes are challenges for viticulture and wine production. It appears that the observed warming over the last 50 years has been largely beneficial for viticulture in many regions through longer and warmer growing seasons with less risk of frost. However, the trends have been shown to be more influential on the poleward fringes by providing more consistent ripening climates for existing cultivars, making warmer climate cultivars more viable or opening up once forgotten regions again. On the other extreme, already hot regions have experienced warmer and generally drier conditions that have produced challenges in ripening balanced fruit. Concomitant with the warming trends have come better technology, better plant material, and better vineyard management, and these adaptations have allowed growers to meet some of these challenges. However, the projections for future climate change will likely be more rapid and to a greater magnitude than our ability to adapt without increased understanding of the impacts and advances in plant breeding and genetics (Bisson et al. 2002; Vivier and Pretorius 2002).

Furthermore, while most of the climate change discussion here and within many other studies has focused on temperature-related impacts, other potential issues affecting grape and wine production and quality include changes in vine growth due to a higher CO_2 concentration in the atmosphere, added moisture stresses in water-limited regions, and changes in the presence or intensity of pests and vine diseases. Even with our current state of knowledge, much uncertainty still exists in the exact spatial and temporal nature of changes in climate; therefore, the wine industry will need to be proactive in assessing the impacts, be ready to implement appropriate adaptation strategies, be willing to alter cultivars and management practices or controls or mitigate wine quality differences by developing new technologies. However, probably the greatest adaptation challenge will be how both growers/producers and consumers respond culturally to changes in regional identities due to cultivar changes or wine style changes.

References

Amerine, M. A. and A. J. Winkler. 1944. Composition and quality of musts and wines of California grapes. *Hilgardia* 15:493–675.

Amerine, M. A., H. W. Berg, R. E. Kunkee, C. S. Ough, V. L. Singleton, and A. D. Webb 1980. *The Technology of Wine Making*, 4th ed., Westport, CT: AVI Publishing Company, Inc.

Andreoli, R. and M. Kayano. 2005. ENSO-related rainfall anomalies in South America and associated circulation features during warm and cold Pacific Decadal Oscillation regimes. *International Journal of Climatolology* 25:2017–2030.

Bindi, M., L. Fibbi, B. Gozzini, S. Orlandini, and F. Miglietta. 1996. Modeling the impact of future climate scenarios on yield and variability of grapevine, *Climate Research* 7: 213–224.

Bisson, L. F., A. L. Waterhouse, S. E. Ebeler, M. A. Walker and J. T. Lapsley. 2002. The present and future of the international wine industry. *Nature* 418: 696–699.

Blanco-Ward, D., J. M. Garcia-Queijeiro, and G. V. Jones. 2007. Spatial climate variability and viticulture in the Miño River Valley of Spain. *Vitis* 46(2): 63–70.

Branas, J. 1974. *Viticulture*. Montpellier, France: Dehan.

Butterfield, R. E., M. J. Gawith, P. A. Harrison, K. J. Lonsdale and J. Orr. 2000. Modelling climate change impacts on wheat, potato and grapevine in Great Britain. In *Climate change, climatic variability and agriculture in Europe: An integrated assessment. Final Report*, ed. T. E. Downing, P. A. Harrison, R. E. Butterfield, and K. G. Lonsdale. University of Oxford: Environmental Change Institute.

Carter, S. 2006. *The projected influence of climate change on the South African wine industry*. Interim Report IR-06–043, International Institute for Applied Systems Analysis.

Carter, T. R., M. L. Parry, and J. H. Porter. 1991. Climatic change and future agroclimatic potential in Europe. *International Journal of Climatology* 11:251–269.

Chabin, J-P., M. Madelin, and C. Bonnefoy. 2007. The Beaune vineyards and global warming: Réchauffement climatique, quels impacts probables sur les vignobles – Université de Dijon. *Proceedings of the Bacchus Symposium* Dijon, France.

Chuine, I., P. Yiou, N. Viovy, B. Seguin, V. Daux, and E. Le Roy Ladurie. 2004. Grape ripening as a past climate indicator. *Nature* 432, 289–290.

Crespin, Y., M. LeBerre, and P. Uvietta. 1987. The system climate-vine-vinegrower: example of a dynamic model. *Bulletin de l'OIV* 60:5–26.

Daly, C., M. Halbleib, J. I. Smith, W. P. Gibson, M. K. Doggett, G. H. Taylor, J. Curtis, and P. A. Pasteris. 2008. Physiographically-sensitive mapping of temperature and precipitation across the conterminous United States. *International Journal of Climatology* 28(15):2031–2064.

de Blij, H. J., 1983. Geography of viticulture: rationale and resource. *Journal of Geography* 112–121.

Dilley, M., 2000. Reducing vulnerability to climate variability in Southern Africa: The growing role of climate information. *Climatic Change* 45:63–73.

Duchêne, E., and C. Schneider. 2005. Grapevine and climatic changes: A glance at the situation in Alsace. *Agronomy for Sustainable Development* 25:93–99.

Easterling, D. R. et al. 2000. Observed variability and trends in extreme climate events: A brief review. *Bulletin of the American Meteorological Society* 81: 417–425.

Esteves, M. A. and M.D. Manso Orgaz. 2001. The influence of climatic variability on the quality of wine. *International Journal of Biometeorology* 45:13–21.

Frauenfeld, O. W., R. E. Davis, and M. E. Mann. 2005. A Distinctly interdecadal signal of Pacific Ocean–Atmosphere interaction. *Journal of Climate* 18(11):1709–1718.

Fregoni, M. 2003. L'indice bioclimatico di qualitá Fregoni. In *Terroir, zonazione viticoltura*, ed. M. Fregoni, D. Schuster, and A. Paoletti, 115–127. Piacenza, Italy: Phytoline.

Garnett, E. R., and M. L. Khandekar. 1992. The impact of large-scale atmospheric circulations and anomalies on Indian monsoon droughts and floods and on world grain yields: A statistical analysis. *Agriculture and Forest Meteorology* 61: 113–128.

Garreaud, R. D., 2000. Cold Air Incursions over Subtropical South America: Mean Structure and Dynamics. *Monthly Weather Review* 128: 2544–2559.

Garreaud, R.D., M. Vuille, R. Compagnucci, and J. Marengo. 2009. Present-day South American climate. *Palaeogeography, Palaeoclimatology, Palaeoecology* 281: 180–195.

Gillett, N. P., T. D. Kell, and P. D. Jones. 2006. Regional climate impacts of the Southern Annular Mode. *Geophysical Research Letters* 33: L23704

Gladstones, J. 1992. *Viticulture and environment*. Adelaide, Australia: Winetitles.

Gladstones, J. 2004. Climate and Australian viticulture. In *Viticulture 1 – Resources*, ed. P. R. Dry and B.G. Coombe. Adelaide, Australia: Winetitles.

Glantz, M., 2001. *Currents of change: Impacts of El Niño and La Niña on climate and society*. Cambridge, UK: Cambridge University Press.

Godden, P. and M. Gishen. 2005. Trends in the composition of Australian wine. *The Australian and New Zealand Wine Industry Journal* 20(5): 21–46.

Gordon, N.D. 1986. The Southern Oscillation and New Zealand weather. *Monthly Weather Review* 114: 371–387.

Grifoni, D., M. Mancini, G. Maracchi, S. Orlandini, and G. Zipoli. 2006. Analysis of Italian wine quality using freely available meteorological information. *American Journal Enology and Viticulture* 57: 339–346

Grimm, A. M., V. R. Barros, and M. E. Doyle. 2000. Climate variability in southern South America associated with El Niño and La Niña events. *Journal of Climate* 13: 35–58.

Grimm, A. M., and R. G. Tedeschi. 2009. ENSO and extreme rainfall events in South America. *Journal of Climate* 22: 1589–1609.

Gu, L., P. J. Hanson, W. Mac Post, D. P. Kaiser, B. Yang, R. Nemani, S. G. Pallardy, and T. Meyers. 2008. The 2007 eastern U.S. spring freeze: Increased cold damage in a warming world? *BioScience* 58 (3): 253–262.

Hall, A. and G. V. Jones. 2008. Effect of potential atmospheric warming on temperature based indices describing Australian winegrape growing conditions. *Australian Journal of Grape and Wine Research* 15 (2): 97–119.

Hall, A. and G.V. Jones. 2010. Spatial analysis of climate in winegrape growing regions in Australia. *Australian Journal of Grape and Wine Research* 16: 389–404.

Hendon, H. H., D. W. J. Thompson, and M. C. Wheeler. 2007. Australian rainfall and surface temperature variations associated with the Southern Hemisphere Annular Mode. *Journal of Climate* 20: 2452–2467.

Hijmans, R. J., S. E. Cameron, J. L. Parra, P. G. Jones and A. Jarvis. 2005. Very high resolution interpolated climate surfaces for global land areas. *International Journal of Climatology* 25: 1965–1978.

Huglin, P. 1978. Nouveau mode d'évaluation des possibilités héliothermiques d'un milieu viticole. *C. R. Acad. Agr. France* 1117–1126.

Huglin, P. 1986. Biologie et écologie de la vigne. (Ed. Payot: Laussane, France).

Hurrell, J. W. 2003. *The North Atlantic Oscillation: Climate significance and environmental impact,* ed. J. W. Hurrell, Y. Kushnir, G. Ottersen, and M. Visbeck. Geophysical Monograph Series 134.

IPCC 2001. Houghton J. T. et al. *Climate change 2001: The scientific basis. Contribution of the working group 1 to the third assessment of the Intergovernmental Panel on Climate Change.* Cambridge, UK: Cambridge University Press.

IPCC 2007. Alley R. et al. Climate change 2007: The physical science basis. Summary for policymakers. *Contribution of the Working Group I to the Fourth Assessment of the Intergovernmental Panel on Climate Change. IPCC Secretariat* (http://www.ipcc.ch/).

Jackson, D. I. and N. J. Cherry. 1988. Prediction of a district's grape-ripening capacity using a latitude-temperature index. *American Journal of Enology and Viticulture* 39 (1): 19–28.

Johnson, H. 1985. *The World Atlas of Wine*. 3rd ed., New York: Simon and Schuster.

Jones, G. V. 1997. A synoptic climatological assessment of viticultural phenology. Dissertation, University of Virginia, Department of Environmental Sciences.

Jones, G. V. and R. E. Davis. 2000a. Using a synoptic climatological approach to understand climate/viticulture relationships. *International Journal of Climatology* 20: 813–837.

Jones, G. V. and R. E. Davis. 2000b. Climate influences on grapevine phenology, grape composition, and wine production and quality for Bordeaux, France. *American Journal of Viticulture and Enology* 51(3): 249–261.

Jones, G. V. and E. Hellman. 2003. Site Assessment. In *Oregon Viticulture,* ed. E. Hellman, 5th edition, 44–50. Corvallis, Oregon: Oregon State University Press.

Jones, G. V., 2005. Climate change in the western United States grape growing regions. *Acta Horticulturae (ISHS)* 689: 41–60.

Jones, G. V., M. A. White, O. R. Cooper, and K. Storchmann. 2005a. Climate change and global wine quality. *Climatic Change* 73 (3): 319–343.

Jones, G. V., E. Duchêne, D. Tomasi, J. Yuste, O. Braslavksa, H. Schultz, C. Martinez, S. Boso, F. Langellier, C. Perruchot, and G. Guimberteau. 2005b. *Changes in European Winegrape Phenology and Relationships with Climate.* GESCO Proceedings, August 2005.

Jones, G. V. 2006. Climate and terroir: Impacts of climate variability and change on wine. In *Fine wine and terroir – the geoscience perspective*, ed. R. W. Macqueen, and L. D. Meinert. Geoscience Canada Reprint Series Number 9, Geological Association of Canada, St. John's, Newfoundland.

Jones, G. V. 2007. Climate change and the global wine industry. *Proceedings from the 13th Australian Wine Industry Technical Conference*. Adelaide, Australia.

Jones, G. V. and G. B. Goodrich. 2008. Influence of climate variability on wine region in the western USA and on wine quality in the Napa Valley. *Climate Research* 35: 241–254.

Jones, G. V., B. Bois, M. Moriondo, A. Hall, and A. Duff. 2009. Analysis of the spatial climate structure in viticulture regions worldwide. *Proceedings of the 7th General Assembly of the International Organization of Vine and Wine*. Zagreb, Croatia.

Jones, G.V. A.A. Duff, A. Hall, and J. Myers. 2010. Spatial analysis of climate in winegrape growing regions in the western United States. *American Journal of Enology and Viticulture* 61: 313–326.

Kandji, S. T., L. Verchot, and J. Mackensen. 2006. *Climate change climate and variability in southern Africa: Impacts and adaptation in the agricultural sector*. United Nations Environment Programme (UNEP).

Karl, T. R., P. D. Jones, R. W. Knight, G. Kukla, N. Plummer, V. Razuvayev, K. P. Gallo, J. Lindseay, R. J. Charlson and T. C. Peterson. 1993. A new perspective on global warming: Asymmetric trends of daily maximum and minimum temperature. *Bulletin of the American Meteorological Society* 74: 1007–1023.

Kayano, M. T., C. P. Oliveira, and R. V. Andreoli. 2009. Interannual relations between South American rainfall and tropical sea surface temperature anomalies before and after 1976. *International Journal of Climatology* 29: 1439–1448.

Kenny, G. J. and P. A. Harrison. 1992. The effects of climate variability and change on grape suitability in Europe. *Journal of Wine Research* 3: 163–183.

Kenny, G. J. and J. Shao. 1992. An Assessment of a latitude-temperature index for predicting climate suitability for grapes in Europe. *Journal of Horticultural Science* 67 (2): 239–246.

Kliewer, W. M. and R. E. Torres. 1972. Effect of controlled day and night temperatures on grape coloration. *American Journal of Enology and Viticulture* 23: 71–77.

Köppen, W., 1931. *Grundiss der Klimakunde*. 2nd ed. Berlin, Germany: Walter de Gruyter & Co.

Le Roy Ladurie, E., 1971. *Times of feast, times of famine: a history of climate since the year 1000.* Garden City, New York: Doubleday & Company, Inc.

Lebon, E., 2002. Changements climatiques: quelles conséquences pour la viticulture. CR 6ième Rencontres Rhodaniennes. pp. 31–36

Lobell, D. B., C. B. Field, K. N. Cahill, and C. Bonfils. 2006. Impacts of future climate change on California perennial crop yields: Model projections with climate and crop uncertainties. *Agricultural and Forest Meteorology* 141 (2–4): 208–218.

Lough, J. M., T. M. L. Wigley, and J. P. Palutikof. 1983. Climate and climate impact scenarios for Europe in a warmer world. *Journal of Climate and Applied Meteorology* 22: 1673–1684.

Mantua, N. J. and S. R. Hare. 2002. The Pacific Decadal Oscillation. *Journal of Oceanography* 58: 35–44.

Mason, S. J. and M. R. Jury. 1997. Climatic variability and change over southern Africa: a reflection on underlying processes. *Progress in Physical Geography* 21(1): 23–50.

Mathieu, P. P., R. T. Sutton, B. Dong, and M. Collins. 2004. Predictability of winter climate over the north Atlantic European region during ENSO events. *Journal of Climate* 17: 1953–1974.

Matthews, M. A. and M. M. Anderson. 1988. Fruit ripening in Vitis vinifera L.: responses to seasonal water deficits. *American Journal of Enology and Viticulture* 39 (4): 313–320.

Mazur, M., Wine Enthusiast's 2002 Vintage Chart. The Wine Enthusiast Magazine (http://www.winemag.com/vintage.cfm).

McInnes, K. L., P. H. Whetton, L. Webb, and K. J. Hennessy. 2003. Climate change projections for Australian viticultural regions. *The Australian and New Zealand Grapegrower and Winemaker* February 2003.

McIntyre, G. N., L. A. Lider, and N. L. Ferrari. 1982. The chronological classification of grapevine phenology. *American Journal of Enology and Viticulture* 33 (2): 80–85.

Menzel, A. and P. Fabian. 1999. Growing season extended in Europe. *Nature* 397:659.

Misra, V. 2003. The influence of Pacific SST variability on the precipitation over southern Africa. *Journal of Climate* 16: 2408–2418.

Moonen, A. C., L. Ercoli, M. Mariotti, and A. Masoni. 2002. Climate change in Italy indicated by agrometeorological indices over 122 years. *Agricultural and Forest Meteorology* 111: 13–27.

Mori, K., S. Sugaya, and H. Gemma. 2005. Decreased anthocyanin biosynthesis in grape berries grown under elevated night temperature condition. *Scientia Horticulturae* 105: 319–330.

Mullins, M. G., A. Bouquet, and L. E. Williams. 1992. *Biology of the Grapevine.* UK: Cambridge University Press.

Nemani, R. R., M. A. White, D. R. Cayan, G. V. Jones, S. W. Running, and J. C. Coughlan. 2001. Asymmetric climatic warming improves California vintages. *Climate Research* 19 (1): 25–34.

Nicholls, N. 2009. Drought in southern Australia: trends, causes, impacts. *Proceedings of the 9th International Conference on Southern Hemisphere Meteorology and Oceanography.* Melbourne, Australia.

Nicholls, N., 1985. Impact of the Southern Oscillation on Australian crops. *Journal of Climatology* 5: 553–560.

Paegle, J. N., and K. C. Mo. 2002. Linkages between summer rainfall variability over South America and sea surface temperature anomalies. *Journal of Climate* 15: 1389–1407.

Peel, M., C. B. L. Finlayson, and T. A. McMahon. 2007. Updated world map of the Köppen-Geiger climate classification. *Hydrology and Earth Systems Science* 11: 1633–1644.

Penning-Roswell, E. 1989. *Wines of Bordeaux,* 6th ed., London/New York: Penguin Books.

Pezza, A. B., T. Durrant, I. Simmonds, and I. Smith. 2008. Southern hemisphere synoptic behavior in extreme phases of SAM, ENSO, sea ice extent, and southern Australia rainfall. *Journal of Climate* 21: 5566–5584.

Pfister, C. 1988. Variations in the spring-summer climate of central Europe from the High Middle Ages to 1850. In *Long and Short Term Variability of Climate*, ed. H. Wanner and U. Siegenthaler 57–82. Berlin, Germany: Springer-Verlag.

Podestá, G. P., D. Letson, C. Messina, F. Royce, R. A. Ferreyra, J. W. Jones, J. W. Hansen, I. Llovet, M. Grondona, and J. J. O'Brien. 2002. Use of ENSO related climate information in agricultural decision making in Argentina: A pilot experience. *Agricultural Systems* 74: 371–392.

Pope, V. D., M. L. Gallani, P. R. Rowntree and R. A. Stratton. 2000. The impact of new physical parameterizations in the Hadley Centre climate model – HadAM3. *Climate Dynamics* 16: 123–146.

Power, S., F. Tseitkin, V. Mehta, B. Lavery, S. Torok, and N. J. Holbrook. 1999. Decadal climate variability in Australia during the twentieth century. *International Journal of Climatology* 19: 169–184.

Prescott, J. A. 1965. The climatology of the vine (Vitis vinifera L.) the cool limits of cultivation. *Transcriptions of the Royal Society of Southern Australia* 89: 5–23.

Ramos, M.C., Jones, G.V. and J. A. Martínez-Casasnovas. 2008. Structure and trends in climate parameters affecting winegrape production in northeast Spain. *Climate Research* 38:1–15.

Reason, C. J. C. and M. Rouault. 2002. ENSO-like decadal variability and South African rainfall. *Geophysical Research Letters* 29 (13): 1638.

Reid, R. and G. V. Jones. 2009. Analysis of the Köppen climate classification characteristics of global wine regions. SOAR: Southern Oregon Arts and Research, 2009 Annual Symposium. Ashland, Oregon.

Renner, B. 1989. The shape of things to come. *Wine and Spirit* December 1989: 55–57.

Rodó, X. and F. A. Comín. 2000. Links between large-scale anomalies, rainfall and wine quality in the Iberian Peninsula during the last three decades. *Global Change Biology* 6 (3): 267–273.

Rodriguez, J. M. et al. 2005. *Main conclusions from the preliminary assessment of the impacts in Spain due to the effects of Climate Change.* Project ECCE. Ministry of the Environment and the University of Castilla-La Mancha.

Rosenzweig, C. and D. Hillel. 2007. *Climate variability and the global harvest: Impacts of El Niño and other oscillations on agro-ecosystems.* London: Oxford University Press.

Roujou de Boubee, D., C. Van Leeuwen, and D. Dubourdieu. 2000. Organoleptic impact of 2-methoxypyrazine on red Bordeaux and Loire wines. Effect of environmental conditions on concentrations in grapes during ripening. *Journal of Agricultural Food Chem*istry 48 (10): 4830–4834.

Schär, C., P. L. Vidale, D. Lüthi, C. Frei, C. Häberli, M. A. Liniger, and C. Appenzeller. 2004. The role of increasing temperature variability for European summer heat waves. *Nature* 427: 332–336.

Schultz, H. R. 2000. Climate change and viticulture: A European perspective on climatology, carbon dioxide, and UV-B effects. *Australian Journal of Grape and Wine Research* 6: 2–12.

Selley, R. C. 2008. *The winelands of Britain: Past present and prospective*. 2nd edition. Surrey, England, Petravin.

Siebert, T. E.; C. Wood, G. M. Elsey, and A. P. Pollnitz. 2008. Determination of rotundone, the pepper aroma impact compound, in grapes and wine. *Journal of Agricultural Food Chem*istry 56: 3745–3748.

Smart, R. E. and P. R. Dry. 1980. A climatic classification for Australian viticultural regions. *Australian Grapegrower and Winemaker* 17: 8–16.

Souriau, A., and P. Yiou. 2001. Grape harvest dates for checking NAO paleoreconstructions, *Geophysical Research Letters* 28 (20): 3895–3896.

Stevenson, T. 2001. *New sothebys wine encyclopedia: A comprehensive reference guide to the wines of the world.* 3rd ed. Dorling Kindersley, London.

Stock, M. 2005. Klimaveränderungen fordern die Winzer – Bereitschaft zur Anpassung ist erforderlich; Geisenheimer Berichte. Band 57: 29–48.

Stock, M., F. W. Gerstengarbe, T. Kartschall, and P.C. Werner. 2005. Reliability of climate change impact assessments for viticulture. *ISHS Acta Horticulturae* 689: Proceedings from the VII International Symposium on Grapevine Physiology and Biotechnology. Davis, California

Tate, A. B. 2001. Global warming's impact on wine. *Journal of Wine Research* 12 (2): 95–109.

Todd, M. C., R. Washington, R. A. Cheke, and D. R. Kniveton. 2002. Brown locust outbreaks and climate variability in southern Africa. *Journal of Applied Ecology* 39: 31–42

Tonietto, J. 1999. Les macroclimats viticoles mondiaux et l'influence du mésoclimat sur la typicité de la Syrah et du Muscat de Hambourg dans le sud de la France: méthodologie de caractérisation. (Thèse Doctorat). École Nationale Supérieure Agronomique de Montpellier - ENSA-M. 233p.

Tonietto, J. and A. Carbonneau. 2004. A multicriteria climatic classification system for grape-growing regions worldwide. *Agriculture and Forest Meteorology* 124: 81–97.

Trenberth, K. E. and J. W. Hurrell. 1994. Decadal atmospheric–ocean variations in the Pacific. *Climate Dy*namics 9: 303–319.

Tyson, P. D. 1986. *Climatic change and variability over southern Africa*. Cape Town: Oxford University Press.

Ummenhofer, C. C., M. H. England, P. C. McIntosh, G. A. Meyers, M. J. Pook, J. S. Risbey, A. S. Gupta, and A. S. Taschetto. 2009. What causes southeast Australia's worst droughts? *Geophysical Research Lett*ers 36: L04706.

Ummenhofer, C. C., and M. H. England. 2007. Interannual extremes in New Zealand precipitation linked to modes of Southern Hemisphere climate variability. *Journal of Climate* 20: 5418–5440.

Valimba P., P. Camberlin, Y. Richard, E. Servat, and D. Hughes. 2006. Influences of ENSO and SST variations on the interannual variability of rainfall amounts in southern Africa. In *Climate variability and change: hydrological impacts*, ed. E. S. Demuth, A. Gustard, E. Planos, F. Scatena, and E. Servat. 362–368, Wallingford U.K. and Paris: AISH, UNESCO.

Vera, C. S., and P. K. Vigliarolo. 2000. A diagnostic study of cold-air outbreaks over South America. *Monthly Weather Re*view 128: 3–24.

Vierra, G. 2004. Pretenders at the table – Are table wines no longer food friendly? *Wine Business Monthly* 11 (7): July 2004.

Vivier, M. A. and I. S. Pretorius. 2002. Genetically tailored grapevines for the wine industry. *Trends in Biotechnology* 20 (11): 472–8.

Webb, L. B, P. H. Whetton and E. W. R. Barlow. 2005. Impact on Australian viticulture from greenhouse induced temperature change. In *MODSIM 2005 International Congress on Modelling and Simulation*, ed. A. Zerger and R. M. Argent, 170–176. Modelling and Simulation Society of Australia and New Zealand, December 2005.

Webb L. B, P. H. Whetton and E. W. R. Barlow. 2008. Climate change and winegrape quality in Australia. *Climate Research* 36: 99–111.

White, M. A., N. S. Diffenbaugh, G. V. Jones, J. S. Pal, and F. Giorgi. 2006. Extreme heat reduces and shifts United States premium wine production in the 21st century. *Proceedings of the National Academy of Sciences* 103 (30): 11217–11222.

White, M. A., G. V. Jones, and N. S. Diffenbaugh. 2009. Climate variability, climate change, and wine production in the western United States. In *Climate Warming in Western North America/Evidence and Environmental Effects*, ed. F. H. Wagner, 77–88. University of Utah Press.

Williams, L. E. 2000. Grapevine water relations. In *Raisin Production Manual,* ed. L. P. Christensen 121–126. Oakland, CA.: University of California DANR Publications.

Winkler, A. J., J. A. Cook, W. M. Kliewere, and L. A. Lider. 1974. *General viticulture.* 4th ed., Berkeley, CA: University of California Press.

Wolf, T. K. 1997. *Site Selection for Commercial Vineyards*, Virginia Agricultural Experiment Station, Winchester, Virginia, Publication Number 463–016.

Wolfe, D. W., M. D. Schwartz, A. N. Lakso, Y. Otsuki, R. M. Pool. and N. J. Shaulis. 2005. Climate change and shifts in spring phenology of three horticultural woody perennials in northeastern USA. *International Journal of Biometeorology* 49 (5): 303–309.

Part IV

Cultural/Economic

8 The South African Wine Industry: Meeting the Challenges of Structural and Ethical Transformation

David Bek, Cheryl McEwan, and Tony Binns

Abstract

Since 1994, the South African wine industry has been challenged to effect a complex transformation which has involved re-engaging with international markets whilst tackling an array of social, environmental and political challenges. The multiple dimensions of this transformation are reviewed within this chapter. The first section provides a geographical and statistical overview of the industry and assesses some of the reasons for its successes in engaging with the demands of international markets. A fascinating aspect of the industry's recent biography has been the emergence of a number of initiatives with ethical dimensions that seek to address environmental and social concerns. A sample of these including Fairtrade, Wieta (the wine and agri-industry ethical trade initiative) and the Biodiversity and Wine Initiative is reviewed. The chapter concludes by considering some of the interrelationships between market and ethical imperatives alongside an overview of the ongoing challenges currently confronting the industry. These include maintaining profitability within the highly competitive global marketplace, retaining value added within South Africa, developing globally recognised brands, gaining recognition for the quality and diversity of higher value wines, streamlining ethical codes and effecting meaningful black economic empowerment.

Introduction

Since 1994, the South African wine industry has been faced with imperatives to effect a dual transformation. On the one hand, the inward-looking nature of the apartheid economy, reinforced by decades of sanctions and boycotts, left a legacy of outmoded production and trade structures. There was, therefore, an urgent need to reorientate the industry so that it could integrate effectively with international markets. On the other hand, South African society was riven by racially delineated inequalities which have rendered the nation one of the most unequal in the world. Nowhere have these injustices been felt more strongly than in the wine industry where workers endured the 'worst working conditions experienced in South Africa' (Brown et al. 2003, 23), whilst the interactions between the wine industry and the natural environment were

D. Bek (✉) • C. McEwan
Department of Geography, University of Durham, Durham, UK
e-mail: dave.bek@durham.ac.uk; Cheryl.McEwan@durham.ac.uk

T. Binns
Department of Geography, University of Otago, Dunedin, New Zealand
e-mail: jab@geography.otago.ac.nz

P.H. Dougherty (ed.), *The Geography of Wine: Regions, Terroir and Techniques*,
DOI 10.1007/978-94-007-0464-0_8,

increasingly problematic. Thus, social, economic, environmental and political imperatives have forced stakeholders in the wine industry to re-evaluate production practices and the industry's institutional organisation whilst seeking to meet the challenges of transforming working conditions and ownership patterns. This chapter first provides an overview of the South African wine industry including data on production and market trends. Secondly, it demonstrates the extent to which the industry has been successful in engaging with the international wine market and outlines some of the reported reasons for this success. Thirdly, the South African wine industry can be seen as something of a hotbed for the initiation of projects and programmes with an ethical component that seek to address social and environmental concerns (Ashton 2009; Bek et al. 2007; McEwan and Bek 2009a). A sample of these schemes are outlined and evaluated here. Finally, some of the complex interrelationships between market and ethical imperatives are considered alongside an overview of the ongoing challenges currently confronting the industry.

Geographical and Statistical Overview of the South African Wine Industry[1]

The South African wine industry is largely located within the Western Cape with some production in the Northern Cape and a very small amount in KwaZulu-Natal (see Map 8.1). The area has a relatively mild Mediterranean climate which is ideal for growing a range of grape varieties. Production during the apartheid era was heavily dominated by white grape production, and planting patterns were driven by the politics of the wine industry's statist structure rather than region's growing potential. Replanting strategies were implemented during the 1990s, thus increasing the proportion of reds in order to provide for the tastes of the international market (see Table 8.1). Thus, internationally popular varietals such as Cabernet Sauvignon and Shiraz account for 20% of all grape production, whilst Pinotage offers a distinctive element to the nation's output (SAWIS 2009). South Africa's white wines have achieved some notable success on the international stage with a wide range of white grapes being grown. White grapes still dominate, to some extent buffering the industry from the worst effects of the 'red glut' which has afflicted the global industry.

The local matrix of topography, soils and mesoclimates varies tremendously creating optimal conditions for different varietals. Ever increasing understanding of the variations in the physical environment, reinforced by scientific data generated by highly sophisticated Geographical Information Systems, underpins Wines of South Africa's (WOSA) current marketing campaign 'Variety is in our Nature' (WOSA 2009c). This campaign draws attention to the diverse terroir of the Winelands and, thus, the potential to produce a range of distinctive high-quality wines. The marketing message is strengthened by the region's stunning landscapes, with its mountain ranges and unique Cape Flora making the area an important wine tourism location.

There are currently more than 100,000 ha under wine grape production, representing 3% of the world's output (SAWIS 2008). South Africa is the world's ninth largest producer and is a key contributor to certain key markets, for example, the UK, where market share is around 9%, and Sweden, where it is the leading exporter (Ponte and Ewert 2007). Although South Africa lacks the extent of brand penetration exhibited by Australia and California, it does have three brands (Kumala, First Cape and Namaqua) in the UK's top 20. Nearly 30% of South African wine exports are directed to the UK market and 17% to Germany (see Table 8.2). Exports to the UK have been growing strongly as have other longstanding markets such as Germany and Sweden. There has been notable growth in expanding markets including Russia.

The industry creates employment for 256,298 people including tourism-related jobs (SAWIS 2004). The industry's contribution to national GDP is around 1.5%, whilst the contribution to the Western Cape's GDP has been estimated at 8.2%. Wine tourism has been growing rapidly, generating R5.5 m in 2005 and 23% of the industry's total jobs (WOSA 2006).

[1] Field-based research which informed this chapter has been conducted on three occasions: March/April 2004, supported by a Nuffield Foundation Grant (SGS/00909/G); August/September 2006, supported by a British Academy research grant (SG-43017) and February/March 2009 supported by Otago University. A wide range of wine industry stakeholders have been interviewed during these visits. Further interviews and desk-based research have been conducted in the UK.

Map 8.1 Wine growing areas of South Africa (Adapted from Wines of South Africa, http://www.wosa.co.za/maps/1_overview.jpg

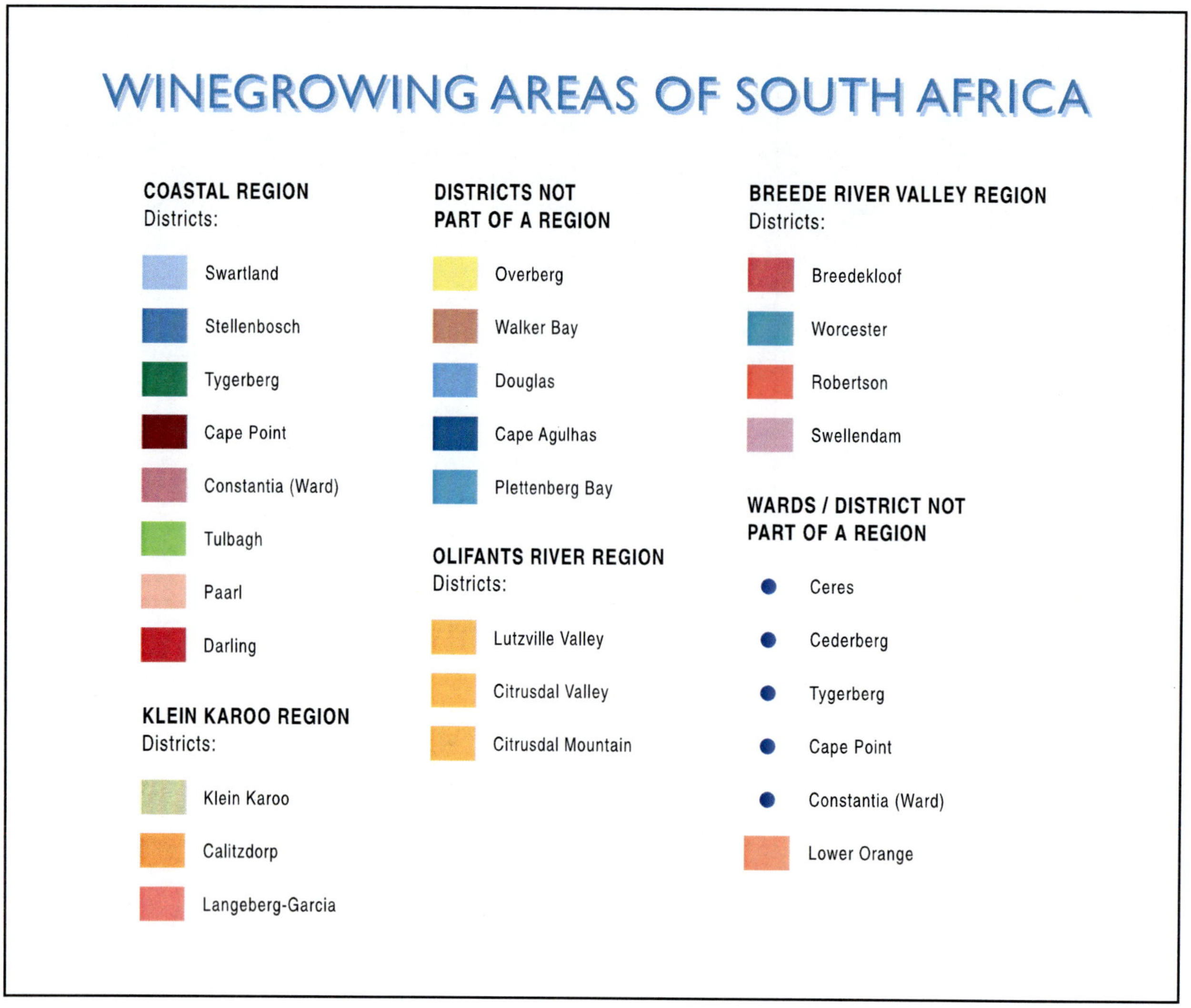

Map 8.1 (continued)

Table 8.1 Production of different grape varieties, 2008

	2008	%	% of all varieties
Red			
Cabernet Sauvignon	13,336	29.1	11.8
Shiraz	9,832	21.4	8.7
Merlot	6,863	15.0	6.1
Pinotage	6,297	13.7	5.6
Ruby Cabernet	2,559	5.6	2.3
Pinot Noir	577	1.3	0.5
Other red	6,333	13.8	5.6
Total red	45,837		40.7
White			
Chardonnay	8,181	12.2	7.3
Sauvignon Blanc	8,406	12.6	7.5
Semillon	1,085	1.6	1.0
Other white	49,208	73.6	43.7
Total white	66,880		59.3

Source: SAWIS 2009

Table 8.2 Destinations of South African wine exports

Packaged and bulk natural wine exports (litres) June 2008–May 2009

	White	Red	Blanc de noir/Rose	Overall	Total exports (litres)	Change over previous 12 months
United Kingdom	33.1	24.3	37.8	28.8	116,479,422	125
Germany	12.6	19.5	28.7	17.1	69,076,824	114
Sweden	7.4	9.8	1.3	8.2	33,316,092	131
The Netherlands	6.5	6.6	16.5	7.1	28,797,436	102
Angola	1.6	8.8	0.0	5.2	21,062,092	145
Denmark	2.3	6.1	0.8	4.2	16,928,629	115
Russia	8.3	0.7	1.2	4.0	16,034,002	243
USA	5.4	2.4	0.6	3.6	14,359,180	114
Canada	3.4	2.9	0.5	2.9	11,873,130	82
Belgium	2.7	1.7	1.7	2.1	8,435,331	93
France	1.7	2.1	3.6	2.0	8,159,229	103
New Zealand	2.4	1.9	0.1	2.2	9,084,876	78
Others	12.7	13.3	7.1	12.4	50,273,521	
Total					403,879,764	118

Source: SAWIS 2008

Overcoming the Apartheid Legacy in the Marketplace

The industry has achieved a remarkably rapid transformation from being backward and inward looking to a state of respected international competitiveness since the end of apartheid. Exports have increased more than 20-fold in 15 years, and markets have been diversified significantly (Boom 2009; Faurie 2008). Indicators of success include the fact that, by 2007, South Africa was the fifth largest exporter by value into the all-important UK wine market, having built upon a period of extended year on year growth of 12%. Furthermore, industry leaders believe that South Africa is a global front runner in promoting ethical and environmental standards, which are not only 'the right thing to do' but also instrumental in providing a marketing edge.

Yet, when sanctions were lifted in the early 1990s, the industry was poorly prepared to exploit newly available opportunities. Whilst overseas consumers were keen to support South Africa's exports for the first time in decades, a rapidly earned reputation for 'mixed' quality wine stymied medium-term efforts to promote export growth. So what factors have driven the remarkable transformation in the fortunes of the industry? The advent of democracy was critical in a number of ways. First, and most obviously, doors were opened in markets that had hitherto been firmly closed. Secondly, the restrictive structures and organisation of the South African wine industry underwent a substantial transformation at the end of apartheid, and the regulatory regime that shaped the industry's activities throughout the twentieth century was dismantled (Ewert 2002). Prior to 1994, production was aimed primarily at the domestic market, with the majority of grapes being sold by producers and their cooperatives to the Co-operative Wine Growers of South Africa (KWV) and other wholesalers (Williamson and Wood 2003). Guaranteed prices protected producers from market variations, creating a distorted market that favoured mass production of low quality grapes, which did little to reflect the unique terroir of the Western Cape. Innovation was effectively stifled, as producers had no real incentive to upgrade technology or skills. Furthermore, restrictions were imposed on planting patterns, preventing producers from matching varietals to suitable areas of terroir. Vineyards have now been completely restructured, from 85% white to 55%, thus enabling the production of quality wines suited to consumer tastes.

Thirdly, there has been a significant growth in intellectual capital within the industry. Decades of international isolation had left producers unable to judge the relative standards of their own product; in the words of Dr. Paul Cluver,[2] 'as with sport you have to play

[2]Dr. Paul Cluver is a leading figure in the South African wine industry, owning his own wine estate in the Elgin region. His wider roles have included acting as Chairman of WOSA and Capespan. Cluver has also played a lead role in the establishment of the BWI, Thandi and the Green Mountain Eco Route tourism initiative.

internationally to know how good you are' (Cluver 23 February 2009). It came as a shock to many to realise how relatively poor the South African product had become. However, individuals responded rapidly to the new opportunities as young winemakers travelled the world to acquire knowledge of the latest technologies and marketing concepts. In return, flying winemakers became part of the local scene, facilitating a churning of ideas. In this sense, 'wine is more international than any other agri-product. There is a unique exchange of people and ideas through networking, which makes a huge difference. It makes a difference to everyone (in global wine production), but it makes a bigger difference (in South Africa) because of the history' (Cluver 23 February 2009).

Ethical Imperatives in the South African Winelands

Labour Standards – The Apartheid Legacy

The history of the South African wine industry is deeply intertwined with the social history of slavery that continued to shape political, economic and cultural power relations for more than 150 years (Kruger et al. 2006). Indeed, the inward-looking and uncompetitive business model underpinning the apartheid-era industry was able to survive due to the appalling employment environments endured by workers. A concept of paternalism defined the relationship between farmer and workers, with the latter receiving 'social dividends' such as housing, electricity and water in addition to wages. However, wage levels were extremely low, and whilst housing was usually available to permanent employees, this was lost if workers were fired (du Toit 1993). Workers were barred from collective organisation and faced punishment if efforts were made to unionise or engage in collective bargaining. Perhaps the most infamous aspect of on-farm labour relations was the 'dop' system, whereby workers were paid a component of their wages in the form of low-grade wine (Brown et al. 2003). As one commentator writes: 'Generations of workers lived and died in shoddy housing, earning very little and slowly drinking themselves to death on 'dops' of cheap wine supplied by the farmers in lieu of cash. All that is changing, but not fast enough' (Anon 2004).

Recent research, however, reveals that the legacy of apartheid-era working conditions still affects labour within the industry. Pay rates tend to be very low, whilst working hours frequently exceed those laid down as acceptable by government legislation (Brown et al. 2003; Ewert and du Toit 2005; Women on Farms Project 2003). Physical abuse of workers is still evident on some farms, and child labour is not uncommon, especially during seasonal peaks of labour demand. Health and safety are frequently a concern, particularly in relation to exposure to agrochemicals. The dop system is still practised on a few estates (Brown et al. 2003). In order to reduce labour costs, many employers are now seeking an increased proportion of contracted off-farm labour. Non-permanent workers are excluded from many basic entitlements and report much higher levels of exploitation and abuse. Such radical shifts in the structure of the division of labour have had profound implications for the broader relationship between workers and employees. Thus, farmers are abandoning many of the 'social dividends' of the paternalist system. As many as a third of farmers, for example, no longer consider the provision of housing to be part of their contract with employees (Ewert and du Toit 2005). The ongoing exploitation of black farmworkers is but one dimension of the considerable, and racially delineated, challenges confronting the transformation process. Land ownership continues to remain almost exclusively in the hands of the minority white population (Kruger 2004; Kruger and Hamman 2004). Whilst wineries are starting to move into black ownership, the rate of change remains slow such that little more than 1% of the industry is under black ownership (Nxumalo 2003; Schweitzer 2008). Employment patterns remain racialised with limited opportunities for black people within more highly remunerated positions. Despite legislation, for example, the 2007 Broad-Based Black Economic Empowerment (B-BBEE) Act, transformation in the wine industry still lags far behind other sectors (Kruger et al. 2006). Groundbreaking initiatives are required to tackle long-entrenched socio-economic, political and cultural legacies.

Worker-Based Empowerment Projects

The imperative to effect transformation has led to the initiation of a number of projects that seek explicitly to empower workers (Bek et al. 2007; WOSA 2004).

Box 8.1 New Beginnings

Amongst the most well known of the first generation of 'empowerment wines' is 'New Beginnings', the first wine made by people of colour from grapes grown on their own land (Bek, et al. 2007). The initiator was Alan Nelson, a Cape Town legal advocate who bought the Nelson's Creek estate (bankrupt under its two previous owners) in the early 1990s (Hiney 2002). As a reward for helping to produce an award-winning wine in 1997 (Anon 2001; Nelson's Creek Wine Estate 2000), Nelson donated 11 ha of the estate to a workers' cooperative on condition that it was used for agriculture (Mohale 2003). Furthermore, by choosing wine production, the co-operative could either have access to Nelson's equipment and hi-tech cellar, or he would purchase their grapes. It was also agreed that the workers would cultivate their land in their spare time, as they remained employees of Nelson's Creek (Anon 2001). The National Department of Land Affairs recognised the initiative as a land reform project and also provided the co-operative assistance which, using income from their initial sales, bought a further plot of land. Profits are used to enable all local children to attend school; paying fees, transport and uniform costs (Botha 2002). Each co-operative member now has a life insurance policy and retirement fund, and each family has received a cash dividend of R6500. In addition, community members have gained viticulture and agricultural training opportunities (Schweitzer 2008). In the longer term, the co-operative is aiming to expand production, create a cellar, construct housing and enhance local education facilities.

These projects have emerged through the initiative of individual estate owners and worker/community organisations. Although business models tend to vary from case to case, such projects have often been categorised as 'worker equity schemes' (Karaan 2004), through which workers have gained legal rights to the means of production, often by redirecting their housing grant entitlement towards the purchase of land (National Library of South Africa 2004). Typically, the workers' group engages in the production of wines under its own label. Examples include the Fair Valley Workers Association that was established to reinvest money into improving education and housing, 'Winds of Change', 'Freedom Road' and New Beginnings, which is profiled in Box 8.1 (SowetoWineFestival.co.za 2009).

The first generation of these projects received mixed reactions, with some commentators praising the innovation inherent in the projects' dynamics whilst others have talked of 'pseudo-empowerment' due to the extent to which the projects have relied upon white farmer patronage (McEwan and Bek 2006). A second generation of projects has emerged which generally involve substantially greater levels of black ownership. These initiatives have often attracted considerable attention both from within the industry and from wider media, and several have achieved notable export success. However, achieving a sustainable marketing edge has often proven problematic, and thus, some projects have aligned themselves to the Fairtrade movement. Thus, the projects are able to benefit from the international recognition associated with the Fairtrade Labelling Organisation (FLO)[3] label, rather than relying upon consumers' abilities to engage with the specifics of South Africa's complex socio-political context.

Fairtrade[4]

FLO-accredited Fairtrade wine entered the marketplace only in 2004. Since then it has grown rapidly within the important UK market, being worth £10 million in 2008, an increase of 22% over 2007 (Fairtrade Foundation 2009). It is predicted that Fairtrade wine will be worth 3% of the total UK market by 2012, more

[3]Bonn based FLO sets the standards for Fairtrade products, which are accredited by FLO-CERT, a company which offers Fairtrade certification services.

[4]Fairtrade is an organised social movement that aims to help producers in developing countries through the payment of minimum prices and a social premium for investment in community projects. Fairtrade sales are growing rapidly year on year with global sales in 2008 representing US$4.1 billion (Fairtrade Foundation 2009) "Facts and Figures on Fairtrade", http://www.fairtrade.org.uk/what_is_fairtrade/facts_and_figures.aspx, accessed 16/07/2009.

than the value of some longstanding producer nations (Ehrmanns Wines 2009). Fairtrade certification on wine was designed and implemented for the first time in South Africa and includes brands that have had notable success. Currently, more than 120 individual Fairtrade wines are marketed in the UK, many of which are sourced from South Africa. Origin Wines, based near Stellenbosch, has a portfolio of 40 wines under its Fairhills brand and is one of the largest Fairtrade projects in the world (Origin Wines 2009). It is important to note that Fairtrade wine, in common with other South African Fairtrade products, can either be sourced from community-based co-operative ventures or from usually white-owned plantation style farms which have successfully met FLO accreditation (Binns et al. 2007; Kruger 2004). The Fairtrade label masks the rather contrasting social relations of production to be found in each setting.

Thandi Wines

The first wine in the world to gain Fairtrade accreditation was 'Thandi' (Xhosa for 'nurturing love') which is sourced from a worker empowerment scheme in the Elgin area. Currently, eight wines are marketed under the Thandi label. The Thandi project,[5] which incorporates both the wine and general farm-based fruit and vegetable production, has its origins in the village of Lebanon situated in the Elgin Valley some 75 km east of Cape Town. Lebanon was founded during the 1960s by the South African Forestry Company Ltd. (SAFCOL) to house its workforce (Dring 2001), but in 1991, SAFCOL restructured its operations and Lebanon's inhabitants were made redundant. After a period of intense negotiations, and, following the intervention of Dr. Paul Cluver, the owner of the neighbouring De Rust wine estate, a solution to the plight of the community was identified. Cluver offered 100 ha of land to the community for the purposes of farming. SAFCOL offered a further 100 ha for rent at a nominal rate. A Farming Trust, a Worker's Trust and a Community Trust were initiated to manage the project (Kruger and Hamman 2004). Profits generated from the farm have been reinvested for educational and other community-related purposes (Ehrmanns 2007). The farm produces a mixture of fruit, vegetables and wine grapes. The wine has been marketed under the 'Thandi' logo. In 2005, 95% of production was exported, mainly to the UK where it was sold by Tesco, Co-op, Sainsbury's and Waitrose. In the 12 months to March 2006, some 42,755 cases were exported. The Thandi Chardonnay won a gold medal at the London International Wine Fair in 2004 (Wardman 2004).

Members of the community have benefited from previously unheralded opportunities. For example, Thandi wines were at one stage made by one of the first black winemakers in South Africa, whose wife is now one of the farm managers and who travels the world as the 'face of Thandi' at international wine fairs (Morris 2003; Smith 2009; Thandi.com 2004a). Members of the Cluver family have acted as mentors to the community since the inception of the project and the Cluver estate made its cellar facilities available to Thandi winemaking staff in the early days of the project (Thandi.com 2004b). The contribution of established farmers is particularly important, as they can offer management and technical expertise, as well as access to equipment and other expensive resources during the period of transition. Otherwise, the market price of land, the high start-up costs associated with land-based industries and a lack of technical experience on the part of new entrants would militate against success. Experts argue that redistribution of land alone is unlikely to provide an economically sound approach for effecting meaningful transformation (Morris 2004).

The integration of Thandi into crucial support networks has been a critical factor in its success. Specifically, Thandi wines are marketed by 'the Company of Wine People', a firm part-owned by the Capespan group. Dr. Paul Cluver played a key role in gaining support from Capespan and Vinfruco, the Company of Wine People's precursor, a process facilitated by his involvement with both organisations. In 2006, the UK government's Department for International Development entered into a 3-year matched-funding arrangement with the Company of Wine People, whereby up to £400,000 has been made available to support Thandi's marketing. By building partnerships with key organisations, Thandi's stakeholders have been able to access business management support and the existing supply chains and marketing

[5]The Thandi project analysed here refers to local community oriented activities occurring near Elgin, although it should be noted that the Thandi banner is used by Capespan to promote Fairtrade fruit production operating more widely within South Africa.

networks of international exporters. Ewert et al. (2006, 2) note, 'in a neoliberal world, with northern protectionism and 'home-made' market driven land reform, partnerships are necessary for small growers to gain access to dynamic markets – especially in export markets where economies of scale, quality of criteria, branding and efficient supply chains are '*sine qua non*''.

The Capespan Foundation has been heavily involved in supporting the development of the Thandi project. Indeed, the Foundation's business plan, which accords with Fairtrade Foundation best practice, provides enhanced training and skill development opportunities to the community, such that they can aspire to higher status positions within the value chain. Furthermore, the Capespan Foundation aims to limit the duration of its involvement and therefore works towards an exit strategy whereby community members become sufficiently 'capacitated' to take on management responsibilities. The Capespan Foundation also ensures that the Fairtrade premium is reinvested in projects which benefit the local community, such as the purchase of a computer and a tractor. By 2006, the social premium received from Fairtrade accreditation had amounted to R114,000. The farm project has been expanded and now contains a restaurant and a farm shop and it also offers wine tastings which had previously taken place in the Cluver cellar. The Capespan Foundation has bought out the rights to the Thandi name which is now used as a unifying brand for a nationwide network of Fairtrade fruit production (Capespan Foundation 2003). The project has not been without its critics who have questioned the nature of the power relations between stakeholders in the project, specifically in relation to shareholdings and decision-making (Anon 18th September 2006). The complexity of the partnership arrangements required to make the project successful is believed to have weakened communication between stakeholders, leading some to question whether they are receiving a fair reward. Efforts have been made to ensure that clarity exists regarding the business and that a reward structure has been implemented (Ewert et al. 2006). It is planned that the Thandi Winery will become the first black-owned winery by 2014 (Huddleston 2009). Currently, one third of the business is owned by the Company of Wine People with the communities of Lebanon being a 45% shareholder and Lutouw, Neitbegin and Paardenkloof possessing shares in the remainder (Huddleston 2009). Thandi will take control over exports with marketing delivered by the Company of Wine People.

Home-grown Certification and Compliance Measures

Compliance with a whole raft of Health, Safety and Ethical codes and protocols is part and parcel of daily existence for producers in agricultural industries across the globe (McEwan and Bek 2009b). The apartheid legacies outlined above have rendered South African suppliers uniquely vulnerable to exposés relating to non-compliance and even abuses. Despite, or maybe because of, this problematic context, the South African Winelands region has been notable for a number of home-grown codes, protocols and initiatives that have emerged in the last decade. These encompass forms of ethics relating to labour standards, health and safety issues (pertaining to sites of production and consumption) and environmental management. Indeed, the ethical innovation demonstrated in South Africa is in some cases considered to be world-class in terms of best practice, such that wider scale adoption is occurring in relation to a number of initiatives. This section provides a sample of the schemes that have emerged in the last decade.

WIETA

The Wine and Agricultural Ethical Trade Association (WIETA),[6] a not-for-profit, voluntary organisation, was formed in 2002. This originated as a UK government Ethical Trade Initiative (ETI) pilot project in 1999 and operates via a tripartite alliance comprising corporations, NGOs and trade unions. WIETA has been described by one ETI representative (ETI representative Simon Steyne, quoted in Distell 2002) as the 'jewel in the crown' of ETI worldwide initiatives. It established a code of good practice based on the ETI base code (ETI 1998: 5–8) and South African legislation (Wieta 2003). Key principles include no child labour, freely chosen employment, the right to a

[6]In early 2009, WIETA formally changed its name to AETI (SA) – Agricultural Ethical Trade Initiative – to reflect the organisation's changing remit, which for a variety of reasons increasingly focuses upon the fruit and flower sectors.

healthy and safe working environment, the right to freedom of association, non-excessive working hours, the right to a living wage, prohibition of unfair discrimination and rights to worker's housing and tenure security. This is monitored through a social auditing inspection process which is complemented by efforts to educate producers and workers about the content of the code and its practical implementation. The imperative is to establish a dynamic upward trajectory of improvement in social conditions on farms rather than to 'fail' or shame producers. WIETA's auditing and support process is perceived by a number of key stakeholders, including some UK retailers, as representing a 'gold standard' in terms of quality. WIETA has 199 members, of whom 103 are wine producers, including major co-operatives (KWV, Distell, the Company of Wine People, Western Wines) and individual farms and estates. Audits began in early 2004, and by mid-2006, 33 producer members had been accredited.

Stakeholder Perceptions of WIETA's Impacts

In terms of the efficacy of its inspection processes, WIETA is generally held in very high regard by the majority of stakeholders. Audits are seen to be of a high standard, comprising good methodology with broad coverage of social, health and safety issues. This contrasts with commercial audits which are considered much more variable and lacking in transparency. Some are tick-box exercises, according to a WIETA executive, 'two hours including tea and a chat with producer' (interview 01/09/06). Another WIETA representative suggests that some retailers organise their own audits, and although they use some 'high calibre individuals who know the process and have the skills', they very often 'don't know the local environment and they come with a westernised mind set' (interview 14/09/06). In contrast to this 'window dressing' (ibid.), WIETA methodology is evidence driven, based on dialogue and representation at various levels, including casual workers, which provides a comprehensive picture of the organisation. As a CEO of a wine industry marketing association stated in a 21 September 2006 interview, WIETA has done really well operating in a difficult environment and has produced material that is very reputable.

Some producers stress the importance of WIETA in raising awareness of new legislation and the extent of non-compliance on their farms, since even progressive producers lack familiarity with the detail of much health and safety and labour law. One producer representative in Stellenbosch recounted his surprise at the audit findings:

> We were audited for the first time in September 2004... After the initial audit we were provided with a report that thick [indicates with both hands] that was structured around the labour law codes and the H&S codes. We were okay on the labour but were big-time non-compliant on the H&S stuff... Out of that process came various outcomes – correct signage, proper chemical storage facilities, and a proper H&S management system... Without WIETA we would never have had a clue about many of the problems that were identified... It really opened our eyes... (Interview 19/09/06).

In addition to raising awareness and attending to areas of non-compliance, producers also referred to the significance of accreditation in terms of the export market and, specifically, in countering the tainted image of exploitation and poor working conditions within the South African wine industry. Another achievement has been to make owners and producers aware of conditions that can be detrimental to the health, safety and labour rights of their workers. For example, many winery owners were unaware that using certain filtration media can cause silicosis and were inadvertently putting their cellar workers at risk. Preventative measures are being taken at WIETA audited sites. In other cases, housing for temporary cellar workers has been improved. However, the costs of achieving compliance are high. Sites are rarely compliant after the initial audit, and implementing corrective actions can be expensive. One producer reported costs of R200–250,000 or £14–19,000/$29–36,000.

UK retailers were instrumental in WIETA's emergence. High levels of retailer buy-in were achieved during the ETI pilot, particularly from supermarkets. WIETA's extension into a locally managed association has been strongly supported by these supermarkets which contribute heavily to WIETA funding by donating their share of the EU Common Customs Tariff rebate back to the wine industry. Retailers play a direct role in demanding that basic material quality is met by suppliers[7] (Ponte 2007, 21), including social and envi-

[7]There are three main types of wine producer in South Africa: estate wineries (producing wine from grapes grown on their own land), co-operatives (processing grapes of farmer member shareholders and pressing around 80% of the harvest) and independent cellars and wholesalers (buying in grapes and/or wine for bottling under their own brand names or third-party brands).

ronmental certification. For retailers, codes of conduct are a necessary business protocol offering a relatively inexpensive measure to protect corporate reputations. There was much initial optimism that WIETA's profile would be driven by strong retailer support. However, the reality has been more ambivalent. Experience has indicated that retailers utilise WIETA in cases where the risks to their reputation are greatest. On the one hand, this indicates that retailers recognise the integrity of the WIETA process, but on the other, risks to corporate reputation are greatest for 'own label' products which encompasses only a small proportion of wines. Thus, retailers portray themselves as providers of shelf space for estate and cellar-produced wines with minimal responsibility for the inherent ethics of these goods provided by such third parties. Fruit, which is usually sold with the retailer's own label, is a different matter as the UK retailer Tesco discovered to its cost following an NGO-led exposé of ongoing practices amongst its suppliers (ActionAid 2005). As a result, Tesco requested that WIETA undertake the audits among its fruit suppliers in order to ensure that minimal standards were being met.

Worker organisations and advocacy NGOs have been heavily involved in WIETA from the start and recognise the quality of the audits and the impact the process has had upon the working lives of employees. However, representatives have expressed frustration regarding WIETA's impact on transformation, with 'WIETA lacks teeth' being a common complaint. Other stakeholders suggest that this miscomprehends WIETA's role:

> At the last AGM, we had a march by trade unions, **'toyi-toyiing'**, all about fighting for visibility… WIETA has no actual power. They want WIETA to be more assertive and aggressive. This is partly about their own frustrations, if truth be told, because the degree to which unions have penetrated is very small. They are very fragmented. We have seen a lot of disillusion on farms. People do not quite understand the role of WIETA. They want WIETA to do their job for them. (Marketing association CEO 21/09/06).

Unfortunately, there are concerns that some labour groups have overplayed their hand and exploited WIETA as a type of political platform, leading producers to be unduly wary of WIETA. Many respondents within the industry are aware that WIETA is not an enforcement agency and is largely market driven, but expectations among workers are often higher than WIETA's remit. Thus, understanding WIETA's role in the context of debates about transformation and empowerment is important in assessing its actual and potential impact within the wine industry.

Assessing WIETA's Contribution to Wider Transformation

WIETA's impact on effecting change on farms might appear to be relatively minor if viewed in isolation from other developments. However, it is also part of networked activism which connects trade unions, grassroots movements, NGOs, producers, government departments, legislative bodies and international retailers. The possibilities for future transformation appear promising. While progressive elements in the wine industry are committed to improving standards and working conditions, WIETA is positioned within a broader set of local and international contexts that places limitations on its potential impact. A key issue is whether it is possible for ethical labour practices to be implemented in the context of enormous downward pressures on labour conditions (Hale 2000). Critical attention should remain squarely on power relations – between supermarkets and cellars, between cellars and farmers and between farmers and workers. It may fall well short of transformative notions of economic justice by focusing on labour conditions, but at least it reveals how poor these labour conditions remain. While WIETA cannot fundamentally alter power relations within the South African wine industry, or within global supply networks, it could bring significant benefits for workers whose lives are often wretched. In normalising the concept of decent working conditions, WIETA thus has the potential to play an important role in transforming the wine industry. As with other ethical trade initiatives, it might be considered an important stepping stone to longer-term, more substantive transformation.

WIETA's role as a local initiative is also significant in blending the ETI code with South African labour laws and adapting normative parameters to local contexts. Unlike some ethical trade initiatives, WIETA relies on high levels of stakeholder involvement and is not driven by decision-making from North to South. It is still part of an ethical complex that is, to some extent, governed by the corporate strategies and management systems of retailers which might blunt its 'ethical force' (Popke 2006). Indicators of achievement will, to some extent, reflect the values and concerns of Northern retailers and consumers because of their significance in the wine industry (Blowfield and

Frynas 2005). However, these also encompass the aspirations of South African workers. By demonstrating its capacity and ability to define its own ethical goals, WIETA thus has the potential to illustrate what might be achieved when Northern retailers, civil society organisations and consumers relinquish some of their control over the ethical trade movement. However, the future challenges for the rebranded WIETA are formidable. These include direct and indirect competition from local and international forms of ethical codes, and the downward price pressures faced by producers which reduce capacity to engage with ethical issues and capacity constraints within the organisation which make it difficult to cope with an increasing workload.

Farm Health Awards (FHA)

RUDNET is the NGO arm of the Black Association of the Wine and Spirits Industry (BAWSI) which aims to make the wine industry fully representative and able to play a meaningful role in empowering black South Africans to become farmers and farm owners. RUDNET was established in 2002 to facilitate socio-economic empowerment and skills training of farmworkers, to promote the social inclusion of farmworkers in civil organisations and to address social problems such as poverty, alcoholism, limited education and poor health on farms. It has run the Farm Health Awards (FHA) for 6 years, successfully recruiting over 100 producers. The scheme has received much positive publicity in the Western Cape, and producers speak highly of it. The FHA, unlike WIETA, is an enterprise initiative and a product of the local industry. It has two elements: the nomination of farmworkers who have acted as a source of inspiration to their communities and an auditing system examining various aspects of working conditions on farms (RUDNET 2005). The audit takes the form of an interview with the producer, completion of a largely tick-box questionnaire and a site visit. Farms are given a score that is not reported publicly except for the winners in each size category and also advice on areas in which they can improve. The winners receive awards at an annual gala event. Reported benefits include (1) producers gain an insight into the needs of workers, (2) good practice is rewarded and (3) positive publicity raises standards more widely. However, critics point out that self-assessment is an inappropriate mechanism for evaluating such complex issues in that paternalist managerial styles are reinforced rather than challenged.

Safeguarding the Natural Environment

The Cape Floral Kingdom (CFK) is a World Heritage Site containing in excess of 10,000 species, more than the whole of the Northern Hemisphere. *Fynbos,* which includes a remarkable diversity of plant species including many members of the Protea family (*Proteaceae*), Heath family (*Ericaceae*) and Reed family of restios (*Restionaceae*), is found widely across the region (Figs. 8.1 and 8.2). Distinctive landscapes are also evident in areas of *renosterveld*, a grassy shrubland dominated by members of the daisy family. Despite only representing 0.05% of the earth's land area, the CFK contains 3% of all species, many of which are endemic to the region (SouthAfrica.info 2009). Around 90% of wine production is located in the CFK, and any expansion could endanger areas of natural habitat, especially in terms of the *renosterveld* and the lowland *fynbos*. Furthermore, the CFK is under increasing pressure from agriculture, urban development and the spread of invasive alien species. By 2004, only 4% of the original *renosterveld* remained. This situation represents an environmental crisis with significant social and economic implications. Yet, with over 80% of the CFK in private hands and with 90% in winemaking areas, the goal of conservation can only be attained by convincing the landowners of the value of biodiversity. The South African wine industry has been at the forefront of the development of regulatory practices to protect biodiversity. This commitment is embedded in the Biodiversity and Wine Initiative (BWI) and in the guidelines of the Integrated Production of Wine scheme (IPW).

Biodiversity and Wine Initiative (BWI)[8]

The Biodiversity and Wine Initiative (BWI), which is co-funded by the World Bank, operates as a partnership between the wine industry and the conservation sector and aims to minimise further loss of this unique and threatened natural habitat and to contribute to sustainable wine production through the adoption of biodiversity guidelines. Producers are encouraged to become members of the BWI. In order to be accepted, 2 ha of natural or restored vegetation must be set aside for conservation and the farm must have an ecologi-

[8] Interviews and site visits were undertaken at the following BWI Champion estates during February/March 2009 at Waterkloof, Oak Valley, Paul Cluver, Vergelegen and Hermanuspieterfontein.

Fig. 8.1 The Spectacular Winelands scenery, viewed from the Devon Valley, Stellenbosch (Source: authors)

Fig. 8.2 Conserved fynbos vegetation on an upland slope on the Hermanuspieterfontein estate in the Overberg district (Source: authors)

Box 8.2 Integrated Production of Wine

IPW was initiated in 1998 as a semi-regulatory system providing guidelines that conform to international standards for 'Good Agricultural Practices' for farms and 'Good Manufacturing Practices' for cellars to produce wines that are 'healthy, clean and environmentally friendly' (IPW 2004, cited in Ponte and Ewert 2007, 45). The scheme is voluntary, but registered IPW actors harvest 97% of South African grapes (WOSA 2009a). The proportion of actually audited farms and cellars is much smaller, however, since the system is regulated via self-monitoring. Members complete a series of self-assessment forms which indicate their current level of attainment. Compliance can be achieved by scoring a minimum of 50% of the total score, a low threshold by international standards. All IPW cellars should be audited within a 5-year period, but farm visits remain at 36 per annum, a tiny proportion of over 4,400 wine farms (SAWIS 2006). In future, it is likely that conformity certificates will be awarded to cellars only if an external audit has been passed (Ponte and Ewert 2007, 45). A recent change in the procedures permits producers to request an audit if they wish. The IPW assessment system is backed by a training infrastructure to support producers. The biodiversity element of IPW has been strengthened considerably following criticisms that the previous version did not 'adequately address issues such as threatened ecosystems such as *Renosterveld* or lowland *fynbos* or the need to remove alien plant species' (BWI 2009a). Chapter 2 of the guidelines is now referred to as the 'Biodiversity Guidelines', following input from conservation experts.

cally viable proportion of natural vegetation still intact, which is at least 10% of the farm's area. Furthermore, members must participate in the IPW scheme and score in excess of 85% on the biodiversity self-assessment section (see Box 8.2 above) (BWI 2009a). Members are encouraged to aspire to achieve the status of BWI Champions – 'exemplary producers who have earned additional recognition and status for their excellent track record of environmental responsibility…These flagship farms have a comprehensive conservation management plan in place with dedicated staff members implementing these plans including extensive alien vegetation[9] clearing, fire management and voluntary conservation agreements committing their natural areas to long term conservation', (BWI 2009e). As of July 2009, there were 15 Champions, 10 producer cellar members and 120 individual members, collectively representing more than 750 farms (BWI 2008, 2009c, d).

Benefits of BWI can be identified in three areas. First, there are clear environmental benefits. The total area given over to conservation is in excess of 115,000 ha representing more than the total vineyard footprint (BWI 2008). Many rare plant species have been discovered on farms and are now protected. Producers are reporting an enhanced appreciation of the uniqueness of the environment within which they work and are recognising the imperative to protect its core attributes. Furthermore, the removal of alien species is reducing the fire risk and improving the availability of water. Second, the wine industry is benefiting financially through its adoption of a proactive stance on the implementation of biodiversity guidelines which acts as a unique selling point to differentiate 'Brand South Africa' (Wineland 2005, in Ponte and Ewert 2007, 69). WOSA's campaign, 'Variety is in our Nature', communicates the direct relationship between South Africa's varied terroir and the distinctive qualities of the wines that emanate from the region. WOSA's decision to market on the basis of biodiversity arises partly from the difficulties in establishing a unique identity for South African wines on the basis of varietals and terroir – in contrast to Old World wines. This is because of the use of modern winemaking techniques in both vineyard and cellar which purists argue obscure or even eliminate the influence of terroir in making different regions unique. The BWI has been an increasingly prominent feature of marketing events

[9] Alien vegetation refers to plants imported from external ecosystems. Many thousands of alien species exist in South Africa, posing severe threats to indigenous vegetation.

such as the Cape Wine Expo. Industry figures believe that South Africa's voluntary focus on environmental issues has placed the nation 18 months ahead of other producer nations in terms of marketing, especially in key markets such as the UK. In 2009, it was agreed that members could use the BWI logo on their product labels as some retailers are explicitly seeking to stock BWI-accredited wines. Related to this, South African producers have been well placed to meet the increasingly stringent environmental conditions imposed by European importers. Third, the BWI presents an opportunity to create employment opportunities for people from historically disadvantaged communities in alien vegetation clearance schemes. Many farms are linking with work groups formed through the government's Working for Water programme (Binns et al. 2001; Barnes et al. 2008). In many cases, existing farmworkers have benefited from the opportunity to work on alien clearing during the quiet periods on the farm when they might otherwise have been laid off. Thus, there are significant livelihood benefits. In addition, some champions have begun to use their engagement with BWI as a catalyst for tourism projects, which, in turn, create some direct on-farm employment. For example, the Green Mountain Eco Route, the world's first biodiversity wine route in the Elgin area, builds upon the environmental credentials of two BWI Champions (GreenMountain.co.za 2009).

Responses to BWI

The response from producers to biodiversity has been good because 'they get it' (wine industry executive, interview 21/09/06). In the early stages of the BWI, this required a long-term vision often working against the aims of marketing at that time. However, as environmental issues have become increasingly prominent within consumer mindsets, so benefits for BWI members are becoming apparent. Interviews suggest that producers are aware of the potential future importance of creating a market niche, generating sustainable farming and water use and developing potential links to tourism. However, a number of respondents were at pains to explain that BWI accreditation might not in itself sell wine or add a premium to the value but it may offer access to shelf space (Interview with Dr. Paul Cluver 23 February 2009). In the medium term, it is likely that environmental credentials will become a sine qua non for access into many markets (Theron 2005). Critics express concerns that BWI is little more than a form of 'greenwash' – allegations that are strongly refuted by the BWI, leading industry figures and participants. Questions have also been raised about the costs associated with achieving champion status in terms of the need to conduct wide-ranging management programmes led by a dedicated member of staff. Given the ever-tightening price pressures experienced by producers, such investments are beyond the scope of many estates. Finally, supporters of social transformation express concern that environmental credentials are distracting consumers from consideration of social issues.

The case study below (Box 8.3) illustrates how one BWI Champion has tackled conservation challenges and the range of socio-economic and environmental benefits that can result.

Challenges, Turf Wars and the Market

The South African wine industry has clearly made a highly successful reintegration into the global trading system. There are, however, a number of challenges ahead. First, whilst export volumes are continuing to grow apace, the value of each litre of wine has fallen sharply in recent years, partly due to the increase in bulk wines exported for bottling elsewhere. Such declines in value have fortunately been offset, for the time being, by the weakening of the Rand on international currency markets. Maintaining profitability, let alone meeting social and environmental objectives, will be a major challenge for many in the industry. Second, there is an economic and social imperative to retain elements of the value chain, such as bottling, within South Africa. Indeed, some Fairtrade producers have noted concerns that intermediaries are exporting their wine in bulk for bottling in Europe. Such producers question whether these practices are within the spirit of Fairtrade. The industry has been responding by initiating the use of 1.5 l pouches which are not only more environmentally friendly than bottles but are produced in South Africa (Park 2009). The need for such innovative value-adding local activities is significant within the broader context of a retailer-driven international market typified by ever-growing downward pressure on prices for the core product.

Third, there are concerns that South Africa lacks sufficient big brands to compete for value on the

Box 8.3 Case Study of a BWI Champion, Vergelegen Wine Estate[10]

Vergelegen, a prestigious wine estate owned by mining behemoth Anglo-American (Fig. 8.3), was nominated as the first BWI Champion in 2005 (BWI 2009b; Theron 2005). The award was the culmination of several years intensive conservation work triggered by a large fire which had devastated the area in 1999. Initially, 316 ha of land were regenerated with a further 2000 of the farm's 3,200 ha the focus for a longer term project. The management plan initially involved creating 26 management units within which a rolling programme of alien cutting, burning and debris removal has been instigated. The alien seedbank is immense with a projected lifespan of more than half a century. Thus, areas have to be continually revisited to remove new alien growth as up to 100,000 stems a hectare can reshoot. These have to be pulled out to allow the indigenous vegetation to grow through. However, the regeneration at Vergelegen suffered a setback in February 2009 when fires (thought to be arson) raged through the region, affecting a large proportion of the Vergelegen estate (WOSA 2009b) (Fig. 8.4). The regenerated *fynbos* areas suffered as insufficient time has elapsed for a new seedbank to become established (normally 10–15 years). The sheer quantity of burnt debris posed logistical challenges as alien vegetation rapidly reshoots among the debris and quickly becomes established. However, the overall severity of the fires may have been reduced by the existence of the rejuvenated *fynbos* areas as fire spreads more slowly within the low lying *fynbos* than the taller canopies typical of the alien vegetation.

[10]Sources: Site visit 11 March 2009, including interviews with Conservation Manager Gerald Wright and winemaker Andre Van Rensburg; interview with Inge Kotze, BWI Co-ordinator, 27 February 2009.

Fig. 8.3 The labour intensive grape sorting process in the Vergelegen cellar (Source: authors)

(continued)

Box 8.3 (continued)

Fig. 8.4 Evidence of fire damage in the Somerset West area, February 2009. Tall alien species contribute to the rapid spread of such fires (Source: authors)

Benefits from the conservation work have included:

1. The number of bird species on the estate has risen from 80 to 120, whilst other species such as antelope are more evident.
2. Wetland areas are rejuvenating as indigenous species use substantially less moisture than aliens. In turn, this is enabling farms such as Vergelegen to become more self-sufficient in their usage of water.
3. Employment has been provided for 50 people on the estate specialising in alien removal. Certified skills training (e.g. in use of power saws and agrochemical management) has benefited 20 individuals. It is estimated that such employment opportunities have benefited upwards of 600 people via extended families.

global stage. Brands are evolving but currently lack the recognition and influence of Australian and Californian brands such as Wolf Blass, Jacob's Creek and Gallo. Fourth, estate owners and winemakers face the challenge of developing higher value quality products which play upon the diverse terroir of the Winelands. Indeed, some observers speculate that the region's true potential has yet to be realised, with areas such as Elgin possessing the ability to produce award winning wines. Equally, WOSA and individual marketing executives face the challenge of successfully marketing such products on the international stage. Fifth, it is clear that the industry has been highly successful in devising and implementing initiatives which support socio-economic upliftment and promote environmental sustainability. While being worthy in their own right, such initiatives convey a series of marketing advantages due to their compatibility with consumer concerns and retailer demands. Furthermore, retailers are raising the bar for market entry and, instead of offering a premium for meeting certain standards, they are increasingly setting these as the minimum. However, the competitive advantage will soon disappear as other nations adopt comparable methodologies. The key for South Africa in the short term will be to devise streamlined but robust, multi-purpose certification systems which will also bring the benefit of reducing the audit burden upon producers.

The relationships between these initiatives and the market are complex. In a highly competitive market place, there is no space for voluntary 'add-ons' that do not enhance the marketability of a wine in some way. In the case of WIETA, some retailers insist upon producers being members if they wish to supply the content of a retailer's 'own brand' wines. By sourcing wines from WIETA members, retailers ensure that they are protected from the criticism that could result from an exposé of labour practices in the industry. South African products remain a source of vulnerability with some consumers likely to be sensitive to revelations that apartheid-era inequalities still remain. In the case of Fairtrade, Thandi grabbed an opening by becoming the first wine to gain such status and has subsequently benefitted from the opportunity to access shelf space via its fair trade accreditation and its unique heritage (Fig. 8.5). BWI offers a point of entry into the fast growing 'green' and 'organic' market segments, whilst some producers are able to use their involvement with BWI to enhance their product's marketing appeal.

The emergence of these various initiatives has meant that South Africa has become a fascinating 'test bed' within which to observe the institutional and moral dynamics of different forms of ethical venture. Some observers talk of 'turf wars' between different codes with each competing for members, who, in turn, are evaluating the perceived costs and benefits of each system. For example, it certainly seems to be the case that the explosion of Fairtrade products has constrained WIETA's dissemination within the wine industry. The rapid development of Fairtrade wines from a standing start in 2004 has

Fig. 8.5 The Thandi Farm stall and restaurant at Grabouw (Source: authors)

taken many by surprise. For retailers, Fairtrade, with its high levels of public recognition in key markets, offers an opportunity to meet their perceived social and moral responsibilities whilst safeguarding their reputations. Thus, the existence of Fairtrade lines within stores can be interpreted as a smokescreen behind which public relations departments can shroud the social conditions existing at the site of production for the majority of their lines which may only have been subject to the more superficial scrutiny of a commercial social audit.

Equally, high profile environmental accreditation such as BWI may compete for consumer share rather than expanding the overall sum of ethical purchasing. Indeed, there is evidence that environmental credentials have more appeal than social ones among consumers. In the same vein, some observers have pointed out that issues relating to social transformation are all but invisible in WOSA's environmentally oriented marketing campaign 'Variety is in our Nature' (Ponte and Ewert 2007). From a producer perspective, the existence of several broadly 'ethical' codes, including the raft of commercially driven systems, is highly problematic since each imposes a cost burden which must be absorbed even though many criteria may actually overlap. As a result, discussions are beginning to take place within South Africa concerning ways in which certification systems may be streamlined or even merged in order to reduce the need for different inspections. If a rigorous, streamlined, multi-criteria certification system can be generated in the near future, then South Africa will gain significant advantages in the global marketplace.

Arguably the greatest challenge confronting the industry lies within the realm of black empowerment. Whilst codes and initiatives such as WIETA and Fairtrade raise the bar in terms of labour standards, Black Economic Empowerment (BEE) legislation has sought to tackle the racially delineated inequalities inherited from the previous regime. The industry itself has taken measures in order to support change such as the creation of the South African Wine Industry Council in 2006 which was set up to deliver the Wine Industry Charter (see also Williams 2005). However, enormous obstacles to transformation persist. A formidable 'wine industry complex' has evolved dominated by white Afrikaner men, many of whom were leading supporters of the National Party under apartheid. To a large extent, these white elites continue to run the industry's major businesses and institutions. It is thus hardly surprising that processes of black empowerment and broader societal transformation are often interpreted rather conservatively. Furthermore, microenterprises, including approximately 80% of wine farms, are exempt from BEE legislation although all producers are expected to strive to comply. However, a survey revealed that 80% felt they lacked the resources to comply, whilst only 5% have become involved in shareholder schemes for BEE or profit-sharing programmes (Wine Industry Council 2007). Without a legislative requirement to adopt BEE codes, therefore, market-driven transformation based on social and environmental codes is likely to remain important in generating change within the industry.

References

ActionAid. 2005. *Rotten Fruit: Tesco profits as women workers pay high price.* London: ActionAid.

Anon. 2001. *Local wines for British labour party function.* Paarl post.

Anon. 2004. *Wine industry faces its regmaker.* www.busrep.co.za/index.php?fSectionId=561&fArticleId=274903# (last accessed 17 April 2004).

Ashton, G. 2009. *Can our wine industry become a model for the rest of our agriculture sector?* http://sacsis.org.za/site/news/detail.asp?idata=250&iChannel=1&nChannel=news&iCat=1415 (last accessed 16 July 2009).

Barnes, A., M. Ebright, and E. Gaskin, W. Strain. 2008. Working for Water, South Africa., http://www.translinks.org/Docustore/tabid/409/language/en-US/Default.aspx?Command=Core_Download&EntryId=145 (last accessed 10 August 2009).

Bek, D., C. McEwan, and K. Bek. 2007. Ethical trading and socio-economic transformation: critical reflections on the South African wine industry. *Environment and Planning A* 39: 301–319.

Binns, T., P. Illgner, and E. Nel. 2001. Water shortage, deforestation and development: South Africa's 'Working for Water' programme. *Land Degradation and Development* 12: 341–355.

Binns, T., D. Bek, E. Nel, and B. Ellison. 2007. Sidestepping the mainstream: Fairtrade. Rooibos tea production in Wupperthal South Africa., In *Alternative food geographies: Representation and practice,* ed. D. Maye, L. Holloway, and M. Kneafsey, 331–349. Oxford Elsevier.

Blowfield, M., and J. Frynas. 2005. Editorial: Setting New Agendas - Critical Perspectives on Corporate Social Responsibility in the Developing World. *International Affairs* 81 (3): 499–513.

Boom, R. 2009. *South African Wine Harvest Report 2009.* http://www.sawis.co.za/info/download/HARVEST_REPORT_2009.pdf (last accessed 16 July 2009).

Botha, S. 2002. Labour Party's wine buys criticized by local estate. www.paarlpost.co.za/archive/170220/business.html (last accessed 27 August 2003).

Brown, M., A. du Toit, and L. Jacobs. 2003. *Behind the label: A workers' audit of the working and living conditions on selected wine farms in the Western Cape.* University of the Western Cape: Labour Research Service, Women on Farms Project and the Programme for Land and Agrarian Studies.

BWI. 2008. *Cape Winelands make conservation history - conservation hectarage exceeds the total vineyard footprint.* www.bwi.co.za/.../BWI%20CAPE%20WINE%20Press%20release%20080919.doc (last accessed 04 August 2009).

BWI. 2009a. *The link with IPW.* www.bwi.co.za/ipw (last accessed 04 August 2009).

BWI. 2009b. *Champions, Vergelegen.* http://www.bwi.co.za/members/member.asp?MemberID=1 (last accessed 17 July 2009).

BWI. 2009c. *Champions, Hermanuspieterfontein,* http://www.bwi.co.za/members/member.asp?MemberID=109 (last accessed 16 July 2009).

BWI. 2009d. *Champions, Waterkloof,* http://www.bwi.co.za/members/member.asp?MemberID=24 (last accessed 16 July 2009).

BWI. 2009e. *How to join. What is the difference between a member and champion?* http://www.bwi.co.za/member-option (last accessed 16 July 2009).

Capespan Foundation. 2003. *Thandi fruit programme. Code of practice.* Capespan Foundation, internal document.

Distell. 2002. *Vineyards on the moral high ground.* http://www.distell.co.za/print.aspx (last accessed 30 September 2003).

Dring, P. 2001. *Grapes of Freedom.* http://www.waitrose.com/food_drink/wfi/drinks/wineproducers/0109076.asp (last accessed 25 September 2003).

du Toit, A. 1993. The micro-politics of paternalism: The discourses of management and resistance on South African fruit and wine farms. *Journal of Southern African Studies* 19 (2): 314–336.

Ehrmanns. 2007. Fairtrade Statement 2007. http://www.ehrmannswines.co.uk/filestore/documents/Ehrmanns_Fairtrade_Statement_2007.doc (last accessed 25 July 2009).

Ehrmanns Wines. 2009. Fairtrade. http://www.ehrmannswines.co.uk/fairtrade/ (last accessed 10 August 2009).

Ethical Trading Initiative (ETI). 1998. Purpose, Principles, Programme, Membership Information. London: ETI, London, p 5–8

Ewert, J. 2002. Co-operatives to companies: The South African wine industry in the face of globalization. In *Globalization, Localization and Sustainable Livelihoods*, ed. R. Almas, and G. Lawrence. Aldershot: Ashgate Publishing.

Ewert, J., and A. du Toit. 2005. A deepening divide in the countryside: Restructuring and rural livelihoods in the South African wine industry. *Journal of Southern African Studies* 31 (2): 315–332.

Ewert, j., Eva, G. and Hamman J. 2006 The inclusion and empowerment of farm workers through partnerships: the case of 'Thandi fruit and wine, http://www.regoverningmarkets.org/en/filemanager/active?fid=574

Fairtrade Foundation. 2009. *Facts and Figures on Fairtrade.* http://www.fairtrade.org.uk/what_is_fairtrade/facts_and_figures.aspx (last accessed 16 July 2009).

Faurie, J. 2008. *South African wines bear export fruit.* http://www.engineeringnews.co.za/article/south-african-wines-bear-export-fruit-2008–11–14 (last accessed 16 July 2009).

GreenMountain.co.za. 2009. *Green Mountain Eco-route.* http://www.greenmountain.co.za/ (last accessed 16 March 2009).

Hale, A. 2000. What hope for 'ethical trade' in the globalised garment industry? *Antipode* 32 (4): 349–56.

Hiney, T. 2002. *The Grape of Good Hope.* http://www.observer.co.uk/Print/0,3858,4368895,00.html (last accessed 21 October 2002)

Huddleston, N. 2009. *Thandi aims for full black ownership.* http://www.harpers.co.uk/news/news-headlines/7647-thandi-aims-for-full-black-ownership.html (last accessed 25 July 2009).

Integrated Production of Wine (IPW) (2004). The South African system of Integrated Production of Wine (IPW). Stellenbosch: IPW.

Karaan, M. 2004. *Overview of BEE intiatives in the wine industry.* Unpublished report, Department of Agricultural Economics, University of Stellenbosch.

Kruger, S. 2004. *Fairtrade in South Africa: Growth, empowerment and fairness in the South African context.* unpublished consultation paper, Programme for Land and Agrarian Studies, University of the Western Cape, Cape Town.

Kruger, S. and J. Hamman. 2004. *Guidelines for FLO's Empowerment Strategy in South Africa.* unpublished consultation paper, Programme for Land and Agrarian Studies, University of the Western Cape, Cape Town.

Kruger, S., A. du Toil, and S. Ponte. 2006. De-racialising exploitatiion: Black Economic Empowerment' in the South African wine sector. *DIIS Working paper 2006/34.* Copenhagen: Danish institute for internatonal Studies.

McEwan, C., and 'D. Bek. 2006. (Re)politicizing empowerment: Lessons from the South African wine industry. *Geoforum* 37(6): 1021–1034.

McEwan, C., and D. Bek. 2009a. Placing ethical trade in context: WIETA and the South African wine industry. *Third World Quarterly* 30 (4): 723–742.

McEwan, C., and D. Bek. 2009b. The political economy of alternative trade: Social and environmental certification in the South African wine industry. *Journal of Rural Studies* 25 (3): 255–266.

Mohale, M. 2003. The Story of New Beginnings,the First Black Winery in South Africa. http://m1.mny.co.za/MbGetM.nsf/UNID/MMOE-5L3FPZ?OpenDocument (last accessed 27 August 2003).

Morris, R. 2003. *Wine cellar secretary organises her casks to toast her heartbeat.* www.busrep.co.za/index.php?fSectionId=553&fArticleId=332905 (last accessed 21 April 2004).

Morris, R. 2004. *Thandi blazes trail for land reform model.* http://www.busrep.co.za/index.php?fSectionId=561&fArticleId=332898 (last accessed 17 April 2004).

National Library of south Africa. 2004. Wine Fram Empowerment-South Africa.

Nelson's Creed Wine Estate. 2000. *Nelson's creed wine estate.* http://www.nelsonscreed.co.za/history,html (last accessed 21October 2002).

Nxumalo, F. 2003. *Wine Trust highlights empowerment needs.* www.busrep.co.za/index.php?fSectionId=561&fArticle=242757 (last accessed 17 April 2004).

Origin Wines. 2009. *Fairtrade.* http://www.originwine.co.za/ (last accessed 16 July 2009).

Park, J. 2009. *Thandi cuts carbon footprint of wine range with PET pouch*. http://www.packagingnews.co.uk/news/910847/Thandi-cuts-carbon-footprint-wine-range-PET-pouch/ (last accessed 25 July 2009).

Ponte, S. and J. Ewert. 2007. *South African wine – an industry in ferment*. Tralac Working Paper 8. www.tralac.org

Popke, J. 2006. Geography and ethics: everyday mediations through care and consumption. *Progress in Human Geography* 30 (4): 504–12.

RUDNET. 2005. *Farm health award: Findings and interpretations*. www.rudnet.org.za (last accessed 20 July 2006).

SAWIS. 2004. Macro-economic impact of the wine industry on the Western Cape. http://www.sawis.co.za/info/download/Macro-eco_impact_study-summary2004.pdf (last accessed 04 August 2009).

SAWIS. 2006. *South African Wine Industry Statistics 2006*. Paarl: SAWIS,

SAWIS. 2008. *World and Local News 2008*. http://www.sawis.co.za/info/download/World_and_Local_News_2008.pdf (last accessed 16 July 2009).

SAWIS. 2009. *Production and Market Estimates 2008–12*. http://www.sawis.co.za/info/download/Production_and_market_forecasts_2008_to_2012.pdf (last accessed 16 July 2009).

Schweitzer, E. 2008. How black farm workers become land and business owners: Actors, resources, contexts and outcomes of black empowerment projects in the South African wine industry. *Vienna Journal of African Studies* 15: 31–53.

Smith, D. 2009. Black winemakers in South Africa discover a nose for overcoming bitter fruit of apartheid. http://www.guardian.co.uk/world/2009/jul/23/south-africa-wine-making-apartheid (last accessed 23 July 2009).

SouthAfrica.info 2009. *The Cape Floral Region*. http://www.southafrica.info/about/animals/capefloralregion.htm (last accessed 16 July 2009).

SowetoWineFestival.co.za 2009. *Empowerment projects in the wine industry*. http://www.sowetowinefestival.co.za/Website_EMPOWERMENT%20PROJECTS%20in%20the%20WINE%20INDUSTRY.pdf (last accessed 25 July 2009).

Thandi.com 2004a. *Our team*. www.thandi.com/wine_team.asp (last accessed 31 October 2004).

Thandi.com 2004b. *Wine associations*. www.thandi.com/wine_assoc.asp (last accessed 31 October 2004).

Theron, K. 2005. *Vergelegen at the forefront of biodiversity*. http://www.wineland.co.za/2005junevergelegen.php3 (last accessed 04 August 2009).

Wardman, J. 2004. *Thandi comes of age*. http://www.wine.co.za/News/News.aspx?NEWSID=5832&Source=News (last accessed 22 October 2004).

Wieta. 2003. *Wine industry ethical trade association: retailer briefing document*. Cape Town: Wine Industry Ethical Trade Association.

Williams, G. 2005. Black economic empowerment in the South African wine industry. *Journal of Agrarian Change* 5 (4): 476–504.

Williamson, K., and E. Wood. 2003. The dynamics of the South African wine industry cluster: A basis for innovation and competitiveness? unpublished research paper, Graduate School of Business, University of Cape Town.

Wineland. 2005. *Vergelegen at the forefront of Biodiversity*. http//www.wineland.co.za/wineland_archive/2005junevergelegen.html (last accessed 16 January 2009).

Wine Industry Council. 2007. *The wine industry transformation charter*. Stellenbosch: Wine Industry Council.

Women on Farms Project. 2003. *Labour wedge campaign: Working conditions in South African agriculture*. Stellenbosch: Women on Farms Project.

WOSA. 2004. *Transformation in the wine industry*. http://www.wosa.co.za/transformation.asp (last accessed 01 November 2004).

WOSA. 2006. The Cape's best of wine tourism winners announced. http://www.wosa.co.za/sa/news_article.php?id=213 (last accessed 05 August 2009).

WOSA. 2009a. Scheme for the integrated production of wine. www.wosa.co.za/sa/sustainable_ipw_info.php (last accessed 04 August 2009).

WOSA. 2009b. Update on forest fires in Cape Winelands. http://www.wosa.co.za/news_wosa_article.php?id=1145 (last accessed 04 August 2009).

WOSA. 2009c. Variety is in our nature. http://varietyisinournature.com.www118.nur4.host-h.net/ (last accessed 10 August 2009).

Competitiveness and Sustainability in Wine Tourism Regions: The Application of a Stage Model of Destination Development to Two Canadian Wine Regions

9

Barbara A. Carmichael, Ph.D. and Donna M. Senese, Ph.D.

Abstract

Recently, both wine production and wine tourism have increased rapidly, particularly in areas outside of traditional old world wine regions. In this chapter, a stage model is presented which outlines the development of a hypothetical wine tourism destination region. Two geographical regions in Canada, the Niagara Peninsula of Ontario and the Okanagan Valley of British Columbia, are the sites for testing the elements of this spatial model. The purpose is to identify their stages of development and to suggest their positions as competitive and sustainable tourism destination regions. Balancing the elements in the internal supply characteristics and in the demand characteristics for wine tourism is important in achieving an optimal state of tourism development.

Introduction

Recently, both wine production and wine tourism have increased rapidly, particularly in areas outside of traditional old world wine regions. The processes guiding this growth are complex and synergistic. The globalized popularity of wine tourism can be linked to the rise in popularity of cultural tourism and fits neatly into the cultural industries (Britton 1991) and experience economies (Pine and Gilmore 1999). Wine tourism provides the ideal setting for Urry's (1990) "post tourists," often represented by an aging, educated, and affluent baby boomer generation. Such tourists are often in search of the education and class attachment associated with epicurean consumption and the discovery of new destinations.

Hall and Macionis (1998, 197) define wine tourism "as visitation to vineyards, wineries, wine festivals and wine shows for which grape wine tasting and or experiencing the attributes of a grape and wine region are the primary motivations for the visitor." A wider definition of wine tourism is "experiential tourism occurring within wine regions providing a unique experience which includes wine, gastronomy, culture, the arts, education and travel" (Dowling 1998, 78). These definitions suggest that visitation may be motivated both by "grape wine" specifically and more generally by the attributes of the "grape wine" region sometimes referred to as "winescapes" (Peters 1997) or wine tourism "terroir" (Hall et al. 2000b). Macionis (1998) has described the importance of the wine region in

B.A. Carmichael, Ph.D. (✉)
Department of Geography and Environmental Studies, NeXt Research Centre, Wilfrid Laurier University, Waterloo, ON, Canada N2L 3C5
e-mail: bcarmichael@wlu.ca

D.M. Senese, Ph.D.
Community Culture and Global Studies, IK Barber School of Arts and Sciences, University of British Columbia Okanagan, Kelowna, BC V1V 1V7, Canada
e-mail: donna.senese@ubc.ca

P.H. Dougherty (ed.), *The Geography of Wine: Regions, Terroir and Techniques*,
DOI 10.1007/978-94-007-0464-0_9, © Springer Science+Business Media B.V. 2012

motivating visitors. Macionis and Cambourne (1998) suggested that while the primary motivation of wine tourists are wine related, there are a number of other motivations that are integral to the total wine experience. These include festivals, socializing, day out, country setting, vineyard destination, other attractions, learning about wine/wine making (education), eating at winery/picnic/barbeque, tour of the winery, meeting the winemaker, and entertainment. These opportunities increase as a wine production region develops into a wine tourism region (Yuan et al. 2006).

Wine tourism destination development is rapidly growing in popularity in areas with favorable growing and marketing conditions. There are numerous new areas and well-established areas of wine production in the world that are developing wine tourism, but within these areas, what is the optimal level of wine tourism development in providing quality tourism experiences and to what extent are these regions competitive and sustainable? This chapter begins with a description of the growth in wine tourism, particularly in North America. The next section outlines the significance of this growth as noted by a number of tourism academics as well as some of the recent theoretical developments in this emerging research area. Next, a stage model outlines the development of a hypothetical wine tourism destination region. The purpose is to identify their stages of development and to suggest their positions as competitive and sustainable tourism destination regions. Balancing the elements in the internal supply characteristics and in the demand characteristics for wine tourism is important in achieving an optimal state of tourism development.

Wine Tourism Growth

The resistance to globalization has secured an important place for the localized consumption of cultural products like wine. Wine regions provide the consummate "destination branding" through designation of appellation and claims of terroir. While the production of wine has long been paramount within many old world cultural economies, it is the experience of discovering wine regions, and the consumption of wine and wine lifestyles, that now push expansion of the wine industry outside of its traditional European home base. The agglomerative appeal of controlled appellations provides a significant boost to regional destination branding, which has contributed to an upward spiral of winery establishments in many parts of the world. This upward spiral has been quite evident in the rural landscape of North America.

There are now more than 100 American Viticultural Areas (AVAs) in the USA, most established between 1983 and 1991 (Robinson 1999). By 2008, there were 4,850 bonded wineries in the USA, approximately 2,687 of them in California (California Wine Institute 2008; Wine Business Magazine 2009), while the 2008 count of wineries in Canada stood at 297 (Wine Business Magazine 2009). There are now wineries operating in all 50 American states and in eight Canadian provinces. As Fig. 9.1 illustrates, much of the explosion of winery establishment in North America has been very recent, which raises concerns about the sustainability of the wine industry and of wine tourism growth.

Industry stakeholders have noted the significance of tourism to the wine industry recently. Publication of wine tourism research in industry journals such as the *Journal of Wine Research* has been eclectic and sporadic, ranging from Hall's (2003) work on wine tourism and risks to biosecurity to Charters and Pettigrew's (2005) discussion of the aesthetic experience of wine consumption. Carlsen's (2004) review of global wine tourism research introduced the first issue of *the Journal of Wine Research* dedicated entirely too wine tourism. Case studies of wine routes (Correia et al. 2004) were presented as well as analysis of wine tourist behavior (Alant and Bruwer 2004; Mitchell and Hall 2004) and demographic profiling of wine tourists (Tassiopoulos et al. 2004).

The growth of interest in wine tourism as a special interest niche in the tourism market has garnered much attention among tourism, business, and hospitality scholars. Much of this work reflects a keen interest in the links between wine tourism, restructuring and development (Hall and Mitchell 2001), development of the wine trail (Jaffe and Pasternak 2004), and further development of wine tourism hospitality and service (O'Neil and Palmer 2004; Brown and Getz 2005; Sparks 2007; Charters et al. 2009). Interest in the area of tourism marketing has also been wide ranging. Williams (2001) and Williams and Kelly (2001) assess wine tourism destination images then link them to cultural tourism. Other research specifies the difficulties of wine tourism development in nascent wine regions such as Chile (Sharples 2002), Michigan (Wargenau and Che 2006), Australia

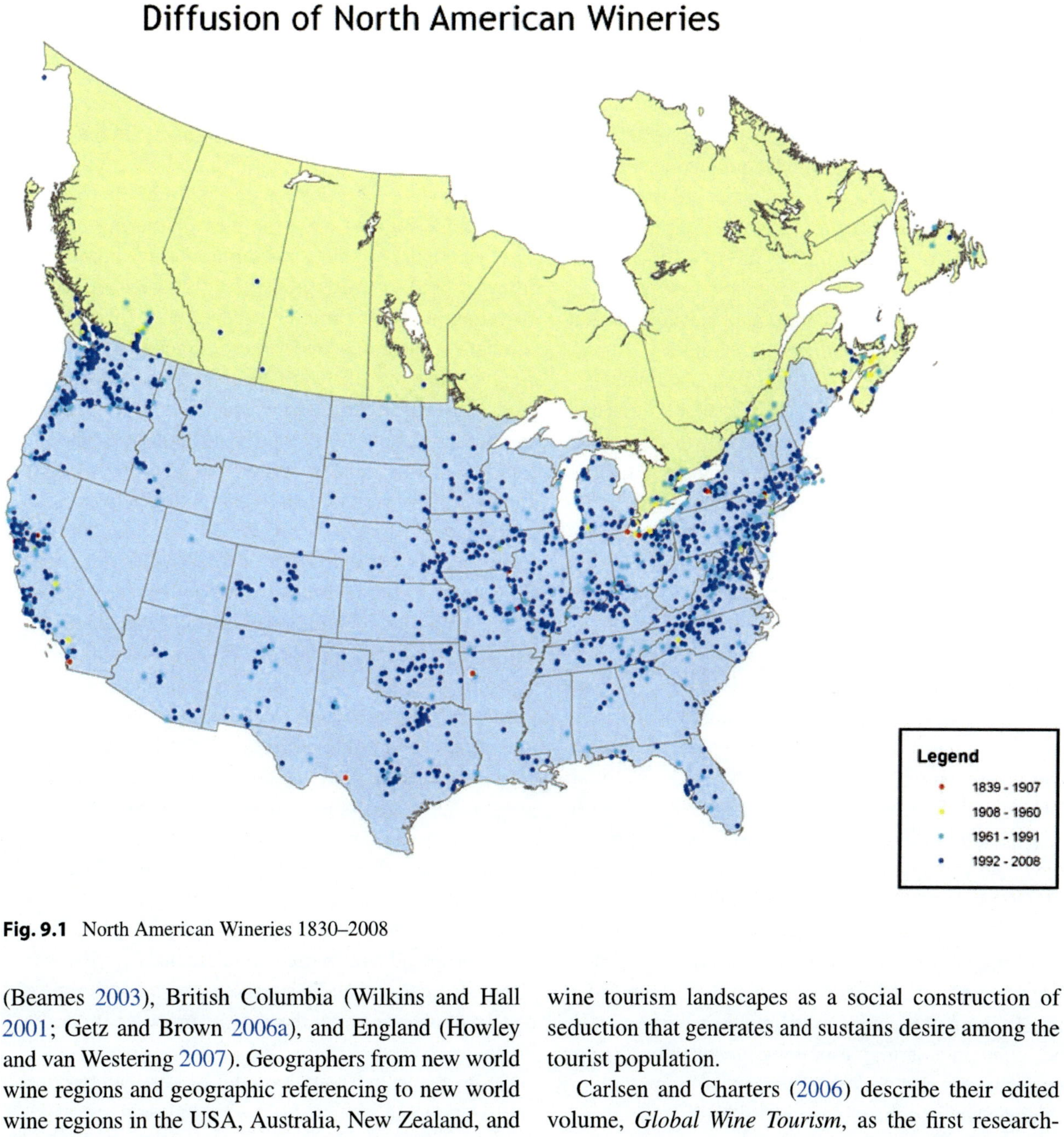

Fig. 9.1 North American Wineries 1830–2008

(Beames 2003), British Columbia (Wilkins and Hall 2001; Getz and Brown 2006a), and England (Howley and van Westering 2007). Geographers from new world wine regions and geographic referencing to new world wine regions in the USA, Australia, New Zealand, and Canada are prominent in many case studies in wine tourism (Carlsen and Dowling 2001; Getz 2000; Dodd and Beverland 2001; Telfer 2000; Telfer 2001a, b; Williams 2001).

Ravenscroft and van Westering (2001) take one of the earliest theoretical positions in the wine tourism literature, offering wine tourism as a bodily practice connected to entry into a social world. More recently, Bell (2008) discusses wine tourism and other forms of alcotourism as a manifestation of place promotion where drinking is a form of consuming place (Bell 2008, 294), while Cartier and Lew (2005) entertain wine tourism landscapes as a social construction of seduction that generates and sustains desire among the tourist population.

Carlsen and Charters (2006) describe their edited volume, *Global Wine Tourism*, as the first research-based volume to approach the phenomenon of wine tourism. The book includes both product approaches (Taylor 2006; Adams 2006) and market approaches (Tassiopoulos and Haydam 2006; Charters and Fountain 2006) to wine tourism with reference to the nature of wine tourism settings and issues of regional development and sustainability, especially in emerging wine tourism regions. *Global Wine Tourism* is now among three edited books amalgamating wine tourism case studies published since 2000 including Hall et al.'s (2000a) *Wine Tourism Around the World* and Getz's (2000) volume *Explore Wine Tourism*.

While regional image and imagery and the distinction of appellation and terroir were always important in the discussion that links wine and tourism, recent research also links these distinct regional images to sustainability of the industry. For instance, both Hall et al. (2003) and Hashimoto and Telfer (2003) have pointed to the reliance of wine tourism, and the wine industries, on regional branding for market leverage and promotion. Johnson and Bruwer (2007), on the other hand, hail regional positioning through appellation or terroir as the key to "sustainable competitive advantage that raises the product out of mere commodity status" (Johnson and Bruwer 2007, 164). With specific reference to sustaining wine regions in Canada, Poitras and Getz (2006, 427) claim, "The appeal of wine regions is based on a difference of place, in which both natural and cultural elements are important, and these differences must be branded." This echoes Williams and Dossa's (2003, 26) claim that "natural and cultural landscapes should be recognized as integral if not central components of the tourism asset base in wine tourism regions."

Getz and Brown (2006a) and Wilkins and Hall (2001) have both attempted to benchmark wine tourism development through destination life cycle concepts and sustainable wine tourism industries in British Columbia. While Wilkins and Hall's (2001) SWOT analysis identified the strengths, weaknesses, opportunities, and threats for wine tourism, Getz and Brown (2006b) went on to use life cycle analysis in order to identify sustainable strategies for wine tourism development in the Okanagan Valley of British Columbia. Gammack (2006) explored the knowledge infrastructure of the wine tourism industry in Australia and found an emerging e-business profile among tourism managers and a close alignment of wine tourism sustainability and the wine industry itself.

A Stage Model for Competitive and Sustainable Wine Tourism Destinations

With reference to tourism destinations in general, Goeldner et al. (2000) defined a tourist destination in its simplest terms as a particular type of geographic region within which a visitor enjoys various types of travel experiences. Gunn (1988) differentiated between destination resort areas and touring supply areas. Both types of destinations require two attributes for long-term success: competitiveness and sustainability (Goeldner et al. 2000; Crouch and Ritchie 1999; Ritchie and Crouch 2000). Competitiveness relates to viability, i.e., the ability to compete successfully in the marketplace. Sustainability is the ability to maintain the quality of its physical, social, cultural, and environmental resources while a destination competes in the marketplace. Wine regions are examples of touring supply areas where the product consumption is at a number of different product levels and scales (Carmichael 2005). Traveling through rural landscapes to experience a number of wineries and other attractions becomes more evident as the wine tourism region develops. Clearly, as wine regions become increasingly dependent on tourism revenues, the need to sustain the economic viability of this industry grows (Skinner 2000). It is important, however, to make a distinction between sustaining the wine industry and developing wine tourism according to sustainable development principles (Butler 1999). Unrestrained development can drastically alter a tourism region and impact experiences for both tourists and residents.

Formal models of tourism development have provided insights into some of the stages a destination may progress through and some of the issues, such as carrying capacity, that are raised by tourism growth. Stage models are not new to tourism and geographical research. Spatial and evolutionary models of tourism development, as reviewed by Dredge (1999), Fagence (1995), Oppermann (1993), and Pearce (1995), have attempted to draw order and spatial patterning from the widespread and varied phenomenon of tourism. However, many of these models have focused on the pattern of change rather than on the processes underlying these changes.

This chapter presents a model of destination competitiveness and sustainability that builds on previous stage models such as the seminal work on the tourist area life cycle model (TALC) of Butler (1980). It also includes a spatial as well as a temporal component. The importance of spatial processes like networking and strategic alliances in wine touring destinations draws on the work of Telfer (2000, 2001) and on the works of Hall et al. (2000b) and Johnson and Bruwer (2007). The proposed model in this chapter outlines the importance of three factors in destination development: internal characteristics (supply), demand characteristics (users), and external forces (wider social, political, economic, and cultural environments).

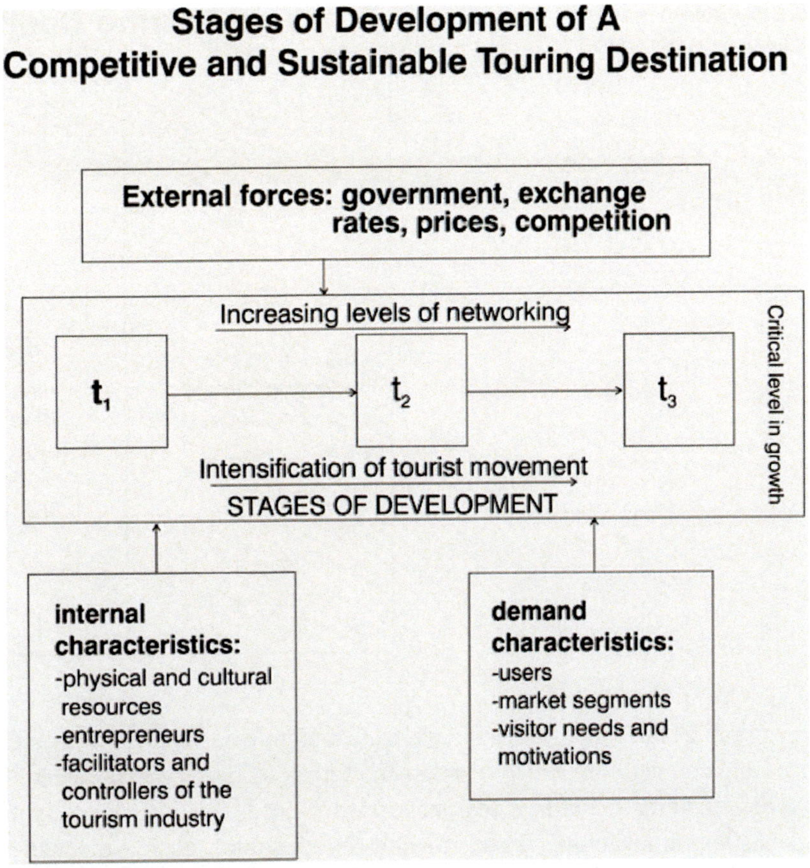

Fig. 9.2 Stages of development of a competitive and sustainable touring destination

Carmichael (2001) first discussed some of the ideas presented in the proposed model.

Other researchers have also applied stage models to wine tourism development. Skinner (2000) applied Butler's TALC model, which incorporates exploration, involvement, development, consolidation, stagnation, decline or alternatives cooperation, and conservation to the Napa Valley wine region. She placed this region in the consolidation stage where carrying capacity limits are being reached through "traffic, noise and overwhelming presence of outsiders" (2000, 294). Application of this model is useful because it draws attention to the need for tighter control over further development. While the application by Skinner (2000) stressed sustainability concepts, research by Dodd and Beverland (2001) took a grounded theory approach to develop an organizational life cycle application to wineries that focused on strategic implications. Their suggested stages were winery establishment, winery recognition, regional prominence, and maturity and decline. At different stages, wineries are likely to have differences in focus, changing from rural tourism to cultural, industrial, special interest, and event tourism – and eventually to urban tourism as the region becomes highly developed. This includes changes in visitor types, tourist facilities, community attitudes, development of networks, sales focus, pricing, and media coverage. The findings of Dodd and Beverland (2001, 20) are based on the New Zealand context, and because of this regional limitation, they call for the development of other models and applications in other regions.

This chapter proposes a three-stage tourism model:
Stage 1, (*t1*), Winery Independence.
Stage 2, (*t2*), Wine Tourism Development.
Stage 3, (*t3*), Wine Tourism Integration.

Figure 9.2 shows the basis of the proposed development trajectory of tourism destination regions. Over time, a touring destination experiences increasing levels of integration and cooperation between businesses, increases in networking, marketing and promotion, and growth in the number of tourists and their spatial flows around the region. This process reflects one of the ways in which touring destinations may develop. Individual trajectories will depend on internal capabilities such as physical and cultural resources, entrepreneurs, facilitators, and controllers of the tourism industry; external

Fig. 9.3 Wine touring destination development

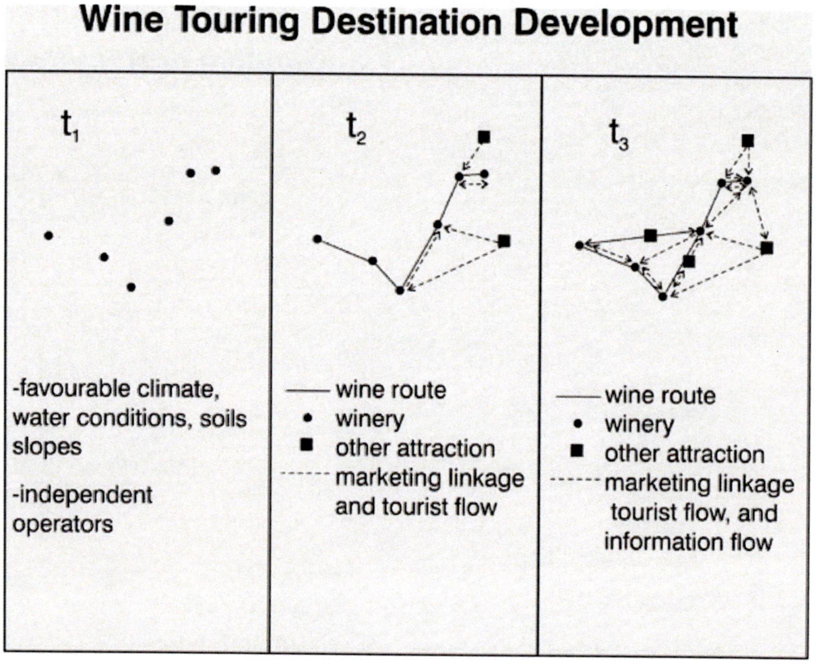

forces such as the role of government, exchange rates, prices and competition; and tourist demand forces such as types of tourists, markets segments, visitor needs, and visitor motivations. This growing momentum should not continue indefinitely, or diseconomies of scale may set in as a critical level of growth occurs. Achieving the optimal level of development and intensification is important in planning and managing a sustainable and competitive touring destination.

Figure 9.3 represents a suggested graph network view of the ways in which the intensity of interaction may increase as the wine destination develops. Stage 1, (t1), winery independence, shows independent operators with no interaction. Stage 2, (t2), wine tourism development, shows the development of marketing linkages between some wineries and attractions creating a wine route linking some of the wineries and tourist flows. In stage 3, (t3), winery integration, there is an increase in marketing linkages between wineries and attractions, greater tourist flows between wineries, and more intensive information flows. This increasing intensity benefits the wine tourism industry until the end of stage 3 when critical capacity in terms of tourism development occurs. Should the area become too crowded and congested, diseconomies of scale may set in and information overload may occur.

Figure 9.4 describes in more detail the stages in the model. Networks change over time from no connectivity through low connectivity to high connectivity. Tourist movements initially rely on a separate winery experience, but with the development of a wine route, there is increasing circulation of tourists dispersed through the region, and the regional experience becomes important. In terms of marketing, there is little awareness of wine tourism products and little investment in marketing in stage 1. As wine tourism develops in stage 2, awareness grows and joint promotion initiatives occur. By stage 3, the wine tourism product is more complex. It features events and festivals and is assisted by formal and informal partnerships and promoted by public and private sector marketing. Regional branding is important in competitive positioning of the wine touring destination. By stage 3, a functional cluster of wine businesses has emerged with linkages to other clusters, food for example. Clusters benefit from information flows in which a wine council or winery institutes, or associations and linkages with university departments occur.

The remainder of this chapter looks at the levels of development in two Canadian wine tourism destination regions – the Niagara Region of Ontario and the Okanagan Valley in British Columbia. Niagara and the Okanagan are two of the oldest and most important wine tourism regions in Canada. While our main analysis relies on data from the late 1990s, we are fortunate to have parallel data on wine tourism dating forward to 2008 in the Okanagan region and from 2005 in the

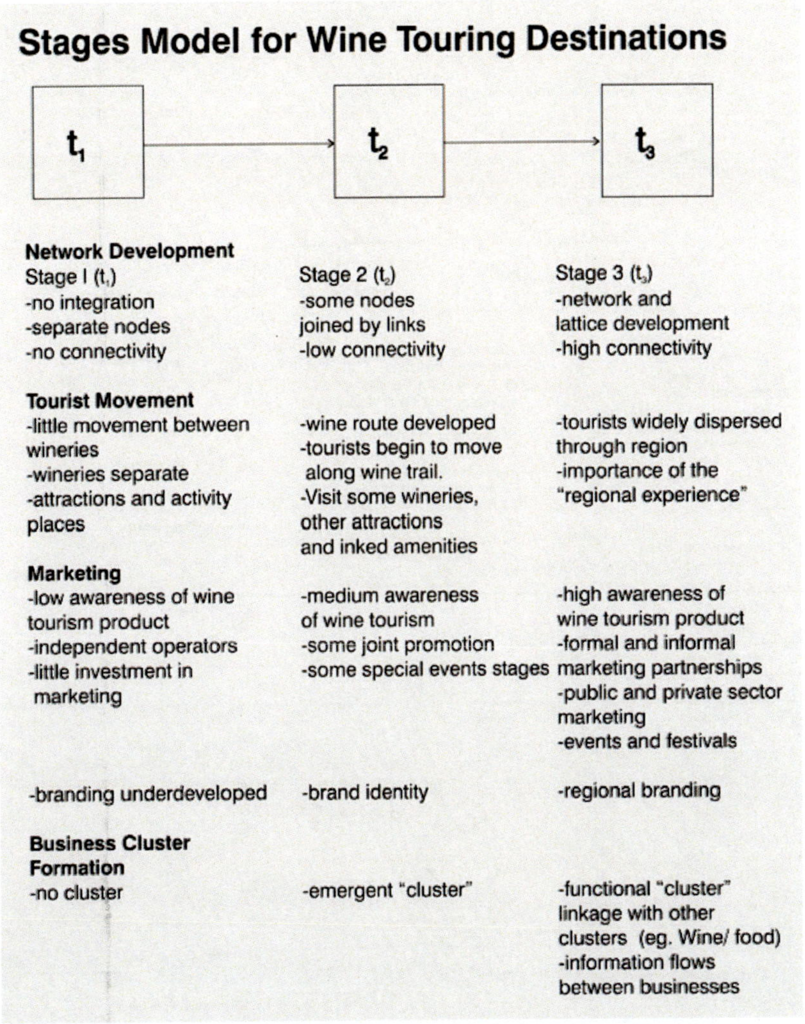

Fig. 9.4 Stage model for wine touring destinations

Niagara region. The longitudinal nature of these data allows us to examine growth and development of wine tourism through stages of the proposed model. The research questions, as far as data limitations allow, are: To what extent is wine tourism competitive and sustainable in these areas? In addition, what level of development is optimal for competitiveness and sustainability?

Niagara Region

The Niagara region in Ontario provides an interesting example of wine tourism. The industry has grown rapidly over the last 20 years. Wine quality has improved considerably with the use of imported European vines and effective management. Furthermore, the wineries have developed marketing linkages with each other and with other tourism initiatives in the region. Wineries are just one element in the attraction mix catering to the touring market. Wineries are located in an attractive agricultural region situated between Lake Ontario and the Niagara Escarpment. This rural landscape is likely to be part of visitor expectation and motivation in visiting the Niagara wineries. A wine route, developed by the Wine Council of Ontario, links some of the wineries.

In 1998, 450,000 people visited the region's wineries (Wine Council of Ontario 1998 factsheet). By 2004, now a designated tourism destination, Niagara attracted over 20 million visitors per year with 750,000 of those visiting wineries (Grape Growers of Ontario 2004 factsheet). In Ontario, the number of visitors to wineries topped one million people by 2007 (Grape Growers of Ontario website). Most visitors come to Niagara, which accounts for 90% of the province's grape growing volume, most of which goes into winemaking, and is the site of the majority of the wineries.

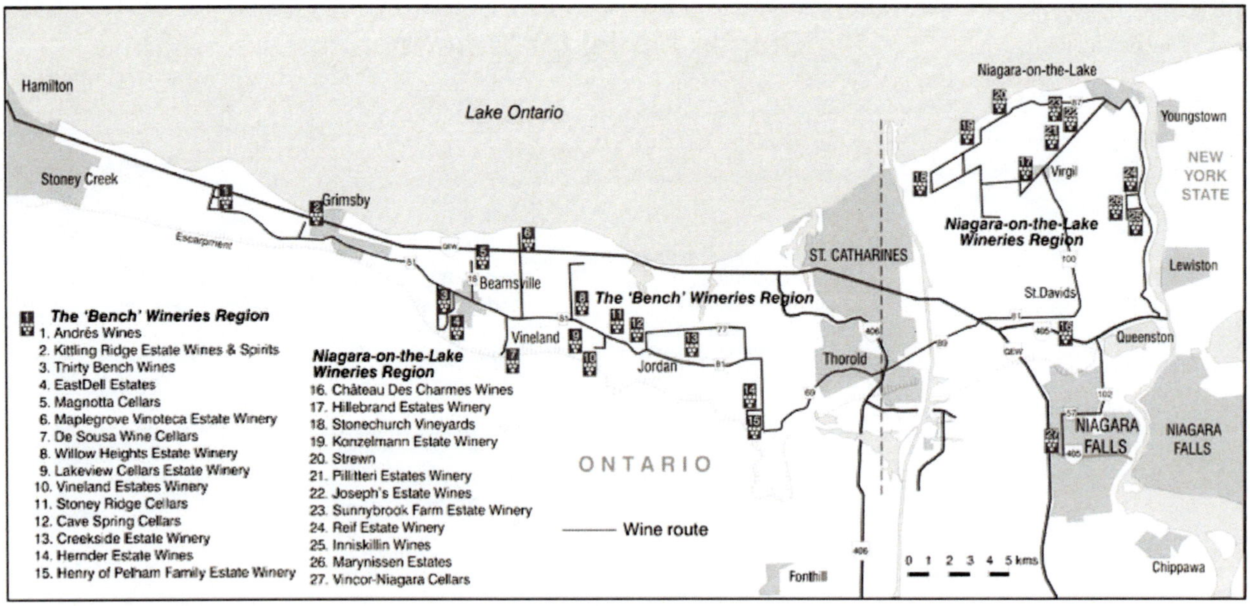

Fig. 9.5 Map of Niagara wineries 2000

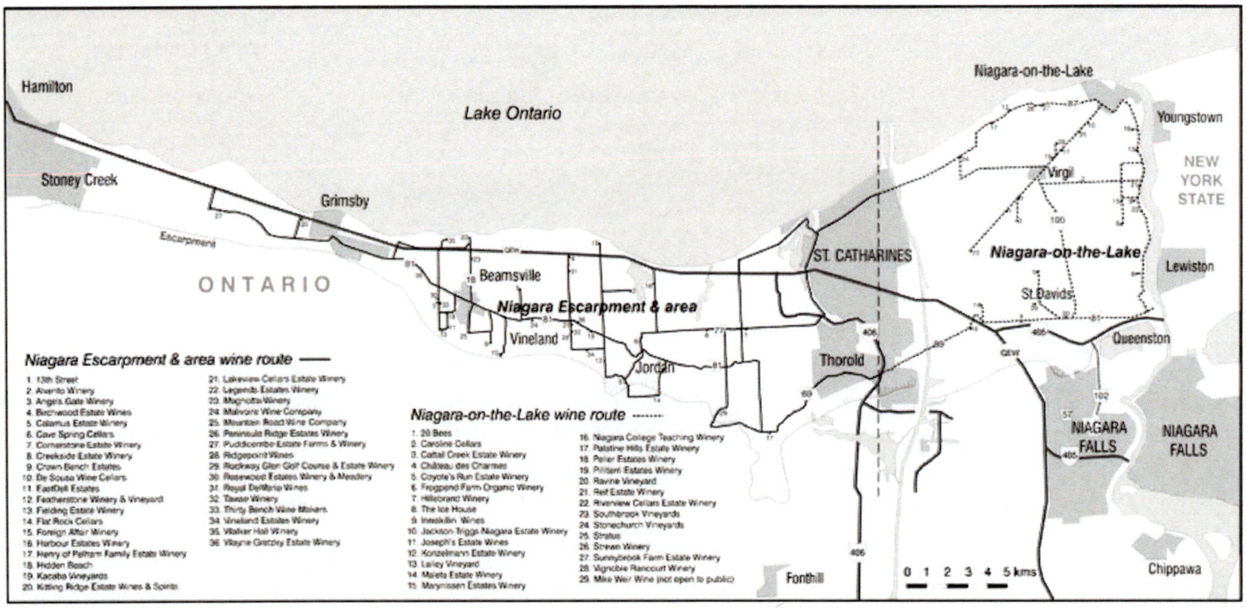

Fig. 9.6 Map of Niagara wineries 2010

There were 13,600 acres under vine in Niagara (Grape Growers of Ontario website). The number of wineries has expanded from 15 wineries in 1988 to over 50 wineries in 1998, with new ones opening every year and improvements made to existing winery infrastructure (Wine Council of Ontario 1998 fact-sheet). In 2000, 27 of the more than 50 wineries were members of the Wine Council of Ontario (Fig. 9.5). By 2007, 64 wineries were members of the Wine Council of Ontario, 36 were on the Niagara escarpment area (the bench) and 29 in the Niagara-on-the-Lake region (Official Travel Guide to Wineries of Ontario 2007). Figure 9.6 shows the Wine Council of Ontario members in 2007. By 2010, this number had increased to 67 wineries, with 37 in the Niagara-on-the-lake area and 30 in the "bench" escarpment area (Wine Country Ontario, Your Official 2010 Year Round Travel Guide).

In 1999, a study of the competitiveness and sustainability of this growth used three main methods for data collection: participant observation, a visitors' exit survey at eight selected wineries, and personal interviews with representatives from eight wineries as well as from related organizations within the region. The survey used a stratified random sample of wineries chosen from a list provided by the Wine Council of Ontario, including four wineries from the "bench" or the edge of the Niagara Escarpment region and four wineries from the area close to Lake Ontario near the small town of Niagara-on-the-Lake. Further stratification of the sample divided wineries into small, medium, and large wineries reflecting the diversity of wineries within the region. A sample of 400 visitors – 50 respondents from each winery – served as a desirable sample size to achieve a 95% level of confidence. The survey occurred on seven randomly selected weekend days during the months of August and September 1999, resulting in a response rate of 71.4%. Data collected included visitor trip characteristics, motivations, demographics, and perceptions of the wine tourism experiences. Where appropriate and indicated in the text, some follow-up interviews with six "bench" winery operators in 2005 indicate important changes in the region since the time of the original survey.

Niagara Area Competitiveness

Competitiveness involves physical attributes, management, and marketing factors. An attractive resource base and the uniqueness of landscape and products differentiate successful touring regions. The Niagara wine region is a distinctive agricultural area with a mix of viticulture, fruit growing, and horticulture. It enjoys a favorable microclimate for growing agricultural produce because of its location between the Niagara Escarpment and Lake Ontario. The Niagara Escarpment designation as a World Biosphere Reserve adds to the uniqueness of the region. Since wineries are only present in a few areas in Canada, visitors in a 1999 survey rated "the importance of a unique experience" highly in their visit to the winery. The region also enjoys a critical mass of attractions including Niagara Falls and the Niagara Escarpment. In addition, many wineries surround attractive towns like Niagara-on-the-Lake and Niagara Falls. The winery region has good accessibility to Toronto and Niagara Falls and high levels of mobility offered by proximity to a major highway, the QEW.

A number of marketing and management factors, including marketing partnerships, target marketing, and visitor and information flows, assist in making the Niagara Region competitive.

Marketing Partnerships

Niagara developed marketing alliances that linked tourism system elements together and offered increased flexibility and choice for the wine tourist. All of the wineries studied were involved in joint promotions. This not only saved money for the individual winery but also helped create a regional image for a number of wineries. The role of the Wine Council of Ontario is to market the province's wine, negotiate prices with grape growers, and work with the provincial government. The Wine Council runs a number of festivals throughout the year including "The Six Unforgettable Weeks of Summer," "The Niagara Grape and Wine Festival," and "The Days of Wine and Roses."

Niagara wineries were also included in the cooperative marketing programs of the Niagara Economic and Tourism Corporation. The corporation promotes agritourism as a whole through combining wine and food in various festivals venues such as the "Taste of Niagara." Some wineries, especially those near Niagara-on-the-Lake, developed partnerships with upscale restaurants, hotels, and the Shaw Festival in Niagara-on-the-Lake. In addition, there were linkages with bicycle tour companies and hiking organizations that mostly involved the "bench" wineries.

Wineries recognize the importance of the Niagara brand and the need for cooperative linkages. As one winery owner stated, "It is important that the winery down the street from you is doing well because if they aren't, it can harm you and the Niagara reputation." From the partnership perspective, Niagara is in stage 3 of the model, although there are two distinct geographical winery subregions.

However, it is worth noting that there are conflicting goals among the wineries that is affecting the efficiency of the winery integration stage. In December 2009, the largest vintners in the province – Vincor Canada Ltd, Andrew Peller Ltd, and Colio Estate wines – withdrew their membership from the Wine Council of Ontario. The cause was a controversial tax increase on wines sold through private wine stores to pay for the wine council's marketing and a VQA subsidy program at the

LCBO. The large wineries believe a new organization can better serve their interests.

Their withdrawal from the Council affected the organization's budget in that the three wineries alone contributed 13% or $500,000 of the budget (*The Standard*, St. Catherine's December 9, 2009). Vincor operates popular wineries including Niagara-on-the-Lake's Inniskillin. Jackson Triggs and Andrew Peller produce Peller Estates and Hillebrand in the Niagara-on- the-Lake region as well as 30 bench wines in the escarpment area. These large wineries started a new association called the Winery and Grower Alliance of Ontario. They are all Cellared in Canada (CIC) members. CIC wines are a blend of foreign wine and local grapes and may be sold through private wine stores – a privilege that predates the 1993 Free Trade Agreement. The majority of Ontario wineries sell only Vintner's Quality Assurance (VQA) wines that use 100% Ontario grapes.

Target Marketing

The existing market is well educated – some 80% graduated from college and university and 44% maintain high household incomes of more than $90,000. Seventy-two percent are professional/skilled workers, and most are between 31 and 70 years old. A third of the respondents are in the age category 41–50. The wineries in the region are targeting both the middle-aged, higher income group as well as the lower income Generation X. Only one winery marketed the seniors' coach tour.

Visitor Flows

Visitor trip characteristics reflect some of the linkages that exist between the wineries and secondary elements like accommodation, catering, shopping, and other natural and cultural attractions. Fifty-nine percent of survey visitors reported visiting other attractions. The most popular combination of attractions was Niagara Falls, the Shaw Festival, and Niagara-on-the-Lake.

Information Flows

Linkages are both formal and informal. The marketing role of the Wine Council of Ontario as a facilitator of formal information flow is pivotal. At the time of the 1999 survey, networking and information flow among Niagara-on-the-Lake businesses was better than in the bench area close to the escarpment. Since then, the bench area has grown rapidly. New wineries opened and became new members of the Wine Council of Ontario. Research conducted in 2005 indicated that marketing partnerships remain important to the winery operators.

Competitiveness requires gaining the optimal state of growth in tourism numbers and in tourism product development. The winery survey in Niagara showed that there were indications of crowding at some of the wineries. In addition, significant differences exist between perceived expectations of service quality and actual experiences for a number of reasons: willingness to help, promptness of service, knowledge, individual attention, and consistency of service (described in more detail in Carmichael 2005).

Niagara Area Sustainability

Sustainable development involves an activity in which jobs and revenues provide for the local economy without jeopardizing resource integrity. At the time of the original survey in 1999, the Niagara wine industry employed almost 5,000 people. At the same time, the average visitor to the wineries spent $38 on wine and a total of $130 in the region. It is interesting to note that when winery representatives reply about the importance of tourism to the sustainability of wineries, they indicated that only livelihoods and profits affected sustainability. The wider aspects of maintaining the quality of the visitor experiences and maintaining rural landscape integrity were not voiced as part of sustainability. However, visitor data in 1999 suggested that the rural landscape was the most important reason for winery visitor enjoyment of the Niagara region, receiving a Likert scale rating of 5.79 out of 7, where 7 is very important. All wineries had plans for expansion and were at different stages in seeking planning permission.

Follow-up research with six winery operators in the bench region near the escarpment in 2005 indicated that winery operators were keenly aware of the importance of the rural landscape, particularly the escarpment lands, to the resource integrity of their tourism product. One respondent mentioned "the need to ensure that our buildings work with the existing visual impression, with a minimalist approach so as not to distract from the natural and rural landscape of the property." Environmental stewardship was very important to one winery respondent in this 2005 research. He did not use insecticides, found natural ways to farm the land that included bringing in predatory insects, and conserved water through trickle irrigation.

Sensitive escarpment lands within the bench region are under stricter control in terms of tourism product development than those in the Niagara-on-the-Lake region. In fact, the Niagara Escarpment Commission (NEC) has now implemented revised and stricter criteria for winery size and the nature of development. Since the time of the survey in 1999, some of the expansions have gone ahead before the new regulations, despite the sensitivity of the environment. Research indicates that in 2005, however, some of the newer wineries found it difficult to expand or even open up a patio restaurant. This has caused some resentment as some of the uses that were restricted or involve a long application process were already in place at older wineries. One winery called the NEC "bureaucratic and difficult to deal with." Nevertheless, such legislation helps in maintaining rural land use and sustainable tourism.

An innovative program to introduce a voluntary initiative aimed at improving environmental stewardship was developed in 2007 by the Wine Council of Ontario. Best practices guidelines and self-assessment criteria were made available to identify problem areas (Martin 2007). The Sustainable Winemaking Ontario initiative covers aspects of viticulture ranging from wastewater management on the farm to efficient energy usage in all winery operations, using an eco-labeling program – a green leaf icon – to market participating wineries (Wine Country Ontario, Your Official 2010 Year Round Travel Guide 2010). According to the Guide, 21 of the area's 67 wineries indicate that they are members of Sustainable Winemaking Ontario.

Achieving an optimal state of competitiveness and sustainability is the goal of a successful touring destination. The development of wine tourism in the Niagara region is at a critical stage in approaching the optimal state. The region has achieved successful marketing partnerships and visitor flows. However, visitor crowding at some of the wineries indicates the need to set limits to growth. As of 1999, the Niagara Region was approaching stage 3 in the model: wine tourism integration. Some of the existing wineries have found it difficult to cope with increasing numbers of visitors. Winery expansion, however, may affect rural image and resource integrity especially on sensitive escarpment lands. Since the time of data collection, the number of wineries has increased and some have expanded. Some of the most recent wineries in the region are organic wineries: Frogpond Farm Organic Winery, Wayne Gretszy Estate Winery, Rockway Glen Golf Course, and Estate Winery (see Fig. 9.6). While stage 3 is in effect in Niagara today, tighter planning and voluntary initiatives may prevent a lapse in the critical limits of planned growth.

Okanagan Region

The Okanagan Valley of British Columbia is one of a growing number of New World wine regions that has experienced recent and rapid growth. Like many New World wine regions, missionaries introduced wine, and European settlers culturally reproduced it. The first Vitis vinifera varieties, including Chardonnay, Csaba, and Muscat, were cultivated in the 1930s despite continued government support for hybrid lambrusca vines (Association of British Columbia Grape Growers 1984).

In the 6-year period following the 1988 North American Free Trade Agreement and the General Agreement on Tariff and Trade between the USA and Canada, grape growers in the Okanagan were paid $8,100 per acre to pull out undesirable grape varieties and replant in vinifera varieties. A total of 2,400 acres of grapes were pulled out and replanted in vinifera varieties (McGuire 1993). In 2009, the Okanagan and Similkameen Valleys contained over 7,000 acres of land cultivated in vinifera varieties, producing 21,647 tons of grapes. This accounts for 97.35% of crop production in British Columbia (British Columbia Wine Institute 2009). In 2009, the number of wineries in the Okanagan totaled over 130, with several startup operations waiting first vintage. Many wineries were established after 1990 (Fig. 9.7) when VQA standards were introduced by the British Columbia Wine Institute (Senese 2008a). This is astonishing growth for an industry that totaled only 14 wineries and 3,500 acres of vineyard in 1988 (Schreiner 2009).

The remarkable growth of the Okanagan wine industry coincides with an equally remarkable shift in the region's tourism industry. For decades, the tourism brand of the Okanagan depended on the unique amenities associated with its warm arid climate, orchard industries, large lakes, and scenic mountains. The current Okanagan tourism brand centers on the wine industry and its associated rural aesthetic that includes the

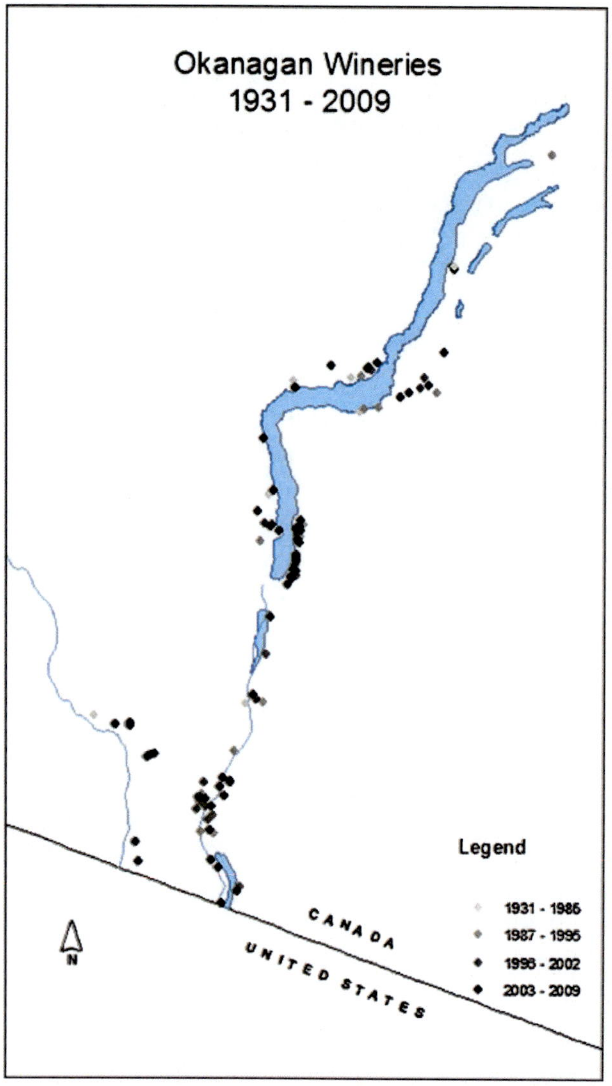

Fig. 9.7 Map of Okanagan wineries

scenic landscapes, recreation facilities, and integrated services and amenities that support the wine industry. Currently, wineries and vineyards provide approximately 80% of the agritourism product in the Okanagan. Of the 53 wineries providing tourism-related products and services in the Central Okanagan alone, 48% provide a gift (non-wine) shopping experience, 22% offer tours, 18% have full restaurant food service, and 12% provide accommodations (Senese 2009).

The production of quality wines competitive on the international stage is an important key to the appeal of the Okanagan as a wine tourism destination. However, the production of wine itself appears secondary to the production of the wine tourist experience in the Okanagan, where winery tours, tasting rooms, festivals, on-site accommodation, dining, theater productions, and concerts have become widespread. The Okanagan Wine Festival recently celebrated its 25th anniversary. While the Fall Wine Festival is the largest in terms of market draw and number of events, the Festival Society now hosts four festivals, spring, summer, fall, and winter with over 100 participating wineries in each festival.

As another illustration of the continued growth of wine tourism in the valley, Mission Hill, one of the largest wine producers in the region, recently announced a massive $10 million expansion to its winery that has little to do with wine production. The expansion includes a 104-room auberge, 30 cottages situated on either side of the auberge, a free-standing restaurant, a wellness center, a small conference center of approximately 1,000 square meters, artisan residences for singers, musicians, and other performers invited to play at the winery, and an art gallery to feature Mission Hill owner Anthony von Mandl's extensive collection of glassworks and tapestries (Luciw 2010).

Application of the Stage Model to the Okanagan Wine Tourism Region

The stage model (Figs. 9.2– 9.4) illustrates the development of wine tourism from the appearance of independent wineries through development of wine tourism to the integration of the tourism industry. The influence of supply and demand characteristics, together with the external forces of political economy and wider social cultural change, provides the context for applying the Okanagan's destination development experience to the model. As this examination of wine tourism in the Okanagan seeks out an indication of the industry's sustainability and competitiveness over time and through stages, it is imperative that a long-term view of the wine tourism industry be captured.

The Okanagan wine tourism region is examined by triangulating three sets of data and three methods of data collection covering the period 1998–2008. A visitor's exit survey was conducted in 1998 at 16 wineries in the Okanagan Valley as part of a comprehensive report for the British Columbia Wine Institute and the Kelowna Museums. Data were collected on visitor trip characteristics, demographics, motivations, wine and tourism experience, behavior, and perceptions. The exit survey was followed up by open-ended interviews with a random sample of 30 wine tourists. The interviews investigate the qualitative nature of trip motivation, satisfaction, and experience among wine tourists. The 1998 exit survey included a geographically stratified sample of wineries, five from the Kelowna-Boucherie wine area, five from the central Peachland-Penticton wine area, and six wineries from the South Okanagan wine area including Oliver, Okanagan Falls, and Osoyoos (Fig. 9.7). Further stratification of the sample into farmgate, estate, and major wineries reflects the diversity of wineries in the region. The sample included 679 surveys collected over the course of 14 days spread throughout the months of July and August of 1998 (Senese and Wilson 1998).

To capture developments in the wine tourism industry since the 1998 survey and apply them to the stage model of sustainability and competitiveness, two separate data sets supplement the original data. The first is a set of open-ended elite interviews conducted in 2007 with eight wine industry stakeholders in the Okanagan. The stakeholders agreed to participate in on-site open-ended interviews to discuss the future of the industry, the roles of networks, and the integration of the wine industry in the regional tourism economy of the Okanagan (Senese 2008b). Secondly, in order to capture the most recent wine tourism scenario in the Okanagan, there is a review of a wine consumer research study published in by Tourism British Columbia (2009) in partnership with the Thompson Okanagan Tourism Association and seven participating wineries. The multi-location, two-stage survey was conducted between August and October of 2008. It resulted in 1,977 surveys of visitors to wineries in the Okanagan. The study was conducted in order to examine wine tourism product development, travel motivations, and trip planning behaviors (Tourism British Columbia 2009).

Okanagan Area Competitiveness: Physical Attributes, Management, and Marketing Factors

The Okanagan Valley has a long history as a tourism region with a varied resource base providing a wide range of landscapes, attractions, and products. Within British Columbia, the Dry Interior Plateau of the Okanagan constitutes one of the most ecologically complex and diverse ecosystems and biogeoclimatic zones in Western Canada. Bordered by the Coast Mountains to the west and the Monashee Range to the east, the Okanagan is the northern point of the arid zone extending southward into the western USA and including the Sonoran Desert. It is a diverse landscape of large lakes, arid grasslands, nationally significant agricultural lands, as well as home to many rare and endangered species.

The Okanagan is also an important agritourism region. It is one of only two soft fruit agricultural areas in Canada. It is the major producer of tree fruits (95% of harvest) and grapes (98% of production) in British Columbia. The remainder of the agriculture sector is diverse and produces commodities from fruits, grapes, and ginseng to cattle, dairy, and forage on approximately 4,600 farms and 170,000 ha (Statistics Canada Agricultural Profile 2001, 2006). Other than agriculture, the Okanagan economy is increasingly dominated by "non-resource" service-producing sectors that capitalize on amenity resources that support migration and tourism arrivals.

The Okanagan tourism region is connected to major population centers in the Lower Mainland of Vancouver and the USA. northwest by the Coquihalla Highway that opened in 1986. It is also serviced by an international airport in Kelowna that is now Canada's tenth busiest airport. As a tourism destination, the Okanagan has historically capitalized on the hot, dry climate, large lakes with accessible beaches, and agricultural landscapes to attract family-oriented, summer visitors from Alberta, the Lower Mainland of British Columbia, and to a lesser extent the northwest states of the USA.

There are more than 130 wineries stretching the 100 km length of the Okanagan Valley on the east and west benches of Okanagan, Skaha, and Osoyoos Lakes. While the historical precedent for cultivation in the region is orcharding and ranching, there has been a significant conversion of agricultural land, particularly orchard land, to vineyard since 1985. This coincides with the post Free Trade wine industry boom in the Okanagan (Schreiner 2009). Conversion of orchard to vineyard and winery growth has been most pronounced in the south Okanagan on medium to small plots of land adjacent to the built up areas of small community service centers of Penticton, Oliver, and Osoyoos. The site, situation, and size of these vineyards and converted orchards represent the ideal geographical conditions for wine tourism. They allow relatively easy access to tourism infrastructure, along with transportation connections to major tourism markets in the Lower Mainland and USA.

According to the 1998 exit survey conducted by Senese and Wilson (1998), landscape amenities including mountain, vineyard, and lake views were rated highly among survey respondents, while the built environment was not rated as highly. The 1998 survey data also revealed that the passing landscape encountered during the wine tour journey was the most important attribute influencing enjoyment and satisfaction with the wine tour. These results provide one indication that at this early stage of destination development, the regional experience and not individual encounters with specific wineries is an important attribute determining wine tour satisfaction. This appears to illustrate the transition from stage 1 to stage 2 in the Okanagan as the attributes of wine routes begin to develop fostered by the early appearance of winery clusters in the initial post 1988 Free Trade burst of winery growth.

Follow-up interviews with tourists from the 1998 study indicated that a comprehensive view of rural amenities including climatic, scenic, and natural amenities formed the basis of motivation to visit the winery and the region. Clearly, natural capital wrapped up in the physical environment is seen as paramount to these wine tourists. However, there was also considerable, direct reference to the desire for more access to the cultural services associated with the rural experience of agriculture, and viniculture usually provided this through education, history, and food (Senese and Wilson 1998). Again, this is indicative of a transition to stage 2 in the model where there is an awareness of resources available to the tourists but little idea about how to access or connect those resources. Sense of a brand identity associated with physical or cultural resources of the destination also appears to be lacking among the tourists surveyed in 1998. Wine tourists themselves could not isolate one motivating factor for visiting the winery or one factor that ensured satisfaction of the visit. These results did vary depending on location, that is, visitors to estate and farmgate wineries differed in their motivations from those visiting large commercial wineries. Visitors to the smaller, more intimate estate and farmgate wineries were more interested in quality of the environment and the evocative acts of discovery, tasting, and learning about wine than were their counterparts in larger commercial wineries where service, staff knowledge, and tours were more important (Senese and Wilson 1998).

Winery stakeholders interviewed in 2006 were highly emotive when asked about the physical resources of the wine tourism region. Collectively, the stakeholders referred to the intrinsic natural value of the land as the backbone to both the wine and the tourism industry associated with it. The links between quality of life, lifestyle, and environment are frequently given as a prime factor by winery proprietors interviewed. The term terroir was also mentioned frequently as an important resource for the wine and tourism industries by winery stakeholders (Senese 2008b).

Travelers to Okanagan wineries surveyed in 2008 were asked to identify the unique characteristics of the Okanagan Valley that made it different from other wine destinations. Half (54%) indicated that the valley's scenery and unique geography make it different from other destinations, while one third (35%) cited its sunny dry weather. Of those surveyed, 79% were satisfied with their wine tourism experience; of those who were highly satisfied, most indicated quality of wine, food, and atmosphere as the reason (Tourism BC 2009). The continued predominance of the quality of the physical environment and atmosphere 1998–2008 is significant. A high-quality natural environment appears to remain in demand by tourists, and proprietors over the 10-year study period have obliged. The emergence and regional adoption of a brand identity, as indicated in stages 2 and 3 of the model, appears to be emerging while the important characteristics of the physical environment are maintained. The association of a quality environment with quality wine and a quality tourism experience certainly factor positively for any notion of a competitive and sustainable wine tourism destination.

Marketing and Management Alliances

Presently, there is a variety of marketing and management alliances in the Okanagan wine region. They include vertical alliances among wineries, horizontal alliances among wineries and the services that support them, and regional alliances or multi-sectoral spatial alliances that appear at different scales throughout the valley. Multi-sectoral alliances are based on aligned niches in the arts and culture, agricultural, culinary, and wine sectors. The British Columbia Wine Institute and the Grape Growers Council of British Columbia promote vertical alliances. Tourism British Columbia promotes "circle routes" of wine tourism and integrated niche tourism regions throughout the province.

The Okanagan wine region also markets with Wines Northwest, an alliance of Pacific Northwest wine regions that targets the western North America touring market. The Okanagan tourism region is divided into regional marketing areas represented, for example, by the South Okanagan Tourism Association, the Thompson Okanagan Tourism Region, and the Okanagan Cultural Tourism Corridor that market wine routes and tours throughout the valley. The Okanagan Wine Festival Society promotes four seasonal wine festivals throughout the region. The fall wine festival occurs over 10 days with 165 events integrating wine and culinary attractions. The American Bus Association consistently ranks the fall festival among the top 100 events in North America.

The wine industry in the Okanagan is encouraged by a marketing management policy referred to as markup free delivery (Belliveau et al. 2006; B.C. Wine Institute 2006). In British Columbia, there is a 100% markup included in the sale price of all wine products. Sales through government-owned liquor stores result in the markup being captured by the government while the winery is able to keep the markup for sales that occur on site. Belliveau et al. (2006) state, "markup free delivery" and the basis in agritourism expose the wine industry and small wineries in particular to the risks and vulnerabilities of the tourism industry in all of its fickleness, vagaries, and cycles.

The expansion of marketing alliances was a factor in the results of the 2009 Tourism BC travelers' survey where 44% of wine tourists surveyed had used local or regional tourism websites to plan their travel (Tourism BC 2009). While the entire region is integrated into the network of marketing and management alliances, stakeholders interviewed in 2006 indicated that most networking with other wineries and cultural stakeholders was informal in nature and facilitated by members of the private sector. Stakeholders noted increasing micro-geographies of market localization within the wine industry and a converse increase in market regionalization of the wine tourism industry (Senese 2008b). For instance, South Okanagan wineries have now aligned themselves into a separate wine trail that is further subdivided into the Golden Mile and Black Sage Benches. The wineries of Penticton/Summerland are subdivided geographically into the Naramata Benches and Summerland Bottleneck, and the wineries of Kelowna are marketed as Kelowna South Slopes, West Kelowna, and Mount Boucherie.

During the same time, the provincial (Tourism BC) and regional (Thompson Okanagan Tourism Association) tourism groups have marketed wine tourism as circle routes containing the Okanagan and Similkameen Valleys in their entirety. These increasingly complicated networks, partnered formally and informally among public and private groups, give strong indication of stage 3 development in the model. Here, networks are latticed with high connectivity between and among groups, and a regional brand is developed through integration of these networks into linkages such as circle tours. These complex webs of networking seem quite distant from the survey results of 1998 that indicated that most tourists depended on word-of-mouth advertising from friends and family and the advice of winery operators to plan their wine tourism experience (Senese and Wilson 1998).

Target Marketing

The origins of wine tourists in 1998 were evenly divided between the Lower Mainland of British Columbia, Alberta, and the local valleys. Visitors were well educated (75% graduated from college or university) and affluent (40% with a household income over $75,000). Most of the respondents surveyed and interviewed were between the ages of 36 and 66 years of age traveling with at least one other person but no children. The market description did not change markedly in the 2008 survey, where 50% were from British Columbia, 26% from Alberta, 13% from other parts of Canada, and 5% from the USA. Almost three quarters (74%) had a college or university degree or diploma, the majority were between 36 and 64 years of age, more than half (52%) had an annual household income of at least $100,000, and the average party size was 2.6 and traveling without children (Tourism BC 2009). Stakeholders interviewed in 2006 noted a creeping elitism in the market, "our visitors are middle aged, well travelled and have money" (Senese 2008a, b). Stakeholders also described their market as comfortable people who like the outdoors and are interested in health and relaxation. According to the stakeholders, a significant growth area in the wine tourism market was among retired visitors from the USA and Europe.

Visitor Flows

According to the 1998 survey, visitation patterns to the Okanagan Wine Region were indicative of the integrative touring market in the area and not a primary attraction destination. Almost 70% of those surveyed indicated that they had arrived by personal vehicle, were in the area for an extended stay of 5–8 days, and would be participating in a variety of activities including water-based recreation or other outdoor recreation such as golf. The majority of respondents (68%) purchased wine at the winery where they were intercepted. Approximately 30% of those surveyed had visited wineries in the Okanagan on prior visits, and close to 50% had participated in a wine tour before. Of those return visitors surveyed, landscape, location, view, and climate were cited most often as the most enjoyable features of the tour that kept them returning to the wine region. However, integrated cultural activities and information were most often viewed as lacking during the visit.

Stage succession between 1998 and 2008 is neatly encapsulated when visitor flows are examined. In sum, winery visitors in 2008 were more interested in wine travel, were more experienced in wine travel, and were more likely to choose the Okanagan wine region as a primary destination. In 2008, tourists stayed in the area longer, bought more wine, and were more likely to return on a wine tour. In the 2008 survey, 79% indicated previous experience wine touring, 94% had used a private vehicle to make the tour, and 76% made the Okanagan Valley their primary destination, and the average time away from home was 9.7 days. The majority (55%) of travelers indicated that wine touring played at least some importance in the planning of their trip. These are considered high-interest wine travelers. Additionally, 71% of those surveyed purchased wine while at the winery where they were intercepted. Travelers to the wineries spent an average of $304.99 per day. High-interest wine travelers reported considerably higher per diem expenditures than low-interest wine travelers, $356.71 and $243.62, respectively. Of those surveyed, 36% were likely or highly likely to return to the Okanagan for a wine tour, while 79% of high-interest wine travelers were likely to take another trip to the Okanagan for a wine tour (Tourism BC 2009). Whereas integrated activities, especially related cultural activities, were seen as lacking in the 1998 survey, the majority of winery visitors in 2008 participated in related and integrated cultural activities while in the region (Tourism BC 2009). The levels of connectivity, integration of experience, and functional clustering appear to place the wine destination firmly in stage 3 of the stage model.

Information Flows

According to the 1998 survey, information linkages were on small regional scales and, especially in the South Okanagan, were very informal (Senese and Wilson 1998). According to some stakeholders interviewed in 2006, more formal information flows from provincial bodies such as the B.C. Wine Institute were regarded as forms of political interference in the operation of private business. This type of public-private discourse was more consistent in estate and farmgate operations than in larger commercial winery operations where formal modes of information networks were encouraged and supported (Senese 2008b). Information networking among Naramata-Penticton and Oliver wineries and wine tourism organizations appeared to be closer than they are in other areas of the valley. Wine subregions such as Kelowna and Osoyoos tend to network more intensely with local and supporting resort, culinary, and cultural businesses than with other winery support networks (Senese 2008b). The level of complexity and subdivision of information flows as well as the level of discontent among some winery operators indicates an expression of diseconomy and unsustainability where levels of interaction are unlikely to support further growth in industry.

Okanagan Area Sustainability

The Okanagan wine industry has become a major economic contributor to the regional economy of the Okanagan as an agricultural activity, a value-added agricultural activity and as a leading figure in the tourism sector. The practice of markup free delivery ensures that the wine industry remains firmly entrenched in the tourism industry here; however, this leaves smaller producers particularly vulnerable to the cycles, stages, and fickleness of the tourism industry. The phenomenal growth of the wine industry from just 14 wineries in 1988 to over 130 in 2009 raises questions about the sustainability of that growth and the ability of the ecological and sociocultural resources of the region to support the industry. Ecological concerns have focused on the arid land base and increasingly competitive demands for limited supplies of water, and this has certainly raised concerns about vulnerability of the industry to climate change (Belliveau et al. 2006).

Growth of the Okanagan wine region has paralleled population growth rates in the area. Population of the Central Okanagan alone grew from 111,846 in 1991 to 176,227 in 2008 (B.C. Stats 2007). Paradoxically, the rural image as branded and marketed to tourists by the wine tourism industry has been a significant catalyst to retirement and other amenity migrants to the Okanagan who compete for water and land resources on the rural-urban fringe. In terms of economic sustainability, wine stakeholders interviewed in 2006 were unanimous in their enthusiasm for tourism industry growth and migration as they increase market demand for their wine and wine tourism products. Stakeholder enthusiasm for tourism development appears grounded in the experiences of their visitors. According to 1998 and 2008 visitor data, the average visitor spent more than $30 at the winery, but desired to spend more on food, recreation, and on-site wine bars. Several stakeholders who acknowledged the dangers of compromising the scenic countryside resources of the region by attempting to accommodate mass tourism arrivals questioned the sustainability of further development of accommodation and dining facilities on site.

Wine tourists to the Okanagan, like cultural tourists in general, are as likely to engage in wine tourism to take in the product itself, as they are to experience the region. This tourist is motivated experientially to discover, to taste, to learn, and to consume wine and wine lifestyles. The landscape as backdrop is important, but it is the experience of the landscape that is essential. Targeting of the Okanagan wine tourist has concentrated on the cultural tourist more generally, and there have been strong information and marketing linkages made between cuisine, gastronomy, aboriginal culture, art, and the wine industry resources. According to visitors surveyed in 1998, more integration of a wider variety of community and cultural resources is desirable. By 2008, a variety of cultural resources had been networked and explored by tourists. With four destination ski resorts and over 40 golf courses in the region, target marketing has also aimed at drawing leisure tourists to winery destinations – especially in off-season periods.

According to Getz and Brown (2006b), the provision of a diversity of quality, cultural, and outdoor recreational products are the most important critical success factors for wine tourism regions. In terms of these observations and the stage model presented, the Okanagan wine region may have reached that optimal state when quality cultural products are integrated into the industry and the natural products of the industry are maintained. Presently, there is a wide array of visitor flows between wineries, service locations, and recreational attractions. A strong variety of partnerships, alliances, and networks exist to maintain those visitor flows. There is also a strong regional reliance on the wine tourism brand for other forms of tourism, significant events, and festivals. In all of these regards, the Okanagan wine tourism region is highly competitive and has evolved to stage 3 of the model. Sustainability of this momentum appears entirely dependent on maintaining the quality of the rural aesthetic and rural environment as its foundation. Given the fragility of that environment, the rapid nature of growth and a lack of cultural integration on a community basis achieving economic, ecological, and sociocultural sustainability will be a challenging task in the Okanagan.

Conclusion

Activity-based approaches to tourism sustainability reflect the resource needs of the industry to sustain economic, environmental, and sociocultural capital. This approach recognizes that there are limits to growth within a dynamic carrying capacity that tend to be cyclical and reactive to the contexts of local communities as well as to wider political economic trends. Stage models, such as the one presented in this chapter that indicates the evolution of tourism destinations, provide a start in the development of a framework and in the search for analytic tools to examine the relative ability to cope with various forms of tourism development. A community-based assessment of sustainability also involves the empowerment of hosts and the protection and preservation of ecological and cultural heritage on a local level. A useful, long-term evaluation of tourism, wine, and sustainability would reflect the dynamic integration of the tourism industry, the wine industry, and the economic, ecological, and sociocultural capital that provides the basis for all concerned. This holistic assessment is beyond the scope of this chapter, but research in both the Okanagan and the Niagara regions clearly indicates the need to understand these dynamic processes more fully.

Both the Niagara region and the Okanagan region are within stage 3 of the model. They have experienced rapid growth in visitor numbers that is now affecting tourism experiences – sometimes in a negative way through crowding and congestion. Both have well-developed marketing partnerships and networking and well-established brands and competitive tourism development. However, in both areas, wineries are planning expansion, and there is encroachment by other forms of development (housing, other industry) that threatens rural aesthetics and agricultural land use. Balancing growth with sustainability is challenging for both areas. Dodd and Beverland (2001, 16) suggest that winery regions develop from rural destinations into urban destinations as the regions become highly developed. It is worrying to think that the Okanagan Valley could become a continuous ribbon development and traffic jam, and that Niagara could become a Toronto suburb. Strict planning regulation is needed to prevent this scenario.

The model presented in this chapter and some of the empirical findings in the two case studies raise interesting questions to be explored in more depth in future research. These include: What are the catalysts that facilitate changes through the stages in the model, and what are the barriers to development and change? Perhaps, the most important applied questions to ask are what level of development is optimal for competitiveness and sustainability in tourism destinations, and how is sustainability linked to competitiveness of such regions? Insights from these two case studies have partially answered some of these questions, but much more work needs to be done.

References

Adams, M. Q. 2006. Nautical wine tourism: a strategic plan to create a nautical wine trail in the Finger Lakes Wine Tourism Region of New York State. In *Global wine tours. Research, management and marketing,* ed. J. Carlsen and S. Charters, 227–241. Wallingford Oxon, UK: Cabi

Alant, K. and J. Bruwer. 2004. Wine tourism behavior in the context of a motivational framework for wine regions and cellar doors. *Journal of Wine Research* 15(1): 27–37.

Association of British Columbia Grape Growers. 1984. *Atlas of suitable grape growing locations in the Okanagan and Similkameen Valleys, British Columbia.* Canada: Agriculture Canada.

B.C. Stats. 2007. *Population Forecasts.*

Beames, G. 2003. The rock, the reef and the grape: the challenges of developing wine tourism in regional Australia. *Journal of Vacation Marketing* 9 (3): 205–212.

Bell, D. 2008. Destination drinking: toward a research agenda on alcotourism. *Drugs: Education, prevention and policy* 15 (3): 291–304.

Belliveau, S., B. Smit and B. Bradshaw. 2006. Multiple exposures and dynamic vulnerability: evidence from the grape industry in the Okanagan Valley, Canada. *Global Environmental Change* 16 (4): 364–378.

Brown, G. and D. Getz. 2005. Linking wine preferences to the choice of wine tourism destinations. *Journal of Travel Research* 43: 266–276.

British Columbia Wine Institute 2009. 2006. *Annual Report.* Kelowna, British Columbia.

Britton, S. G. 1991. Tourism, capital and place: towards a critical geography. *Environment and Planning D: Society and Space* 9: 451–78.

Butler, R. W. 1980. The concept of a tourism area cycle of evolution: Implications for the management of resources. *The Canadian Geographer* 24: 5–12.

Butler, R. W. 1999. Sustainable tourism: The state of the art review. *Tourism Geographies* 1 (1): 7–25.

California Wine Institute. 2008. *California wine industry statistical highlights.*

Carlsen, J. 2004. A review of global wine tourism research. *Journal of Wine Research* 15 (1): 5–13.

Carlsen, J. and S. Charters. 2006. *Global wine tourism research, management and marketing.* Wallingford Oxon, UK: Cabi

Carlsen, J. and R. Dowling. 2001. Regional wine tourism: a plan of development for Western Australia. *Tourism Recreation Research* 26 (2): 45–52.

Carmichael, B. A. 2001. Competitive and sustainable wine tourism destinations. *Proceedings of the Travel and Tourism Research Association Canadian Chapter Conference.* Niagara Falls, Canada October 14–16, 245–249.

Carmichael, B. A. 2005. Understanding the wine tourism experience for winery visitors in the Niagara Region, Ontario, Canada. *Tourism Geographies* 2: 185–204.

Cartier, C. and A. Lew. 2005. *Seductions of place. Geographical perspectives on globalization and touristed landscapes.* New York: Routledge.

Charters, S. and J. Fountain. 2006. Younger wine tourists: a study of generational differences in the cellar door experience. In *Global wine tourism. Research, Management and marketing,* ed. J. Carlsen and S. Charters, 153–160. Wallingford Oxon, UK: Cabi

Charters, S., J. Fountain and N. Fish. 2009. You felt like lingering…Experiencing "real" service at the winery tasting room. *Journal of Travel Research* 48 (1): 122–134.

Charters, S. and S. Pettigrew. 2005. Is wine consumption an aesthetic experience? *Journal of Wine Research* 16 (2): 121–136.

Correia, L., M. J. Passos and S. Charters. 2004. Wine routes in Portugal: A case study of the Bairrada wine route. *Journal of Wine Research* 15 (1): 15–25.

Crouch, G. I. and J. R. B. Ritchie. 1999. Tourism competitiveness and social prosperity. *Journal of Business Research* 44 (3): 137–142.

Dodd, T. and M. Beverland. 2001. Winery tourism life-cycle development: A proposed model. *Tourism Recreation Research* 26 (2): 11–21.

Dowling, R. 1998. Wine tourism: Perfect partners. *Australian and New Zealand Wine Industry Journal* 13 (3): 307–309.

Dredge, D. 1999. Destination planning and design. *Annals of Tourism Research* 26 (4): 772–791.

Fagence, M. 1995. Changing paradigms of orthodoxy: The case of spatial models in tourism planning. *Les Cahiers De Tourisme Aixen-Province*: Centre Des Hautes Etudes Touristiques.

Gammack, J. G. 2006. Wine tourism and sustainable development in regional Australia. In *Global wine tourism, research, management and marketing,* ed. J. Carlsen and S. Charters, 59–66. Wallingford Oxon, UK: Cabi.

Getz, Donald. 2000. *Explore wine tourism: management, development and destinations.* New York: Cognizant Communication Corporation.

Getz, Donald and G. Brown. 2006a. Benchmarking wine tourism development: The case of the Okanagan Valley, British Columbia, Canada. *International Journal of Wine Marketing* 18 (2), 78–97

Getz, D. and G. Brown. 2006b. Critical success factors for wine tourism regions: A demand analysis. *Tourism Management* 27: 146–158.

Goeldner, C. R., J. R. B. Ritchie and R. W. McIntosh. 2000. *Tourism - principles, practices, philosophies.* New York: Wiley.

Gunn, C.A. 1988. *Tourism planning.* 2nd Edition. New York: Taylor & Francis.

Grape Growers of Ontario. 2004 factsheet. St. Catharines, Ontario: Grape Growers of Ontario.

Grape Growers of Ontario Website (accessed May 27 2010).http://www.grapegrowersofontario.com/thevineyards/niagara.html.

Hall, C.M. 2003. Biosecurity and wine tourism: Is a vineyard a farm? *Journal of Wine Research* 14 (2–3): 121–126.

Hall, C. M., E. Sharples, B. Cambourne and N. Macionis (eds.). 2000a. *Wine tourism around the world: Development, management and markets*. Oxford: Butterworth-Heinemann.

Hall, C. M., G. Johnson and R. Mitchell. 2000b. Wine tourism and regional development. In *Wine tourism around the world: development, management and markets*, ed. C. M. Hall, E. Sharples, B. Cambourne and N. Macionis, 196–225. Oxford: Butterworth-Heinemann.

Hall, C. M. and N. Macionis. 1998. Wine tourism in Australia and New Zealand. In: *Tourism and recreation in rural areas*, ed. R. Butler, M. Hall and J. Jenkins, 197–224. Chichester: Wiley.

Hall, C. M. and R. Mitchell. 2001. Wine tourism in the Mediterranean: a tool for restructuring and development. *Thunderbird International Business Review* 42 (4): 445–465.

Hall, C. M., L. Sharples, and R. Mitchell. 2003. Consuming places: The role of food, wine and tourism in regional development. In *Food Tourism*, ed. C. M. Hall, L. Sharples, R. Mitchell, N. Macionis and B. Cambourne, 25–59. Melbourne: Butterworth Heinemann.

Hashimoto, A. and D. Telfer. 2003. Positioning an emerging wine route in the Niagara region: Understanding the wine tourism market and its implications for marketing. *Journal of Travel and Tourism Marketing* 14 (3/4): 61–76.

Howley, M. and J. van Westering. 2007. Developing wine tourism: A case study of the attitude of English wine producers to wine tourism. *Journal of Vacation Marketing* 14 (1): 87–95.

Jaffe, E. and H. Pasternak. 2004. Developing wine trails as a tourist attraction in Israel. *International Journal of Tourism Research* 6: 237–249.

Johnson, R. and J. Bruwer. 2007. the balancing act between regionality and American Viticultural Areas (AVAs). *Journal of Wine Research* 18 (3); 163–172.

Luciw, J. 2010. Public sneaks peek at massive Mission Hill expansion. *Kelowna Capital News,* March 19 2010.

Macionis, N. 1998. Wineries and tourism: perfect partners or dangerous liaisons? In *Wine tourism perfect partners, Proceedings of the First Australian Wine Tourism Conference*, ed. R. Dowling and J. Carlsen, 35–44. Canberra: Bureau of Tourism Research

Macionis, N. and B. Cambourne. 1998. Wine and food tourism in the Australian capital territory: Exploring the links. *International Journal of Wine Marketing* 10 (3): 5–16.

Major Wineries pull out of Ontario Wine Council. *The Standard*, St. Catherine's December 9th 2009 accessed May 28th 2010 http://www.stcatharinesstandard.ca/PrintArticle.aspx?e=2182324

Martin, N. 2007. Sustainable winemaking Ontario –Environmental best practices for wineries. In *Sustainable winemaking Ontario: An environmental charter for the wine industry.* St. Catharines, Ontario: Wine Council of Ontario.

McGuire, D. W. 1993. *The political economy of the grape and wine industry in British Columbia and the impact of the Free Trade Agreement.* Kelowna, British Columbia: Okanagan University College, Department of Economics.

Mitchell, R. and C. M. Hall. 2004. The post-visit consumer behavior of New Zealand winery visitors. *Journal of Wine Research* 15 (1): 39–49.

Official travel guide to wineries of Ontario. 2007. St. Catharines, Ontario: Wines of Ontario

O'Neil, M. A. and A. Palmer. 2004. Wine production and tourism. Adding service to a perfect partnership. *Cornell Hotel and Restaurant Administration Quarterly*. 45 (3): 269–284.

Oppermann, M. 1993. Tourist space in developing countries. *Annals of Tourism Research* 20: 535–556.

Peters, G. L. 1997. *American winescapes. The cultural landscapes of America's wine country*. Boulder, Colorado: Westview Press.

Pearce, D. 1995 *Tourism today: A geographical analysis.* New York: Longman.

Pine, J. and J. Gilmore. 1999. *The experience economy*. Boston: Harvard Business School Press.

Poitras, L. and D. Getz. 2006. Sustainable wine tourism: The host community perspective. *Journal of Sustainable Tourism* 14 (5): 425–448.

Ravenscroft, N. and J. van Westering. 2001. Wine tourism, culture and the everyday: A theoretical note. *Tourism and Hospitality Research* 3 (2): 149–162.

Ritchie, J. R. B., and G. I. Crouch. 2000. The competitive destination: A sustainability perspective. *Tourism Management* 21: 1–7.

Robinson, J. 1999. *The Oxford companion to wine*. Second edition. New York: Oxford University Press.

Schreiner, J. 2009. *The wineries of British Columbia*. Vancouver: Whitecap Books.

Senese, D. 2008a. Turismo enogastronomico Canadese, la nuova frontiera del turismo combinata all'ecoturismo e al turismo culturale. *Biteg*: *Congresso Internatzionale Sul Turismo Enogastronomico*. Italia: Saint Vincent.

Senese, D. 2009. Agritourism, amenity resources and organic growth poles. *Canadian Association of Geographers, Annual Meeting*. Ottawa, May 2009.

Senese, D. 2008b. *The proposed national park in the South Okanagan and Lower Similkameen: integrating regional tourism into national parks*. Penticton BC: Parks Canada.

Senese, D. and W.Wilson. 1998. *Wine tourism in the Okanagan*. Kelowna: British Columbia Wine Institute.

Sharples, L. 2002. Wine tourism in Chile…A brave new step for a brave new world. *International Journal of Wine Marketing* 14 (2) 43–58.

Skinner, A. M. 2000. Napa Valley California: A model of wine region development. In *Wine tourism around the world: development, management and markets*, ed. C. M. Hall, E. Sharples, B. Cambourne and N. Macionis, 283–296. Oxford: Butterworth-Heinemann.

Sparks, B. 2007. Planning a wine tourism vacation? Factors that help to predict tourist behavioural intentions. *Tourism Management* 28 (5): 1180–1192

Statistics Canada Agricultural Profile 2001, 2006.

Tassiopoulos, D., N. Nuntsu and N. Haydam. 2004. Wine tourists in South Africa: A demographic and psychographic study. *Journal Wine Research* 15 (1): 51–63.

Tassiopoulos, D. and N. Haydam. 2006. Wine tourists in South Africa: a demand-side study. In *Global wine tourism. Research, management and marketing,* ed. J. Carlsen and S. Charters, 141–152. Wallingford Oxon, UK: Cabi.

Taylor, R. 2006. Wine festivals and tourism: Developing a longitudinal approach to festival evaluation. In *Global wine tourism. Research, management and marketing,* ed. J. Carlsen and S. Charters, 179–195. Wallingford Oxon, UK: Cabi.

Telfer, D. 2000. Tastes of Niagara: Building strategic alliances between tourism and agriculture. *International Journal of Hospitality and Tourism Administration*, 1 (1): 71–88.

Telfer, D. 2001a. Strategic alliances along the Niagara wine route. *Tourism Management* 22: 21–30.

Telfer, D. 2001b. From a wine tourism village to a regional wine route: An investigation of the competitive advantage of embedded clusters in Niagara, Canada. *Tourism Recreation Research* 26 (2): 23–33.

Tourism British Columbia, Research and Planning 2009. *Okanagan Valley wine consumer research study 2008 results.* Victoria: Tourism British Columbia

Urry, J. 1990. *The tourist gaze: Leisure and travel in contemporary societies*. London: Sage.

Wargenau, A. and D. Che. 2006. Wine tourism development and marketing strategies in southwest Michigan. *International Journal of Wine Marketing* 18 (1): 45–60.

Wilkins, M. and C. M. Hall. 2001. An industry stakeholder SWOT analysis of wine tourism in the Okanagan Valley, British Columbia. *International Journal of Wine Marketing* 13 (3): 77–78.

Williams, P. W. 2001. Positioning wine tourism destinations: An image analysis. *International Journal of Wine Marketing* 13 (3): 42–58.

Williams, P. and K. Dossa. 2003. Non-resident wine tourist markets: Implications for BC's emerging wine tourism industry. *Journal of Travel and Tourism Marketing* 14 (3/4): 1–34.

Williams, P. W. and J. Kelly. 2001. Cultural wine tourists: Product development considerations for British Columbia's resident wine tourism market. *International Journal of Wine Marketing* 13 (3): 59–70.

Wine Business Magazine. 2009. Wines and Vines Directory 2009.

Wine Council of Ontario. 1998 factsheet St. Catharines, Ontario: WCO.

Wine Country Ontario, Your official 2010 year round travel guide. St. Catharines, Ontario: Wines of Ontario.

Yuan, J., S. C. Jang, A. C. Liping, A. M. Morrison and S. J. Linton. 2006. Analysis of motivational and promotional effects of a wine festival. In *Global wine tourism. Research, management and marketing,* ed. J. Carlsen and S. Charters, 196–208. Wallingford Oxon, UK: Cabi.

Northern California Through and Economic Geographer's Lens

10

Greig Tor Guthey

Abstract

Natural endowments, entrepreneurship, and numerous watershed moments have all been important dimensions of the Californian wine industry. Less acknowledged, but of equal importance, is the broader context of industrial development and regional change that has been necessary for the growth of the industry. From what would be regarded today as industrial best practice to land conservation planning, a host of institutional, industrial, and economic shifts undergird this industry's development. The challenge has been to capture the effect of these changes. It has long been a principle of economic geography that proximate interaction of producers in industrial districts generated innovation and economic growth. A general theory of regional development reinforced this view with an understanding of untraded interdependencies of producers in regional economies. That shift in economic geographic theory enabled more refined understanding of the industrial practices of subsectors of the food system including wine production. Consideration of California's experience provides additional insight into how regional innovation systems can support and encourage wine producers to make changes to their regional conventions, and as a result, to the organization of production and the quality of their products.

Introduction

To explain the success and growth of the northern California wine industry, observers often turn to nature, entrepreneurship, and significant historical events. Scholars have long suggested that nature gifted the region with an ideal climate or geology for making dry wines, e.g., Winkler (1938), and Swinchatt and Howell (2004). Other observers point to significant individuals such as Robert Mondavi, among others (Emert 2003). Some firms may have imported particularly good rootstock. And popular culture often focuses on the 1976 Paris Tasting in which two northern California wineries won accolades from a French tasting panel. These are all important dimensions of the Californian wine industry.

Less acknowledged is the regional economic context considered in this chapter. From winemaking best practice to land conservation planning, a host of institutional, social, and economic shifts undergird the growth of the industry. Similar shifts in other industries have generated renewed interest in the region as a

G.T. Guthey (✉)
Department of Liberal Studies, California State University San Marcos, San Marcos, CA, USA
e-mail: gguthey@csusm.edu

P.H. Dougherty (ed.), *The Geography of Wine: Regions, Terroir and Techniques*,
DOI 10.1007/978-94-007-0464-0_10, © Springer Science+Business Media B.V. 2012

unit of analysis in economic geography. But attention to the wine industry as a regional economic phenomenon is fairly recent and growing. A theoretical discussion tracing recent debate about regional development and conventions in economic geography to wine research appears in section one. Section two considers the development of the California wine industry in long-term regional perspective, thereby building on recent work in the field. While recent analysis, e.g., Sanchez-Hernandez et al. (2010), examines innovation in wine industries as shifts between "worlds of production," section three argues that similar consideration of the California industry reveals a great deal of complexity and dynamic tension between productive worlds, rather than shifts from one to another.

Wine, Conventions, and Economic Geography

Few wine geographers address the role of regional economic forces; nature figures more prominently. Exceptions include Dion (1959) who suggests that politics and culture shape our understandings of wine, and Moran (1994) who argues that supportive legislation points to the power of producers in shaping our understandings of terroir and of nature. Likewise, both Barham (2003) and Gade (2004) argue that terroir is a social construction of nature. And Guthey (2008) argues that industrial practices of producers are the main stimulus to industrial upgrading in northern California. Nevertheless, as Moran (1994) argues, geographers and viticulturalists have often turned to nature to explain the spatial location of vineyards and wineries reflecting a clear environmental determinism. In California, the state's flagship wine region has largely been understood as the result of nature (Swinchatt and Howell 2004). This emphasis is apparent in both general understanding of terroir as soil and in the use of mesoclimate zones as the basis for deciding where to plant different varieties of grapes. As Fig. 10.1 illustrates, there are three main mesoclimatic zones that correspond with different varieties of grapes in both Napa and Sonoma Counties.

For a number of years, the challenge for economic geographers has been to theorize the process of regional development in any industry, let alone the wine industry. Making connections to Joseph Schumpeter, Karl Marx, and David Ricardo, Page and Walker (1991) provided an argument for the regional development of the Corn Belt of the United States and identified an expanding regional "division of labor" as a necessary causal factor influencing the growth of Midwest agriculture. As they write, "Midwestern industries developed themselves through an evolution of productive capabilities that owed nothing to nature and everything to regional social arrangements, human capabilities, technological advances, and divisions of labor yielding powerful external economies" (Page and Walker 1991). In other words, industrial growth is conditioned by increasing productivity, building home markets, improving techniques across a range of industries, and thereby reorganizing production.

Another thread in the analysis of regional development is the interaction and knowledge exchange which occurs in specific places, as in, e.g., Annalee Saxenian's (1995) research in Silicon Valley. As knowledge spreads through a region, producers adopt best practices and learn from one another. Saxenian calls this sort of interaction industrial culture and finds that the structure of the firm as well as the absence of previous industrial models in the state enabled an innovative industrial cultural in Silicon Valley. Additionally, culture spills over into the creation of a place broadly recognized for its products and able to generate its own cultural production alongside its products. Thus, Saxenian found that engineers in Silicon Valley felt they were working for the region, not any particular firm.

Research on the mechanisms for the emergence of this sort of regional industrial culture does not completely take off until Michael Storper (1997) applies French conventions theory to economic geography. Developed by French economists interested in "economies of quality" and partially based in research on the French wine industry (Touzard 2000), conventions theory seeks to understand the economic practices of firms as a collective process. Storper's book *The Regional World* provides a theoretical basis for differentiating what he calls "worlds of production." In conventions theory, conventions are broadly shared social practices and agreements. Worlds of production are distinct combinations of conventions conditioned both by endogenous factors and by the particular markets with which a region is integrated. Thus, different constellations of conventions create different opportunities for growth and innovation in different sorts of industries.

Storper proposes that there are four essential types of production worlds shaped by varying degrees of

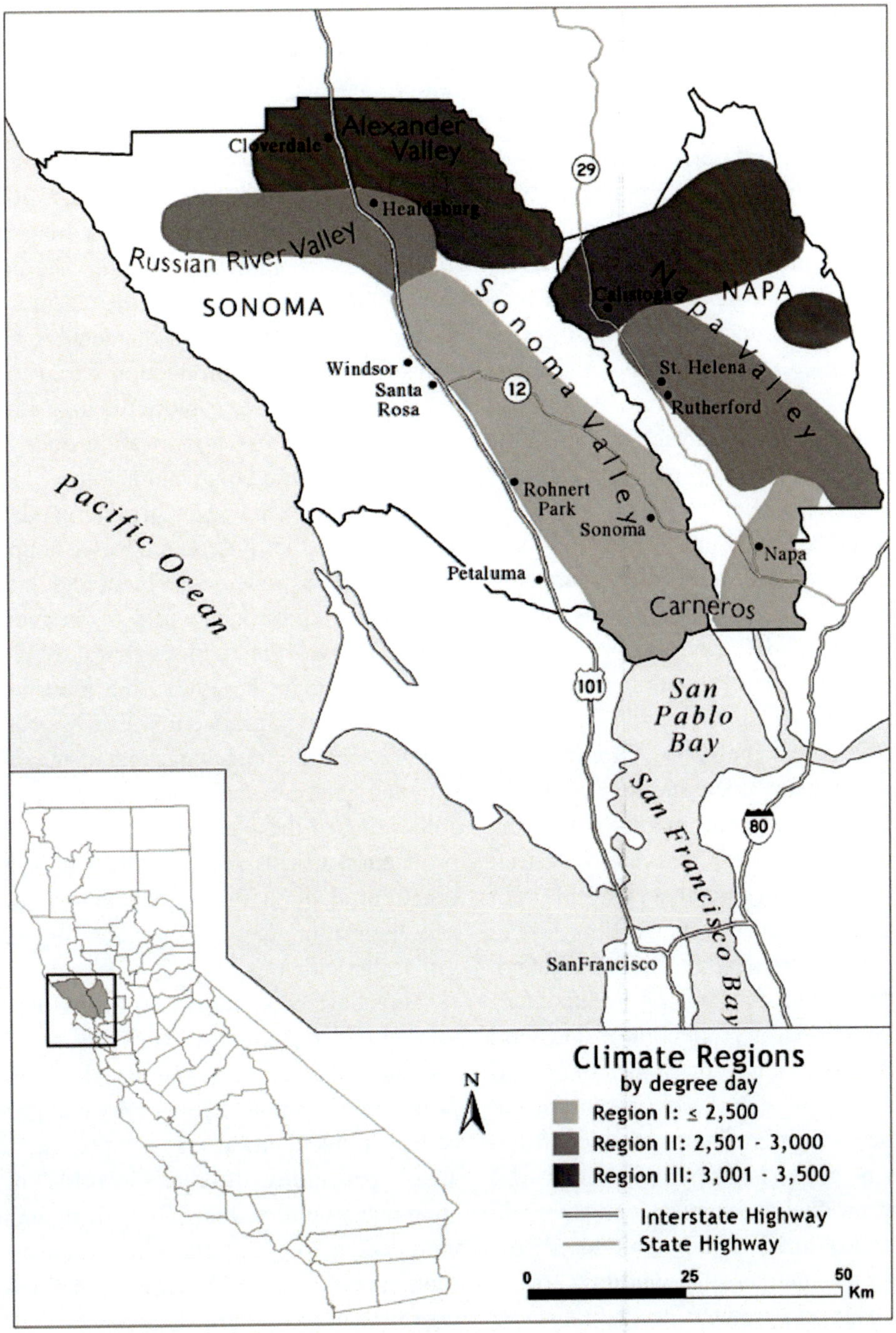

Fig. 10.1 The main climate regions of Napa and Sonoma counties. Cartography cartography by Roseli Ilano, Department of Geography, University of California, Berkeley (Source: Johnson 1994)

uncertainty. In the Marshallian or Interpersonal World, producers and consumers rely on close coordination to produce specialized products and convey their particular qualities. In the Market World, producers make small runs of standardized products. In the world of innovation and intellectual resources, producers face high levels of uncertainty regarding price and quality, and thus, close coordination among and between consumers and producers is necessary. Finally, in the Industrial World, standardized products are built according to established criteria and sold in mass markets with very little uncertainty.

Uncertainty lies at the heart of Storper's worlds of production approach because producers attempt to manage uncertainty by developing different industrial conventions. In general, there are six sorts of conventions. These include commercial conventions which involve price determinations, domestic conventions which involve attachments to place, public conventions which involve trademarks and brands, industrial conventions which concern efficiency and reliability, and civic conventions which involve health and safety concerns (Murdoch and Miele 1999; Sanchez-Hernandez et al. 2010). Murdoch et al. (2000) further suggest that there appear to be ecological conventions which involve environmental concerns. Assembled into worlds of production, these conventions constitute "action frameworks" that support production and lead to innovation.

Drawing on Storper's framework and related work with the theory of conventions, significant research has taken place on the agro-food system starting with a group of researchers in the UK who mapped portions of Storper's framework onto the food sector in a "worlds of food" approach (Murdoch and Miele 1999; Murdoch et al. 2000; Parrot et al. 2002; Morgan et al. 2006). As in Storper's theory, there are four worlds of food: the interpersonal world, the market world, the world of intellectual resources, and the industrial world. And, their analysis suggests that conventions in the food system are a form of collective action, e.g., in a commodity chain or regional innovation system, attached to particular understandings of product quality. Thus, different worlds of production in the food system involve different kinds of conventions and product qualities as a result of the kinds of actors active in each one and the sorts of uncertainty they confront. Corporations may be seen as occupying the industrial world of food and producing for mass markets while also often undermining smaller, more craft-based producers operating in the interpersonal world by appropriating their language and products as alternatives to industrial food become more popular, and thus, ready for scaling up. There is necessarily some overlap and middle ground between food worlds that have become a site of conflict in the food system.

Storper's work has also inspired a number of economic geographers studying wine industries both as part of the food system and as a separate category of analysis. Gade (2004) used elements of conventions theory to understand what he calls a process of patrimonialization in southwestern France while Guthey (2004, 2008) used Storper's framework to describe and understand agro-industrial conventions and their relationship to industrial upgrading and innovation in the California wine industry. Ponte (2009) applied conventional analysis to the South African wine industry's insertion in global value chains. And, most recently, Sanchez-Hernandez et al. (2010) push the debate further by developing "a long-term perspective" on conventions. By tracing the conventions of particular regions overtime, they argue, scholars may be able to understand and characterize the reasons for transition from one production world to another and, more importantly, to capture the long-run evolution of regions as a process of innovation since changing conventions constitute innovation.

In their case study, Sanchez-Hernandez et al. (2010) find that Castile y Leon in western Spain has gone through three transitions. First, the industry shifted from an interpersonal world to an industrial world. With the development of the first designation of origin in the region in the 1960s, the industry shifted to a market world of production leading it to become the second most popular wine region in Spain. Today, a growing interpersonal world is increasingly evident. According to these authors, the above shifts between production worlds are contingent upon the organization of production, the regional innovation system, and new legislation, namely, an increasing number of origin designations.

Thus have economic geographers moved to a more refined understanding of worlds of production within specific industries and regions. At the outset, it was a principle of economic geography that proximate interaction of linked producers in industrial districts generated innovations. Storper's worlds of production approach amplified this view with an understanding of untraded interdependencies of producers in regional economies generally. Then more refined understanding of the industrial practices of subsectors of the food system became possible. Now, we find economic geographers developing research on the economic practices of wine producers in specific places where production organization, legislation, and a regional innovation system support wine producers in a process of regional change. The next section seeks to add to the debate about wine and worlds of production by exploring and developing a long-term perspective on the Californian wine industry. Data come from archival sources, the secondary literature, and interviews that were written up in Guthey (2004).

Towards a Long-Term Perspective on California Wine Production

The story of this industry's transformation in the late twentieth century often includes a transition in the 1960s. For example, Robert Mondavi retells his view of the mid-twentieth century industry as backwards and industrial prior to his epiphany about the American style of winemaking while on vacation in Europe; he establishes his own winery with a new emphasis on smaller scale and varietal grapes in 1966 (Mondavi 1998). The argument in this section is that a long-term perspective on the northern California industry reveals a more complicated industrial evolution. Production, markets, and institutions appear to have been configured in particular ways at different periods of time throughout the history of the industry. However, we find a good deal of dynamism between production worlds so that innovations sometimes emerge out of what might be termed an industrial world of production; while in others, they emerge out of an interpersonal world. Today, we find firms occupying quite complex positions vis-à-vis the worlds of production approach.

The Nineteenth Century Californian Wine World

Wine production in the nineteenth century California was a complex mixing of Storper's ideal typical productive worlds. Wine historians point out that producers were selling wine in Japan, England, Germany, Denmark, Mexico, Central America, and Canada which suggests a certain external demand and industrial scale. At the same time, the wine industry operated on the basis of several endogenous factors including significant local capital formation, favorable regional social relations, supportive state institutions, and a diverse and numerous set of grape growers and winemakers.

The initial extractive economy brought on by the Gold Rush provided an immediate and sizable amount of local capital that was then invested in a range of industries and regions within California (Brechin 1999; Walker 2001). Napa and Sonoma County were in the thick of this investment pattern. Sonoma was the leading agricultural county in the state in the nineteenth century. Wheat, wine grapes, stone fruits, and truck crops were all locally important crops in Napa County. Wine grapes were well established by the 1860s in Sonoma County, mainly driven by local consumption as suggested by Hutchison (1984) who writes that "the boisterous, open-handed new Californians of the Gold Rush era were drinkers."

It is true that the wine industry began in Southern California, particularly in the San Gabriel Valley and near Anaheim where Charles Kohler and John Frohling started their vineyard in 1854; nearly every farmer and landowner was said to have caught "vine fever" during this period (Carosso 1951). But southern California vineyards were decimated by a vine disease known as Pierce's disease in the 1880s, while northern California was relatively unaffected. Thus, Sonoma County became the industry's center. Napa County increased in importance later as growers spun off from Sonoma County wineries, eventually surpassing Sonoma in the early twentieth century.

We often remember the signature wineries of this period: The Buena Vista Viticultural Society and Italian Swiss Agricultural Colony are perhaps the most famous because they exist to this day, though under different ownership. Started by Agoston Haraszthy in 1857, Buena Vista was underwritten by San Francisco banker William Ralston along with nine other silent urban partners. San Francisco banker and grocer Andrea Sbarbaro created the Italian–Swiss Agricultural Colony through his banking connections in 1881 (Pinney 1989). De Turk winery, started by Isaac de Turk in the 1860s, became the leading producer in Sonoma County. However, wine production involved hundreds of rural producers and urban investors. There were between 278 and 500 other grape growers and winemakers in Sonoma County in the 1880s, according to the San Francisco *Merchant* and the state board of viticulture commissioners. Of these, 24 are listed as also winemakers; 23 as winemakers and distillers (Peninou 1998).

In short, both grape growers and wineries clustered in the Napa and Sonoma region. These producers included prominent and lesser citizens, émigrés, senators, and dairy ranchers. George Hearst had a vineyard in Sonoma in 1885. Nevada Silver King and Senator James Graham Fair owned 9,000 acres of land near Petaluma including a vineyard and winery in 1891. By 1893, there were approximately 932 individual vineyards in Sonoma County (Lemmon 1893). Starting in the 1860s, producers like Charles Krug, Jacob Schram, and Gustav Niebaum followed suit in Napa County, where there were 175 wineries by 1886 (Heintz 1999).

One could attribute the proliferation of wineries and grape growers to individual initiative, but the federal government and the state of California supported these efforts in a number of ways. In 1861, as part of a broader state effort to encourage agriculture, the state legislature authorized the governor to appoint three commissioners to investigate the best ways to "promote and encourage the growth of the grape vine in California" out of which the infamous "Father of the California Wine Industry," Agoston Haraszthy, embarked on his tour of European wine districts and imported many wine grape cuttings (Carosso 1951). The California Assembly instituted a State Board of Viticultural Commissioners to conduct research and promote the state industry fervently across the nation. In the same legislative bill, the assembly established a new department of viticulture at the University of California, itself a land grant university established under the federal Morrill Act of 1862. UC Berkeley professor and dean of the College of Agriculture Eugene W. Hilgard lobbied the state legislature to form the board of viticultural commissioners and "instructed the Board of Regents to provide instruction and research in viticulture and enology" (Carosso 1951; Amerine 1981). A number of state institutions supported the nineteenth century wine development.

Upstream from the wineries and grape growers, these public institutions quickly set about to improve industry practices. Historical accounts suggest that most wine was of questionable quality in the early years. One wonders how the state's wineries won any international competition; they lacked refrigeration, knowledge of sanitation, and climate, for example, but many did. Yet, quality concerned Hilgard. The son of a wine entrepreneur in Missouri, Hilgard arrived at UC Berkeley in 1875 and quickly set about to rationalize wine production. As he wrote:

> among the first necessities of the present situation in California wines in the world's market, is the establishment of more definite qualities and brands, resulting from the definite qualities of each of the prominent grape varieties, and of their influence upon the kind and quality of wine in blending before, or as the case may be, after fermentation; of the treatment required by each in the cellar, during the time of ripening; and finally, of all the differences caused by difference of location, climate, etc., as well as by different treatment of vines themselves (Amerine 1981).

His words apply just as well today. Indeed, the oft-heard phrase today that "quality begins in the vineyard" was exactly what Hilgard had in mind in the nineteenth century. The other half of his thinking was that quality emerged through processing which is the basis of much agro-industrialization and a big driver of regionalization. Throughout his tenure, he aggressively and publicly campaigned for improvements in grower and winemaker practices. He conducted and disseminated research on phylloxera which plagued the industry in the 1880s and 1890s. He studied grape varieties in California to determine their qualities and characteristics for winemaking and made his findings available to the industry (Pinney 1989). He examined the optimal temperatures for fermentation of white wine. His research was part of the mission of the first College of Agriculture, a mission continued by the professors of University of California professors during the twentieth century, some of whom can trace their intellectual lineage straight back to Hilgard (Walker 2004).

Downstream from grape growers and wineries, a web of processors, shippers, merchants, and distributors provided essential services. Producers grew grapes on estate and contracted vineyards, processed the juice, and usually shipped in bulk and not under refrigeration to wine merchants in San Francisco and other cities; the distribution side was well developed in San Francisco where between 50 and 60 wine merchants were storing millions of gallons of wine in the city's wine vaults and cellars by the late 1880s. These merchants maintained the wine vaults, bottled the wine themselves, or sold it directly to consumers out of wine casks. Major producers worked both sides of the chain making wine in the countryside, contracting with smaller producers, and handling distribution in major cities as independent wine merchants of sometimes adulterated wine (Pinney 1989). Southern California producer and former Los Angeles mayor Benjamin Wilson provides an example. He used wine agents in the early 1860s in San Francisco, but in 1865, as markets for his wine improved, he established his own branch house in the city to better control his products (Pinney 1989).

While some supplies had to be imported, such as corks and eastern oak used for cooperage, a wider division of labor supported northern California producers. At the start, the merchants sold wine out of the cask, while some producers packaged wine in used imported wine bottles. Yet producers were not just growing grapes in isolation, they were collaborating and, thus, generating new conventions, organizations, and techniques. For example, by 1863, Kohler and Frohling, and Taylor & Co. started the Pacific Glass Works to supply wineries' container needs (Berger 1990; Carosso 1951). In addition, "by the 1870s, there was a specialized branch of the state's cooperage industry devoted [to making barrels for] the wine trade" (Pinney 1989). Machine makers developed "improved grape crushers and stem separators." By the mid-1870s, continuous feed stemmer-crushers became available and, later, the continuous screw press. In 1887, J. M. Curtis introduced a type of wine pasteurizer. There were other specialized tools made for vineyards such as the Napa Plow Company's Vineyard Plow, and specialized service providers like J. L. Heald in Crockett who advised, designed, and provided equipment for many leading wineries of the time.

In addition to this secondary economic activity, wine production spun off new ventures, a classic practice associated with innovative clusters. Scatena & Company was started by Martino Scatena in Healdsburg in Sonoma County. In 1902, he apparently left this business to start his own winery and establish a wine depot in San Francisco. The company changed names around 1906 to the Roma Wine Company and relocated to Lodi (Peninou 1998). His cousins apparently continued their own winery. In Cloverdale, Felicien Vadon started up his own winery after working for Leland Stanford's winery in Palo Alto. J. Gundlach & Co. in the Sonoma Valley started out as a partnership between Jacob Gundlach and Emil Dresel. The two of them implemented advice they received from Agoston Haraszthy who encouraged them along with a number of other early producers to begin making wine. Thus were there important producer interactions and mutual assistance during this period.

Finally, while many producers were small, suggesting a sort of interpersonal wine world, by the end of the century, wine production also became big business and wine quality improved measurably. Started in 1880 as a merger of seven of the state's largest wine dealers, the California Wine Association absorbed many smaller wineries in the late nineteenth century; its market power spurred on similar consolidation among grape growers who started the California Wine Makers' Association. The two eventually collaborated and CWA became the leading producer in the state until Prohibition; CWA processed grapes at its Winehaven facility in Richmond after the 1906 San Francisco earthquake and fire, a facility considered as the largest winery in the world at the time. Its production practices increased product quality through evaluation of wine as part of wine purchases, by establishing specific blending programs for particular brands and through chemical evaluation of its wine. Its wine was so prominent that glasses in London were etched with a redwood tree, the insignia of "the Big Tree" label (Pinney 1989).

In short, we find nineteenth century evidence for a variety of productive worlds described in the literature. While CWA signals a significant industrial world of production in northern California – it controlled 84% of state wine production by 1918 (Peninou and Greenleaf 1954 cited in Amerine 1983) – smaller producers continued to attempt to produce their own brands of wine, which suggests that both interpersonal and market worlds of production persisted as well. About 187 producers made wine in the state by 1900, not including the smaller farm producers which are more difficult to count, according to Pinney (1989). Thus, characterizing the Californian industry in the nineteenth century in terms of one productive world or another is a complicated matter as the variety of productive worlds seems to be an important dimension of Californian production at the time.

Dry Year Dynamism

Whether the industry was industrial, interpersonal, or some combination of productive worlds, as argued above, the nineteenth century wine industry as a whole contracted dramatically with the passage of temperance laws in the early twentieth century. The development of that particular civic convention typically signals the end of winemaking in historical accounts (Pinney 1989). At the same time, a second period of development arose largely geared to shipping grape production with some continued experimentation in alternative forms of wine production allowed under Prohibition-era regulation, so a related industry continued in California albeit without a significant commercial winemaking dimension and lacking a specific understanding of region.

During this period, we continue to find elements of various productive worlds. The 18th Amendment to the Constitution banned the manufacture, sale, import, and export of any "intoxicating liquors" between 1919 and 1933 (Teiser and Harroun 1984). This was an enormous setback for the wine industry not because it banned wine consumption – which continued – but because it cut off the important productive element of agro-processing; this value-added portion of the production process is central to all dynamic agricultural regions, and by limiting the kinds of processing, the Volstead Act dampened the elaboration of new conventions in the Napa–Sonoma region.

At the same time, the Volstead Act enabled some commercial production because it contained a waiver that allowed people to make "non-intoxicating cider and fruit juices exclusively for use in his (sic) home." In practice, there were few ways for authorities to confirm the alcoholic content of homemade products. And while intoxicating liquor was defined in the law, the definition of non-intoxicating liquor was never clearly established. This discrepancy gave home producers broad leeway to make their own wine, beer, and other beverages. Combined with increased immigration from southern Europe, the waiver enabled wine consumption to increase from 50 million gallons in the highest pre-Prohibition year to an average of 76.5 million gallons during Prohibition. Wine consumption outpaced population growth (Teiser and Harroun 1984). These numbers suggest that a certain industrial scale continued in the Californian industry.

But if the numbers suggest industrial scale, an interpersonal world seems to have continued for a time. In Napa and Sonoma, grape growers used the federal waiver to circumvent laws prohibiting the manufacture of alcoholic beverages by providing grapes, grape juice, and winemaking services to home winemakers. One example is the Asti Grape Products Company in Sonoma County – the Prohibition era name for the Italian Swiss Colony winery – which delivered grape juice to consumers' homes for fermentation. The company provided follow-up visits to handle bottling, shipped grapes to eastern markets, and sold a juice concentrate known as "Moonmist." In the worlds of production framework, such new practices and products are evidence of changing conventions, and thus indictors of innovations, all of which suggest that while the Volstead Act was the signature convention of the period and placed significant constraints on commercial wine production, an industry nevertheless continued and people experimented with alternatives throughout these "dark years."

The new world of production had a different organization of production and a different stock of grapes. Some 90% of wine production shifted between 1919 and 1933 away from commercial wineries, and into private homes, garages, and clandestine cellars (Teiser and Harroun 1984). Immigrants preferred purchasing whole grapes to certify their quality leading to a planting boom in California focused on grapes that could travel well to eastern markets. (Sullivan 1994). In New York and other eastern cities, local dealers purchased grapes at rail yard auctions. The dealers then diverted the shipments to immigrant neighborhoods where residents carted grapes home in wagons, wheelbarrows, and baby carriages. In Chicago, middlemen negotiated prices directly with grape farmers who followed their freight shipments to the city on passenger trains. Buyers then carted off the grapes once the price was set (Teiser and Harroun 1984).

Neither grape growing on its own, nor the various related processing efforts encouraged the same level of secondary economic activity as commercial wine production had previously. By the mid-1920s, the grape boom withered due to over planting and a consequent softening of grape prices. In the late 1920s, the federal government declared semiprocessed grape products in violation of the Volstead Act. Many processors subsequently closed all together. There were more than 700 wineries in California before Prohibition – 256 in Sonoma County and 100 in Napa County – but only between 130 and 140 at Repeal (Teiser and Harroun 1984). Only a handful of serious operators remained in either Napa or Sonoma County, those with special licenses from the federal government. Beaulieu Vineyard – the most prominent among them – was sustained by a steady business in sacramental wine, as were Christian Brothers and Beringer Brothers.

One consequence of this decline was that the region not only lost many of the conventions that had been established during the previous period of industrial development, but in addition, it failed to sustain new ones. Much of the winemaking experience, skills, and associated supply infrastructure – the regional innovation system which had been constructed over decades following the Gold Rush – was gone. In theoretical terms, Prohibition transformed the world of production by smashing the region's division of labor – the propulsive element to the region's growth – and with it the important interaction that occurred among linked producers. Some people left entirely. Most facilities were idled; many were dismantled. Wine presses rusted, and cooperage was sold off or dried out and rotted. Prohibition did not end wine production or consumption, but it did place limits on the regional context, change the varieties of grapes available in California, and shift the focus of productive activity to the Central Valley and consumers' homes.

The Post-repeal Industrial World

The shift to shipping grapes had long-term ramifications for consumer and producer understandings of quality, and regional vitality. After Prohibition, the remaining wineries had, at their disposal, very few of the resources needed to make dry wine commercially, even though many former and prospective winemakers restarted winemaking. Home winemaking made recovery that much more difficult. Post-Repeal, this grape trade continued for some time. Repeal occurred in 1933, but home winemakers produced 31 million gallons of wine in 1935 and likely averaged around 20 million per year (Heintz 1999; Lukacs 2000). At the same time, wineries began processing grapes anew and started shipping bulk wine to regional bottlers across the country under a new regulatory system. The bottlers sold this wine in jugs to individual consumers and in barrels to restaurants and retailers who then filled their own glasses and jugs much like pre-Prohibition years. The storage of wine in large barrels invariably let dry wine spoil, so sweet wines were preferred.

Since the overall quality of the wine was poor, there was little need for growers to change the variety of their grapevines or their farming practices; they had two relatively stable markets for grapes: home winemakers and regional bottlers. In this context, the farmers in Napa and Sonoma County were most concerned with protecting themselves from large price swings. Napa and Sonoma County each had two cooperatives for that purpose: The Sonoma County Cooperative Winery and Northern Sonoma Wines, and the St. Helena Cooperative Winery and Napa Cooperative Winery. The cooperatives had their own wine brands, but they shipped much of their production to regional bottlers across the country. The Sonoma County Cooperative Winery and the Napa County cooperatives both sold their bulk wine to Fruit Industries for a period. Fruit Industries was a cooperative sales agency with blending facilities in New York, Chicago, New Orleans, Los Angeles, and San Francisco. Roma Wine Company quickly became the largest California wine producer after Repeal and maintained branches in other cities. Other competitors were United Vintners, the processing side to the Allied Grape Growers Cooperative in the Central Valley; Italian Swiss Colony, owned at this time by Edmund Rossi; and Sunny St. Helena Winery, owned by a partnership that included Cesare Mondavi, who later started Charles Krug Winery as part of C. Mondavi & Sons. More premium producers included Beaulieu Vineyard, Larkmead, and Inglenook. The latter were important because they developed some of the key innovations of the time – varietal wines, vintage bottling, and the like – but generally the industry produced poor to average quality wines and suffered from overproduction until World War II.

Most grapes went into sweet or fortified wines which did not require special care in cultivation or reflect to any significant degree their alleged terroir. Some of this production was by design; a special program to convert some wine into brandy, known as the Prorate, was used to stabilize the wine market by reducing supply. Most premium North Coast grapes were blended with Central Valley grapes to add improved characteristics to bulk wine. This fact is reflected in the evolution of the cooperative wineries' markets. Over time, the Napa cooperatives became associated with one growing bulk wine producer: Gallo. Located outside of the North Coast counties in Merced County, Gallo purchased all of Napa Valley cooperative wine in the mid-1950s. Both cooperatives in Napa Valley signed long-term contracts with Gallo which provided them with an immediate stable outlet for their production through the 1950s and 1960s. The Sonoma County Cooperative Winery started its own label but eventually closed and leased its facilities to Gallo (Peninou 1998). Gallo grew so rapidly; it soon dominated the jug wine market and claimed to be the biggest winery in the world. Their contracts accounted for 65% of Napa County's wine production in 1956 and 50% of California output (Heintz 1999).

Through cooperative action and contracting, growers increased their returns but had limited incentive to change even when Gallo, the University of California, and others urged growers to plant better varieties of grapes (Heintz 1999). Cooperative wine remained a bulk commodity, as did most of California's commercial wine. This state of affairs reflected a short-term vision of grape growing focused on annual swings in grape prices and improving returns above the average price for bulk grapes, but it also reflected the general state of the wine industry which targeted its production towards sweet wines for much of the mid-twentieth century. The stability of the wine market and production strategies locked both types of producers onto a low road for decades. There was some replanting. James Lapsley (1996) tells of 1,000 acres replanted in the 1940s which led to some improvements and reinforced a brief 1950s wine boom. But from a grape supply standpoint, even in 1961, the entire state had only 800 acres of Cabernet Sauvignon, 600 acres of Pinot Noir, and 300 acres of Chardonnay, a paltry amount of "noble vines," which make the best dry red and white wines. In fact, county government records at the time did not even record the varieties of grapevines planted in the region; they simply tracked black and white grapes, reflecting a view of grapes as undifferentiated commodities.

New Wine from Old Producers

The classic story of the wine industry is that post-Repeal wine industry constituted a backwater, and it does seem to be the case on a number of levels. It also appears to be a predominantly industrial world of production with some evidence of market and interpersonal worlds as well. However, while the immediate post-Repeal period was a different time with different markets, innovation proceeded apace. The region's challenges lay in rebuilding the division of labor, in developing the right agricultural resources and regional vision of what the district could become, and in making stable, palatable wines.

A "worlds of production" approach would suggest that effort had to be placed in recreating a dynamic world of production in some shape or form. And history suggests that is where much effort went. Producers developed their own measures of quality between the 1930s and the 1960s. The early version of quality was first and foremost a processual quality. Their innovations focused on figuring out how to make drinkable table wine that did not spoil; how to market it to existing wine drinkers, and how to build future markets. Some wine producers pursued niche markets such as Martin Ray, who sought to make dry red wines for wealthy customers. In that sense, conventions continued to change albeit more slowly in the post-Repeal era.

Bulk wine quality arose out of the practical experience of wine makers after Repeal. In the 1930s and 1940s, producers faced significant product failures and poor public image because they did not deploy the proper techniques to make drinkable dry wine. The lack of knowledge and technique among these producers is why Lapsley (1996) writes that, "In the first years following Repeal, California winemakers were untrained and unscientific, often not understanding the science of fermentation, and they lacked the technology to control it even if they did. As a whole, the California wine industry was primitive and backward." Even the premium wine companies were more often simply trying to get by with old methods, luck, and difficulty.

Several issues faced mid-century producers in California. Winemakers could not control the fermentation process, did not understand microbiology, and did not store wine properly. People are said to have had better refrigerators in their kitchens than the wineries had in their cellars, suggesting that homemade wine was better in some respects than commercial wine. But more importantly, for commercial producers, uncontrolled fermentation led to excessively strong products, spoilage in some cases, and continued fermentation after bottling which caused some bottles to literally explode on store shelves and also caused off colors and odors in the final products (Lukacs 2000). A similar problem affected Paul Masson's champagne production during Prohibition. Production was often lost as bottles exploded. Masson treated these problems as part of the process of production, but faced federal questioning because they suspected that the producer was illegally selling champagne.

A larger issue similar to the distribution problems faced in the nineteenth century was that producers continued to have little control over the treatment of their wine between the winery and the consumer as most wine was still sold in bulk to regional bottlers who bottled and cellared wine themselves and continued as before to sometimes extend or otherwise adulterate wines before sale. There were few quality control practices – such as wine tastings, for example. Producers simply made the wine and shipped it out in bulk or in bottles without knowing what they were shipping (Amerine 1983).

Production problems like these stimulated some smaller producers to tinker with process innovations such as mechanical refrigeration and winery bottling from the very beginning. Louis M. Martini was among these earliest of innovators in Napa County. Martini built the first new winery in the state in 1932 in anticipation of Repeal. Similar techniques were subsequently adopted at Beaulieu, Greystone Cellars (a part of Roma Wine Company based in Lodi), and others but apparently not to the same degree. By 1947, both Beaulieu and Charles Krug Winery were using stainless steel coils inside wood fermenters to control temperatures.

On-site bottling began in earnest during the 1940s. At this time, a large infusion of capital, talent, and technology flooded the wine district as whiskey distillers and regional bottlers invested heavily in wine production to find products for the wartime domestic market. Roma Wine Company was the largest, purchased by Schenley Distillers. National Distillers purchased a number of wineries and began distributing under the Cresta Blanca brand. Regional bottlers joined in the shopping spree after the government seized the country's remaining tanker freight cars for the war effort; the cars had been used to ship bulk wine around the country. This development is often considered a bad turn of events because it associated wine with whiskey, but the move into the wine district led to the integration of fermentation, bottling, and cellaring, and to the development of national brands with stable quality. It was a very mixed bag.

The wines were still only of average quality, and the markets were still largely for low-priced sweet wines. But the roots of the late twentieth century transformation in Californian winemaking can be found here. Mondavi's premium wine production was based in wartime consolidation of the industry. Cesare Mondavi purchased Sunny St. Helena Winery in Napa County after selling Lodi-based Acampo Winery to a regional bottler whose markets were located in Indiana, Ohio, and Kentucky. Sunny St. Helena winery was a smaller bulk producer at the time which Mondavi developed into a successful business based on marketing networks he had previously established. When he and his sons wanted to develop their own brands rather than sell bulk wine, they spun off from Sunny St. Helena to resurrect the defunct Charles Krug Winery.

These pioneering wineries benefited from proximity. They watched each other closely, learned from observation, talked to each other, plundered each other for new talent, and slowly began to spin off their own firms. BV hired Russian winemaker Andre Tchelistcheff, who was trained in France, and later hired Leigh Knowles away from E. and J. Gallo in Modesto (Curry 1994; Conaway 1990). Gallo focused on catering to the jug wine markets through sales, but the company also was an early leader in scientific production techniques. Thus, hiring one of their employees would have meant improving one's practices; Knowles learned Gallo's system and expanded Beaulieu's sales significantly. Cesare Mondavi hired regional talent when he made C. J. Rossi of Napa Valley Grape Products the director of sales for Charles Krug Wineries, Inc. In fact, Mondavi was processing wine for Napa Valley Grape Products, suggesting that there was a good bit of practical crossover among wineries through contracting as well. Thus, by interacting, contracting, and hiring away employees, wineries, large and small, benefited from other firms' expertise and networks.

Producers organized a number of associations to support their efforts and lobby on their behalf at the state and federal level. The Wine Institute is one example, organized by the California State Chamber of Commerce in 1935 to bring together leading producers to discuss how to "win back the home winemaker" (Heintz 1999). More important was the Technical Advisory Committee inside the Wine Institute and the American Society of Enologists (Pinney 2005). The Napa Valley Vintners Association, started with a handful of producers by John Daniel and Louis Martini, met in Sonoma County starting in the early 1930s to discuss common production problems over lunch and wine tastings and later to promote Napa Valley wines (Heintz 1999). Other examples of collaboration include Napa Valley Enological Research Laboratory funded by Charles Krug Winery, Inglenook, Souverain Cellars, and Beaulieu Vineyard (Lapsley 1996; Sullivan 1998). A coalition of wine producers known as the Napa Valley Technical Group periodically met to discuss their production techniques.

These interactions were crucial to the performance and improvement of wineries and winemakers. The most successful among them were those who developed the practices that would ensure a consistent product. Though Beaulieu had some production problems, Tchelistcheff kept that winery among the leaders in Napa Valley by carefully cleaning the winery's equipment and eliminating contact of the wine with iron, among other practices. Tchelistcheff reached out to other wine producers; he employed other current and prospective vintners, such as Lee Stewart at Souverain Cellars, who later hired Warren Winiarski (Conaway 1990; Lukacs 2000). This layering of practical experience with individual entrepreneurship was and continues to be frequent in the California industry.

The Mondavis themselves were both products of and consummate operators in the regional innovation system. Robert Mondavi was trained at UC Berkeley. He learned a lot from close observation of Inglenook's John Daniel, whom he is said to have followed around "like a puppy" (Conaway 1990). He hired Tchelistcheff as a consultant to C. K. Mondavi & Sons. But his main innovation was implementing leading edge research by the University of California that was not initially followed by other wineries and growers after Prohibition. From a worlds of production standpoint then, the post-Repeal industry is a rather interesting period of time to view the interplay between productive worlds in California as the interpersonal world strengthened over the period and an industrial world refined practices with the goal of capturing mass markets.

The Proliferation of the Boutiques

If the transformation of the California wine industry began in the 1950s, it proliferated and deepened over the next two decades as increasing numbers began to think of boutique winemaking as a growing and more promising business model. New wine producers developed a wide range of new resources and conventions equally industrial, commercial, public, civic, and ecological as the industry became not just a more learning focused one but also arguably an early green business. These conventions emerged out of significant investment by winemakers and grape growers, large and small, and from long-term efforts to change regional understandings of place, to reexamine expectations of growth, and to calibrate industrial practices with a changing regional context under pressure from urban residential growth.

Three key dimensions of the industry stand out. First, the initial convention used by Mondavi to build markets among the managers of major corporations in the 1950s – wine tasting – became common practice for most wineries. Wine tastings are not only an important quality control mechanism, they are a primary marketing technique through which it is possible for small wineries to gain customers for life by providing unique experiences for tourists at their production facilities. The challenge for producers of fine wine was in part in building connoisseurship among Americans; tasting and social networks provided the avenues for amplifying levels of wine knowledge among Americans.

Secondly, the industry became increasingly tied to the local context. On the one hand, the wine industry needed a level and certain playing field which was not possible in the context of agricultural land conversion at the time. The floor of the Napa Valley had been planned for housing development in the 1960s. As a result, boutique producers allied with land conservation activists to protect signature portions of Napa County's agricultural resources, known as the agricultural preserve, through tax easements and county planning. These initial changes in the civic conventions of Napa County induced other conventional changes as the shift to protecting Napa County's agriculture base led to a string of institutional changes over a period of years, each with a focus on protecting environmental resources. The land protection movements spill over into Sonoma County about a decade later (Guthey and Whiteman 2009).

Third, while the local context of land use and regulation has been critical to the industry's persistence in the region, the industry has also created ties to nature through the development of public and ecological conventions. Federally recognized areas known as American Viticultural Areas, the US version of designation of origin, began in 1978 with strict rules regarding labeling. Today, these domestic connections proceed through additional schemes connecting agricultural practices and ecological sustainability through labeling, codes of conduct, and additional local policy development. In some global wine regions, the certification of industry practices is a sine qua non for market access. For example, the South African industry has witnessed a proliferation of codes of conduct and like schemes in order to hedge against the legacy of apartheid (see Bek et al., in this volume). Likewise, New Zealand producers are developing rules regarding sustainable winegrowing practices as a precondition for participation in export promotional schemes.

But in northern California, the push for a code of conduct seems less a requirement for export markets than it is a part of a regional politics of place (Guthey 2004). The initial stimulus to adopt codes of conduct emerged out of significant local activism concerning on the one had the development of "industrial vineyards" in Sonoma County and on the other hand pesticide spraying to protect the industry from a new vector of Pierce's disease, the glassy-winged sharpshooter.

Citizens of Sonoma and Napa County along with a similar movement in Santa Cruz County protested the transformation of their landscape due to the combined effects of expansive vineyard development during the tech boom of the 1990s and the potential for state enforced pesticide spraying for the glassy-winged sharpshooter. It became an explicit strategy of activists to "kick them [the industry] in the image" (Guthey 2004). In turn, the industry's major trade groups the California Association of Wine Grape Growers and the Wine Institute developed the California Code of Sustainable Winegrowing, a voluntary set of practices intended to promote sustainable agricultural practice in the industry. Later, Napa County government responded by including incentives for land owners to adopt sustainable agricultural practices reflecting the emergent dimension of social and ecological transitions that are ongoing in the region (Guthey and Whiteman 2009). Thus, while the activists characterized the problem as corporations and industrial vineyards, the cumulative impact of the wine industry as a whole was having an impact on the region at the turn of the millennium leading to policy responses on a number of levels and changes in production practices.

Interweaving Worlds of Production in the California Wine Industry

While the history of California's wine industry is often told through a transition from bulk to premium wine production, this case study of northern California suggests that wine development in the state confounds the worlds of production approach in many ways. Rather than transitioning from one world of production to the other, California's wine worlds involve more closely interwoven linkages, frequent involvement of the national and local citizenry, and thus, an often dynamic tension over the definition and operation of conventions. Prohibition certainly placed limits on commercial winemaking and forced a new round of industrial development in the post-Repeal period. What might explain the interweaving of productive worlds in the state's wine industry?

The proliferation of boutique producers starting in the late 1960s shares a role. There have always been small producers in the state amid the larger firms in the industry, but they multiply rapidly after the 1960s. With this shift in productive scale and consequent changes in how consumers understand wine, even large firms attempt to mimic small-scale production as an economic strategy to attract consumers with affordable brands and lead them to more expensive brands as their tastes and knowledge develop. Thus, frequently large industrial and small firms follow a tiered approach by making wine for multiple segments of the wine market from the lower priced bulk end to the more expensive brands.

The economic synergies between productive worlds are also a product of the region itself. The interaction of producers and the amount of spin-offs that occur among large and small producers alike ties many firms together in a variety of ways. And, the regional system of innovation supports both large and small producers albeit perhaps in different ways. Since the nineteenth century, University of California scientists have conducted foundational research in viticulture, fermentation science, and related fields which they distributed widely among the state's producers regardless of markets or the scale of the firm. University of California scientist Maynard Amerine writes from experience in *The Book of California Wine* when he states that "Before there can be any wine, there must be suitable grapes. Before there can be suitable grapes, there must be a climate." He neglects to say that before any of this can happen, someone had to figure out that vinifera grapes would do well in the climates of Napa and Sonoma County and learn what to do with them once they were harvested.

A student of E. W. Hilgard's, Frederic Bioletti, was instrumental in building this basic knowledge. Two of the state's most important enology professors, Maynard Amerine and Albert Winkler, worked under Bioletti. Despite Prohibition, there was a direct lineage by way of the regional innovation system between the district's early prominence and its later rebirth. Maynard Amerine, Albert Winkler, and other UC researchers published pamphlets, held seminars and taught short courses on winemaking techniques, the importance of sterilization, and various other related topics throughout the 1940s, 1950s, and 1960s. In the early 1940s, UC Davis professors published a series of papers on how to determine the maturity of grapevines, and the best time to pick grapes (Amerine 1983). Their experiments with grape varieties continued into the 1980s under various scientists. Warren Winiarski took short courses at UC Davis as did most wine entrepreneurs of the late twentieth century.

This research initially benefited smaller producers. These scientists' accounts indicate that much of their work went unacknowledged for a long time. The industry did not quickly adopt the university's teachings or its students. Producers did not want to hire university grads, believing that one season's experience prepared one for the job of winemaker (Amerine 1983). But in 1960s, boutique producers began to heed UC advice. The latecomers – the Mondavis and others after them – benefited most from UC because they were hungry for advice and competitive advantage, and in a position to profit by investing in the latest technology and science and in new facilities that placed them at the leading edge (Amerine 1983).

Production problems and product failures stimulated innovation through a complex interaction of private investors, speculative entrepreneurs, and state stimulus led by UC researchers. This is exactly the sort of interaction one would expect to find in an industrial region. In this sense, new types of wine quality emerged out of existing conventions and agreements of wine producers working in close proximity to each other. Producer understandings of quality benefited in important ways from large infusions of capital, state funded research and supportive legislation, and a dense social network which gradually thickened. Through this regional process, they developed increasingly refined understandings of grapes, wine, and natural environment – all necessarily new conventions and, thus, innovations from a worlds of production perspective.

At the same time, the local population came to understand the role of the wine industry in maintaining rural lifestyles and cultures. Producers often responded to the local context in a supportive manner attempting to build on their necessary connections to place through community activism and integration. Voters supported restrictive land use laws intended to protect agricultural land for wine growing and to focus population growth in established cities and towns. They elected green candidates in support of the industry and created new public bodies to protect agricultural land.

In conclusion, a long-term perspective on the California wine industry reveals a complicated interplay of productive worlds in contrast to the case of Castile y Leon by Sanchez-Hernandez et al. (2010). Historically, the northern Californian industry involved multiple production worlds. Today, bulk wine still exists alongside and often includes within the same firm boutique wine brands. Industrial, Market, and Interpersonal Worlds are thus found in overlapping configurations. Behind the apparent binary divisions between large and small producers, understandings of terroir, small batch production, wine tastings, and like issues which would typically be associated with interpersonal worlds of production are today seen as important dimensions of all of California's wine worlds.

Acknowledgment Special thanks to Percy Dougherty for inviting me to contribute to this special volume on geographies of wine, to Roseli Ilano for her excellent map making, and to one anonymous reviewer for their helpful comments and feedback.

References

Amerine, M. 1981. *Wine Production Technology in the United States*. Washington D.C.: American Chemical Society.

Amerine, M. 1983. Oral History of Maynard Amerine: Interviewed by Bob Travers for the Napa Valley Wine Library. 9 July.

Barham, E. 2003. Translating terroir: The global challenge of French AOC labeling. *Journal of Rural Studies* 19 (1):127–38.

Berger, D. 1990. Gallo's Sonoma Secret: It is quietly buying acreage and moving mountains. *The Los Angeles Times* 19 September: B3.

Brechin, G. 1999. *Imperial San Francisco: Urban Power, Earthly Ruin*. Berkeley: University of California Press.

Carosso, V. 1951. *The California Wine Industry: A Study of the Formative Years*. Berkeley: University of California Press.

Conaway, J. 1990. *Napa*. Boston: Houghton-Mifflin Company.

Curry, J. 1994. "Agriculture under late Capitalism: The structure and operation of the California Wine Industry." Ph.D. dissertation, Department of Rural Sociology, Cornell University.

Dion, R. 1959. *Histoire de la vigne et du vin en France des origines au XIX siècle*. Paris: Les Belles Lettres.

Emert, C. 2003. Mondavi Helped Make Napa What it is Today. *San Francisco Chronicle,* 18 June: A1.

Gade, D. 2004. Tradition, Territory, and Terroir in French Viniculture: Cassis, France, and Appellation Contrôllée. *Annals of the Association of American Geographers* 94(4): 848–867.

Guthey, G. 2004. Terroir and the Politics of Agroindustry in Northern California's Wine Industry. Ph.D. Dissertation, Department of Geography, University of California, Berkeley.

Guthey, G. 2008. Agroindustrial Conventions: Some Evidence from the Northern California Wine Industry. *The Geographical Journal* 174 (2): 138–148.

Guthey, G. and G. Whiteman, 2009. Social and Ecological Transitions: Winemaking in Northern California. *Emergence: Complexity and Organizations* 11(3): 37–49.

Heintz, W. 1999. *California's Napa Valley: One Hundred Sixty Years of Winemaking*. San Francisco: Scottwall Associates.

Hutchison, J. 1984. Northern California from Haraszthy to the Beginnings of Prohibition. In *The Book of California Wine,* eds. D. Muscatine, M. Amerine, and B. Thompson. Berkeley: University of California Press and Southeby Publications.

Johnson, H. 1994. *The World Atlas of Wine*. Fourth Edition. New York: Simon and Schuster.

Lapsley, J. 1996. *Bottled Poetry: Napa Winemaking from Prohibition to the Modern Era*. Berkeley: University of California Press.

Lemmon, A. 1893. Report of Andrew Lemmon. In E. Peninou (ed), *History of the Sonoma Viticultural District*, 356–358. Santa Rosa, CA.: Nomis Press.

Lukacs, P. 2000. *American Vintage: The Story of American Wine*. Boston: Houghton-Mifflin Company.

Mondavi, R. 1998. *Harvests of Joy: My Passion for Excellence: How the Good Life Became Great Business*. New York: Harcourt Brace & Company.

Morgan, K., T. Marsden, and J. Murdoch. 2006. *Worlds of Food: Place, Power and Provenance in the Food Chain*. Oxford: Oxford University Press.

Moran, W. 1994. The Wine Appellation as Territory in France and California. *Annals of the Association of American Geographers* 83 (4): 694–717.

Murdoch, J. and M. Miele, 1999. 'Back to Nature:' changing 'worlds of production' in the food sector. *Sociologia Ruralis* 39: 465–483.

Murdoch, J., T. Marsden, and J. Banks. 2000. Quality, nature, and embeddedness: Some theoretical considerations in the context of the food sector. *Economic Geography* 76(2):107–125.

Page B, Walker R 1991. From Settlement to Fordism: The Agro-Industrial Revolution in the American Mid-West. *Economic Geography* 67(4): 281–315.

Parrot, N., N. Wilson, and J. Murdoch. 2002. Spatializing Quality: Regional Protection and the Alternative Geography of Food. *European Urban and Regional Studies* 9(3): 241–261.

Peninou, E. P. and S. S. Greenleaf. 1954. *Winemaking in California*, San Francisco: Peregrine Press cited in M. Amerine 1981. *Wine Production Technology in the United States*. Washington D.C.: American Chemical Society.

Peninou, E. P. 1998. *History of the Sonoma Viticultural District: The Grape Growers, the Wine Makers and the Vineyards*. Vol. 1, *History of the Viticultural Districts of California*. Santa Rosa, CA.: Nomis Press.

Pinney, T. 1989. *A History of Wine in America: From the Beginnings to Prohibition*. Berkeley: University of California Press.

Pinney, T. 2005. *A History of Wine in America: From Prohibition to the Present*. Berkeley: University of California Press.

Ponte, S. 2009. Governing through Quality: Conventions and Supply Relations in the Value Chain for South African Wine. *Sociologia Ruralis* 49(3): 236-257.

Sanchez-Hernandez, J.L., J. Aparicio-Amador, and J.L. Alonso-Santos. 2010. The shift between worlds of production as an innovative process in the wine industry in Castile and Leon (Spain). *Geoforum* 41(3):469–478.

Saxenian, A. 1995. *Regional Advantage: Culture and Competition in Silicon Valley and Route 128*. Cambridge, MA.: Harvard University Press.

Sullivan, C. 1994. *Napa Wine: A History from the Mission Days to Present*. San Francisco: The Wine Appreciation Guild.

Sullivan, C. 1998. *A Companion to California Wine: an Encyclopedia of Wine and Winemaking from the Mission Period to the Present*. Berkeley: University of California Press.

Storper, M. 1997. *The Regional World: Territorial Development in a Global Economy*, New York: Guilford Press.

Swinchatt, J. and D. G. Howell 2004. *The Wine maker's Dance: Exploring Terroir in the Napa Valley*. Berkeley: University of California Press.

Touzard, J. M. 2000. Coordinations Locales, Innovation, et Régulation: L'exemple de la Transition «Vin de Masse» - «Vins de Qualité» en Languedoc-Roussillon. *Revue d'Économie Régionale et Urbaine* 3: 589–604.

Teiser, R., and C. Harroun. 1984. The Volstead Act, Rebirth, and Boom. In *The Book of California Wine*, eds. D. Muscatine, M. Amerine, and B. Thompson. Berkeley: University of California Press and Southeby Publications.

Walker, R. 2001. California's golden road to riches: natural resources and regional capitalism, 1848–1940. *Annals of the Association of American Geographers* 91(1): 167–99.

Walker, R. 2004. *The Conquest of Bread: 150 Years of Agribusiness in California*. New York: The New Press.

Winkler, A. 1938. The Effect of Climatic Regions. *Wine Review* 6: 14–16.

The Origin, Diffusion, and Globalization of Riesling

11

Robert Sechrist

Abstract

The Riesling grape variety first appeared in the Rhine Valley about 1350 AD. It is the only noble grape variety not native to France. Beginning in 1386, Riesling was repeatedly selected over other varieties, first by Cistercians, then by nobles, and finally by commoners after 1750. The variety spread to Alsace before 1477 and to Austria before 1700. European emigrants brought Riesling with them to the Eastern United States before 1650 where the vines failed. During the mid 1800s, Riesling was successfully planted in three far-flung regions: Chile, Australia, and California. Since 1955, Riesling further diffused to New York, New Zealand, Oregon, Washington, and Canada. Riesling is the only variety to have its own international association of advocates and promoters. Today, Riesling grows on over 90,000 acres across six continents.

Introduction

Riesling is a variety of *Vitis vinifera*, the wine grape. Its parents were Gouais Blanc and Sauvignon Blanc crossed with an as yet unidentified, possibly wild, variety (Regner 1998). The Riesling grape sprouted in the Rhine drainage basin no earlier than 1350. Spreading across the Rhinelands of southwest Germany and Alsace, Riesling first captured the attention of religious orders and nobles. Repeatedly selected over other field cuttings, Riesling came to dominate Rhineland landscapes during the Little Ice Age (1300–1850).

The Little Ice Age's climatic and catastrophic geologic-induced weather events played a pivotal role in Riesling's acceptance (Lamb 1995). The Riesling vine, like many others, is hardy and cold resistant. Riesling produces a successful crop where other varieties cannot because of its unique annual growth profile. Of all varieties extant during the Little Ice Age, Riesling alone produced so much at such high sugar levels and at a very low pH (2.9–3.2) in such extreme growing conditions.

The Riesling is well suited to continental climates. Its thick, dark bark is the source of its name *russ*, German for dark wood. The bark is deeply grooved, permitting expansion and contraction to accommodate rapid water uptake. This capacity may also assist with cold tolerance. The Riesling leaves are compact and rounded. The crenulations separating the three lobes are shallow. The small, round, ripe berries are green-gold flecked with brown freckles. A peachy to lavender sheen of yeasts on unwashed fruit is commonly seen in photos. Bunches are usually less than 6 in. in length.

R. Sechrist (✉)
Department of Geography and Planning, Indiana University of Pennsylvania, Indiana, PA 15705, USA
e-mail: rpsecrst@iup.edu

P.H. Dougherty (ed.), *The Geography of Wine: Regions, Terroir and Techniques*,
DOI 10.1007/978-94-007-0464-0_11,

Fig. 11.1 The Rhinelands and Alsace, the home of Riesling

Equally important to the widespread viticultural acceptance of Riesling is its flavor profile. Riesling is intriguing. It presents a different aroma when grown on different soils and bedrock. Riesling is vinified dry to very, very sweet. It readily succumbs to Botrytis which adds honey and honeysuckle flavors. The practices of late and selective harvesting add further complexity to the wines. The cooler fermenting temperatures of late fall preserve more volatile organic chemicals than wine fermented during hotter weather.

The Riesling, always at the poleward fringe of grape production, thrives in two narrow bands around the world. In the southern hemisphere, Riesling vines grow in South Australia, Western Australia, New Zealand, South Africa, and Chile. In the northern hemisphere, Riesling grows in Germany, Austria, British Columbia, Washington, New York, Oregon, and China.

The international diffusion of Riesling began in the mid 1800s. During the contraction of the international wine trade between 1870 and 1970, Rhineland Riesling production fell sharply, more sharply than did any other region or variety during the same period. Since 1970, Riesling has had its ups and downs in the public's mind. "In my life in the wine industry, I think I am on my third Riesling phenomenon…it comes along every so often, but it will plateau and fade away, and then it will spike again" (Tinney 2008). Freddy Price's *Riesling Renaissance* (2004) and Schildknect's "Riesling on a Roll" (2008) also refer to this phenomenon. Over the past decade, the grape has again been rising in popularity and taking its rightful place among the noble white wine varieties.

Place of Origin

The Rhinelands, see Fig. 11.1, are underlain by a sedimentary plain of non-marine origin laid down in the Devonian Era. In some localities, metamorphic schist,

Table 11.1 German Riesling wine-producing regions comparing soil and resulting wine style (Deutches Weininstitut 1997: 10). Saale-Unstat, Ahr, and Franken have no Riesling

Region	Soil type	Wine style	% Riesling	Riesling hectares
Mosel-Saar-Ruwar	Slate	Piquant, racy, light body	**69.3**	5,219
Pfalz	Loam, limestone	Round, full-bodied, aromatic	28.6	5,066
Rheinhessen	Loess, limestone, sand, schist	Mild fruitiness, round, full-bodied	15.6	3,194
Rheingau	Loess, loam, slate	Elegant, fruity, racy	**84.6**	2,422
Wurttemberg	Limestone, marl, loess	Robust, powerful acidity	21.9	2,086
Baden	Loess, loam, volcanic	Full-bodied whites, velvety and fiery reds	8.6	1,193
Nahe	Loess, loam, quartzite, porphyry	Subtly racy, fragrant, fruity	**50.9**	1,061
Mittelrhein	Slate	Racy, fruity acidity	**81.8**	380
Hessiche Bergstrasse	Loess	Elegant, fruity, good acidity	**73.2**	213
Sachsen	Sand, porphyry, loam	Dry, fruity acidity	31.8	62

slate, and gneiss predominate but, for the most part, the bedrock is sedimentary shale. The Devonian Era Rhineland sedimentary plain is heavily eroded today. Current surface features are rounded hills and deeply incised river valleys. The hills are rounded because these stones readily crumble, creating grades in excess of 50%. Hillsides rise 500 ft above the rivers in many places. Were the rocks harder, this would have be a region of canyons and spectacular cliffs. The crumbly nature of the rocks probably prevented the people from building terraces.

Rivers cut deep twisting courses through the region's soft rocks. The rivers, flowing across the post-glacial rebounding land, gouged steep-sided, deep, broad valleys. The rivers create the steepest and longest slopes on their cut banks. Riesling produces its best wines on slate-covered slopes. The level fill banks on the inside curve are often town sites.

Low sun angle and the twisting river valleys create an environment with sharp temperature distinctions between sun and shade. The south-facing slopes on the incised meanders receive the highest volumes of perpendicular summertime solar radiation and are regarded as the best locations for growing grapes. Sites angled toward the sun receive intense, long hours of bright sunlight during the long summer days. Solar rays bounce off the river below, amplifying the sun's intensity. The broken shale reflects light as well. For 6 weeks, from the beginning of June to mid-July, the days are over 16 hours long. A short distance away, in the deep shade of a north-facing slope, the day feels short and cool.

The Rhineland soil residing on the slopes surrounding the Rhine is rocky. Depending on the substrate, the rocks are finger-sized shale, dinner plate-sized slate, or fist-sized gneiss and schist. Through centuries of cultivation, smaller particles of soil eroded downslope, leaving crumbled rock fragments. For centuries, people have hauled fertilizers up the slopes to compensate for the effects of gravity. The crumbly rocks made terracing the slopes impractical. Without the possibility of horizontal terracing, rows were oriented vertically, a choice that facilitates cold air drainage.

Rows of vines are planted vertically up the slopes. Each vine is anchored to its own pole at the crown (3–4 ft off the ground). When trained in the spring, all but two shoots are trimmed from the crown. The shoots are bent downward, forming two large loops. The shoots are tied off below the crown. The result is a heart-shaped loop.

Riesling is affected by soil composition more than any other grape with the possible exception of its parent, Sauvignon Blanc. That "Riesling, give[s] fine wines, though of a different character, wherever they are planted" (California State Board of Agriculture 1915: 136) has been known for centuries. Riesling "wines differ substantially regarding their sensory appearance within the same vineyard designation" (Fischer 1999: 281). Differences in the wine result from viticultural practices, winemaking techniques, and climatic variability. Fine, intraregional distinctions are discernable within the resulting wines. In the Mosel, "the slate has different colors, for example, brown at Brauneberg, black at Bernkastel, blue at Whelen, and pink at Erden, and there are recognizable differences in the taste of the wines" (Price 2004: 22).

The pattern of dominance and quantity (Table 11.1) reveals that Riesling production concentrates along the

Rhine from Koblenz at the mouth of the Mosel upstream to Lake Constance (Bodensee) in the Baden region. Loess soils are an acceptable medium for growing Riesling vines as it provides deep rooting opportunities. Riesling grown on loess lacks the intense acidity apparently derived from slate. The terms limestone and full-bodied occur together frequently.

Climatic temperature and precipitation regimes define the potential for vine growth. Precipitation deficits can often be overcome with irrigation. Fans and heaters can save a small area from frost, but they cannot overcome major oscillations. There are three measures of annual temperature that directly relate to vine survival: (1) length of the growing season, (2) sustained lowest winter temperature, and (3) growing degree days. The length of the growing season is defined as the period between the last frost of spring and the first frost of fall. Too cold for too long and the water transport system explodes – think of it as pipes bursting – and kills the vine. Growing degree days are the annually accumulated average daily temperatures above 50°.

A wide assortment of vines can grow in places when these three measures are consistent from year to year. The real measures of plant viability in a place are irregular weather events. Irregular events, an extreme cold event, a late frost, or a hot summer, kill young plants of pioneering varieties. A killing event with a recurrence interval of 10 years would effectively keep vines from colonizing a region. Often, the worst of these events can be tied to tectonic activity.

In the Rhinelands, at the northern fringe of viticulture, these factors are of utmost importance when selecting grape variety. A single late frost can kill thousands of tender young shoots if buds have already burst. Frozen shoots produce no grapes. An unusual intense winter cold snap freezes vine trunks. Insects and pests reduce yields. Diseases reduce yields and inject off-flavors into the wine. In cool wet summers, grapes do not ripen. Moisture encourages molds and fungal infections. Whenever dieback occurs, the varietal composition of the vineyards necessarily shifts toward surviving varieties.

Riesling is also sensitive to higher temperatures. When exposed to warmer environments, "Riesling can be harvested earlier, but the more rapid ripening there tends to make for wines with less aroma, elegance, and finesse. Here a big problem is that the hot ripening conditions result in grapes with thick skins…contain bitter-tasting tannins…In the Barossa Valley of South Australia…the skin…can be up to seven times as thick as in Germany's Rheingau region!" (Pigott 1991: 13).

Riesling takes full advantage of the prevailing climatic conditions of its homeland. The region receives about 30 in. of rain, spread throughout the year. June, July, and August each has mean daily temperatures over 60°F. It tolerates the winter cold very well. It buds late, avoiding late spring frosts. The clusters hang long into the fall; frost rarely occurs before November. Unlike other grapes, Riesling retains acidity as sugar accumulates, making it extremely tasty. The differences between land and water temperatures cause morning fogs in the late summer and early fall. These fogs create conditions suitable for the sprouting of Botrytis, a beneficial mold that withers and sweetens the berries.

Historical Background

Viticulture was an integral part of Roman culture. When they conquered the Rhinelands, the Romans brought wine with them. Slowly the Romans tamed the land and the people. The stability of the *Pax Romana* brought vineyards north of the Alps. Ausonius of Bordeaux, c 370 AD, refers to the importance of the Rhine's steep slopes for ripening grapes. After decades of trial and error, the Romans identified locations receiving adequate direct solar radiation to allow some grape varieties to ripen. After hundreds of years, each vineyard consisted of multiple, successful varieties. One of these was the Croatia import Gouais Blanc, a Riesling parent.

Attempts to fit grapes to the German *terroir* continued for over 1,000 years before the appearance of the Riesling. Charlemagne, it is claimed, ordered planting on the steepest slopes of the Rheinhessen because he believed the grapes would ripen best there. For the next 600 years, Rhinelanders intercropped red and white wine grape varieties. Peasants routinely vinified them together.

In the Rhinelands, red grapes were already on their way out when Riesling was born. In 1386, Cistercian monks at Eberbach, Rheingau, decided their pale reds could not compete with the intensely colored French reds. They removed all red vines from their 300 ha (750 acres) of vineyards and replanted with white, probably Elbling, grapes. It was a good decision. The 1390s experienced cool, wet summers that would have made their reds even thinner, paler, and weaker. From "1300 - [to] 1430 the upper limit of vineyard cultivation in Baden in southwest Germany was brought down by 220 m" (Lamb 1995: 205).

From 1386, one monastic landlord after another ordered the eradication of all red grape and then all non-Riesling white grapes as climatic conditions continued to turn against other varieties. "In 1672, the St. Clara convent in Mainz instructed its tenants in Geisenhiem, in the Rheingau, to replace their red vines" with Riesling (Price 2004: 17). "In the same year, the Bishop of Speyer specified Riesling for his vineyards at Deideshiem in the Palatinate. The great Benedictine abbey of St Maximin at Trier…in 1695 its Abbot Whitman had planted over 100,000 vines which would certainly have been principally Riesling" (Johnson 1989: 288).

German Wine Categories

Germany currently classifies its wines based on sweetness. Several of the terms used to describe German wine categories have a story with special meaning. Wines fermented dry are called Kabinett. The term Kabinett derives from a story where the good wine was kept in a cabinet, out of sight and reach of the servants and low visitors.

From near the end of the Little Ice Age come tales of how the Rhinelanders learned to deal with Riesling in the irregular freezing events. Each response represented the culmination of 1,500 years of grape-growing experience. Three events secured Riesling's place as Germany's premier variety: the discovery of Spatlese in 1775, Auslese in 1787, and Eiswein in 1794. Curiously, each is portrayed as an accidental discovery.

During those 1,500 years, early frosts, Botrytis, and late harvesting must have occurred and the grapes vinified. The properties of the resulting wines were certainly widely known. Before then, however, wines were made from field blends, not varietals. The fabled discoveries led to the legendary codifications of harvesting, vinification techniques, and defined new high-value products.

The fable for the discovery of Spatlese and, ultimately, the current German wine quality categories highlights the orderly side of German culture. The story begins on the steep slopes of Schloss Johannesburg. By 1700, grapes had been growing on this slope for 1,000 years. By law, established in 1718, Johannesburg growers were forbidden to commence harvest before the Prince-Bishop of Feuda[1] decreed it. The decree kept growers from making wine from under-ripe grapes. More importantly, it kept anyone from getting a jump on the other growers of the region.

In 1775, some 54 years after Schloss Johannesburg was planted exclusively in Riesling, the Schloss Johannesburg-bound courier bearing the "Pick the grapes!" decree was waylaid. Couriers to other communities were, apparently, not so delayed. With great concern, the people of Schloss Johannesburg awaited the courier. As they waited, they watched their ripe grapes succumb to a rot that spread across the vineyards as quickly as the morning fog burned off. When the poor courier arrived 14 days later, the people frantically picked what remained of their harvest. In desperation, they even pressed the rotted fruit, while no doubt cursing the courier. Several months later, their curses supposedly turned to praise when they discovered that the wine from the rotted Riesling grapes was exceptional. Today, there is a statue of the fabled courier in the Schloss Johannesburg courtyard which bemuses the large number of tourists who visit the site annually. The story of this statue is also why Johannesburg is regularly associated with Riesling.

Auslese produces "select harvest wines made from very ripe hand-picked clusters that are usually affected by Botrytis mold" (Wines 2007: 15). The decision to cluster select late harvest Botrytis-infected grapes for a specific wine came 12 years after the "discovery" of Spatlese. After the success of Auslese, Germans developed two additional categories: Beerenauslese and Trockenbeerenauslese in the 1800s. Beerenauslese is made from individually selected Botrytis-infected grapes. Trockenbeerenauslese is made from selected, dried Botrytis-infected grapes.

The first tale of ice wine comes from the late eighteenth or mid-nineteenth centuries. Undocumented ice wine harvests probably occurred several times previously (Schriener 2001). The story is very similar to Spatlese. After a cold, wet summer, some growers near Wurtzbergen, Franconia (Franken), left their grapes on the vine late into the fall to achieve maximum sweetness. In early November, the first hard freeze of the season occurred weeks earlier than usual. To save the vintage, they picked and pressed the grapes in the bitter cold. They fermented this separately, lest it spoil the rest of the vintage (Schriener 2001). The product was remarkable for its sweetness and balanced acidity.

[1]The local political and religious ruler.

Ice wine production was intermittent, and before 1960, production was erratic. The proper environmental conditions to produce it occur naturally only a few times each decade. Germans defined and codified ice wine vinification requirements and production methods in 1961 and modified them in 1971. Under those rules, the ice wine production was encouraged. It has been a successful policy as demand regularly exceeds supply. In 1983, the German government created the first regulatory procedures and classifications for ice wine.

Thus, we have the definitions and history of the German wine categories: Kabinett, Spatlese, Auslese, Beerenauslese, Trockenbeerenauslese, and Eiswein.

Riesling Diffusion

Riesling Diffusion occurred in three phases. The first occurred when Riesling was a hot commodity spreading through the Rhinelands and spread along the northern limit of European vine production. The second phase began with European colonization in 1500. On every continent, early European colonists attempted to grow wine grapes, some of which may have been Riesling, soon after their arrival. Failure was their common experience.

Successful commercial Riesling plantings, however, did not occur outside Europe until the mid-1800s. The continued growth of the wine industry was interrupted by a series of calamities beginning with Phylloxera's arrival in Europe in the 1850s and ending in the 1950s. In between, Europe experienced three major wars which destroyed vineyards and decimated the labor force. Wine grape acreage was most severely reduced in the United States by Prohibition, 1920–1934. The third phase, 1955 to the present, began with the application of plant science to winemaking and continues with an industry-wide drive to improve both grape and wine quality.

Diffusion Within Europe

One spring day, a Gouais Blanc flower was pollinated by a strangely crossed pollen grain somewhere in the Rhinelands. The resulting seed grew. Perhaps it was the leaves, differently shaped than the others, or the yellow-green color of the grapes, or the sweetness of the fruit, or the resulting aromatic and vibrant wines that perhaps caught the attention of some medieval monk. The discoverer's identity, as well as the reasons for propagating Riesling vines, is lost to us. Whatever the rationale, he and his fellow monks liked it so much they propagated and sold cuttings while promoting its merits with the wine as the sales pitch. Their efforts succeeded for they offered something that satisfies.

The original Riesling probably sprouted within 30 years of 1350, the height of the Black Death. Europe was in social, political, and economic turmoil following the Black Death. With so few survivors and with ownership unclear, many vineyards went untended. In the untended fields, grape seeds sprouted. There is suggestive evidence that the Riesling was growing in Kintzheim, Alsace, in 1348, but this report is widely discredited. The first reliable reference to Riesling vines comes from the Rheingau region in 1435. This reference, as are most fifteenth century documents referring to Riesling, is a receipt for the sale of cuttings.

Within 20 years, written references to the Riesling abounded in the Rhinelands, showing its popularity with the literate few. Records show Riesling vines were sold to growers in Trier in 1464. Its presence in the Alsace was documented in 1477. A Riesling vineyard was depicted on a 1490 map of Worms. Another vineyard was identified in the Rheinhessen in 1511.

Based on these documents, we can deduce several facts about who grew Riesling and about its perceived desirability. Ordinary vine cuttings from widely available varieties were freely available. That Riesling cuttings were sold and not given away tells us that they were regarded as special. The purchasers' identities tell us that initially only those with wealth and connections could obtain Riesling cuttings. The written records tell us that those involved with the transactions were literate at a time when few people could read. They further tell us that the receiver required that documents of authenticity accompany the vines. Taken together, they tell us that for about 50 years, Riesling cuttings were in high demand and limited supply. Afterward, second-generation Riesling cuttings made the variety more available. In 1577, Hieronymus Bock reported that "Riesling was growing in the Mosel, the Rhein, and the environs of Worms."

Nobles and religious orders owned large tracts of land and had wealth and connections. It was they who adopted Riesling first. Fashionable among the elites, Riesling spread down the geo-social hierarchy. The initial adopters spread Riesling cuttings to their distant social superiors and locally to their vassals. Vassals of elite adopters contagiously adopted Riesling, often by coercion. In less than 150 years, Riesling was widespread in the Rhinelands and upper Danube basin.

Before the Thirty Years War (1618–1648), the Rhineland vineyard area was triple today's area. The Rhinelands, the core of Charlemagne's crumbling empire, were the scene of the most intense fighting during the Thirty Years War. Military campaigns devastated the Rhinelands. Before the Thirty Years War, the population of Europe was 30 million; at its close, the population was 20 million. Many of the dead fell in the Rhinelands. During the war, and for years after, vast vineyards lay untended. When replanted in the 1670s and later, it was colder than before and during the war. The red wines became paler and weaker; planting white varieties became more imperative. Riesling grapes planted in select locations by monks in the French Alsace and German[2] Rhinelands thrived.

From the close of the Thirty Years War until the mid 1700s, there was a 50% tax in kind on German wines. As a result, the peasants grew Elbling. Elbling is a large producer, yet its wines are both low in acidity and alcohol, giving them a strong tendency to go bad in the barrel. "Only after this tax was removed did Riesling begin its takeover of the Mosel" (Clarke 2007: 201). Monastaries and the vineyards of nobles were exempt from the tax and could afford to grow Riesling.

In 1716, Riesling was so commonly planted and praised that its viticultural practices and production techniques were codified. In 1721, Schloss Johannesburg – the Castle on St. John's Mountain – became the first estate to plant only Riesling, planting nearly 300,000 vines that year. Speyer, in the Pfalz, followed in 1744. In 1787, the Elector of Trier in the Mosel region ordered the eradication of lesser varieties and the planting of Riesling.

Beginning no later than 1500, Riesling cuttings spread outside the Rhine River Valley and into the northern tier of Central Europe's vineyards. "The bishops of Saltzburg planted Riesling in the 17th century in the Sausaler Valley" (Price 2004: 118). The grape gained minor acceptance in the Wachau and Styria regions of Austria. From there it was passed, hand to hand, across the Austro-Hungarian Empire. In the 1830s, Riesling vineyards were scattered across that Empire. Today, Riesling vineyards are found in small pockets in Switzerland, northeast Italy, Slovenia, Romania, Bulgaria, and Hungary. The French banned Riesling beyond the Alsace, preventing its western diffusion. The preceding innovations took place under church auspices in the face of climatic change. Much of the land was in church hands, and the political leaders were also clergy, hence the title prince-bishop.

This changed in the early 1800s with Napoleon's conquest of the Rhinelands. In 1803, Napoleon ordered the confiscation and sale of church lands. As in Burgundy, this put thousands of acres into the hands of private individuals and fragmented vast church vineyards. By 1815, when the Napoleonic Wars concluded, the region was once again in ruin. Slow rebuilding, the potato famine, and political turmoil led to massive German emigration during the mid 1800s. Those emigrants who knew wine planted Riesling wherever they went. In some of the new places, Riesling and other varieties thrived. In other places, conditions required them to adopt other varieties or crops.

Rhineland suffering began anew with the Prussian War and Phylloxera in the 1870s and 1880s. Wars, economic depression, and bad viticultural decisions continued to suppress the German wine industry.

Riesling Diffusion 1632 to 1880

When Rhinelanders left Europe, Riesling cuttings accompanied them. In most of the new locations, Riesling, like most other grapes, grew poorly. The Riesling's first unsuccessful rooting outside Europe may have occurred in the 1630s. Many vine cuttings were sent to the English Colonists in Virginia, and Riesling might have been among them. It is also possible the Dutch planted Riesling at their New Amsterdam (New York) colony. If so, they quickly died. Further attempts, and there were many, in eastern North America failed for the next three centuries.

Three men are responsible for Riesling's initial global diffusion during the mid 1800s. They were James MacArthur of Australia, Don Silvestre Ochagavia Echazarreta of Chile, and Agoston Haraszthy of California. Each acquired a broad assortment of cuttings from Europe and accompanied them to their new destinations in Australia, Chile, and California. Each actively campaigned for the expansion of wine production in their respective lands. Each man started a national industry and is considered the local "father of wine."

[2] German is a misnomer because at that time Germany consisted of approximately 2,000 separate feudal states.

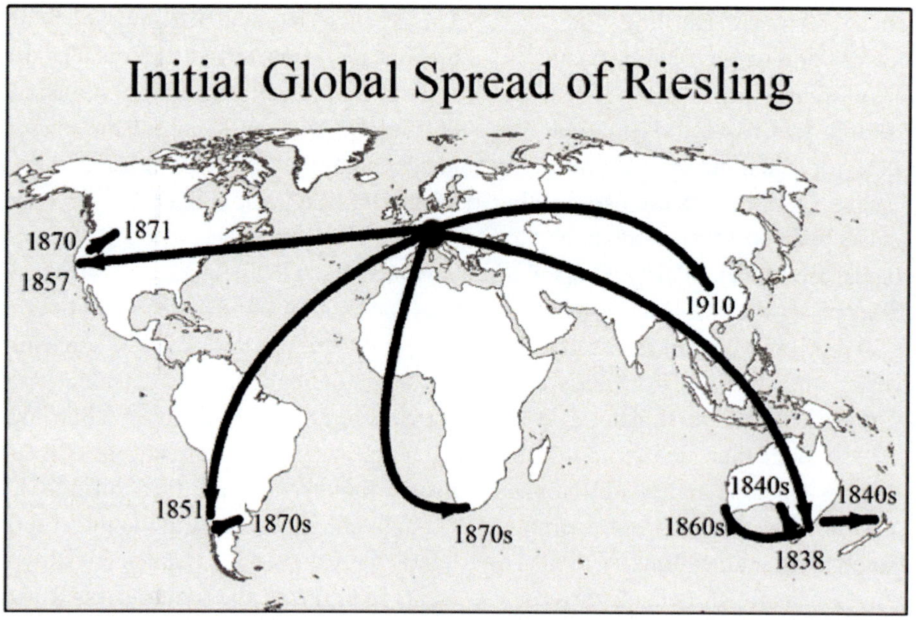

Fig. 11.2 Map showing when Riesling vineyards were planted around the world

In New South Wales, Australia, James MacArthur first planted Riesling in 1838. Riesling wine production began in the 1840s. MacArthur invited Rhinelanders to work in his vineyards. The success helped a wave of Silesian immigrants seeking religious freedom decide to settle in Australia. Most of the immigrants settled in South Australia near the city of Adelaide. Many workers were inexperienced with grapes and planted the vines too close to the coast to produce good wines. Sheep graze in those abandoned vineyards now (Beeston 2001).

The first vineyard in Western Australia was planted near Perth at Olive Farm in South Guildford in 1829. This predates Riesling's arrival in Australia. By 1861, there were 457 acres of vines in Western Australia, and Riesling was certainly well represented (Coghlan 1896). New Zealand was planted with Riesling grapes with the earliest settlement in 1819 at the northern end of North Island. Since then, the viticultural focus has moved southward. The first South Australian vineyards were planted in the mid 1840s. Gilbert at Pewsey Vale and Gramp at Rowland Flat each established vineyards in 1847. In South Australia, wine boomed and then broke. In 1866, there were 6,629 acres of vineyards. Production peaked in 1870 with nearly 900,000 gallons. Supply exceeded demand. Eradication followed and, in 1879, there were only 4,114 acres producing 313,000 gal (Conigrave 1886: 90). The abandoned land reverted to sheep pasturage. When acreage increased again, it was in the more temperate Barossa, Eden, and Clare valleys. The Eden and Clare valleys are famous for their dry Rieslings.

Across the Pacific, Don Silvestre Ochagavia Echazarreta brought Riesling cuttings to Chile in 1851 as part of his mammoth effort to expand the Chilean wine industry. Argentina received its Riesling cuttings from Chile as well as German immigrants. South American Riesling acreage is small and naturally concentrated near the southern limits of production (Fig. 11.2).

Riesling successfully came to North America even later. Successful is the key word. Of the many shipments of cuttings sent to eastern French, English, and Spanish colonies, Riesling must have been tried. For a number of reasons, these early colonial vineyards failed – mostly through ignorance and intervening economic opportunities. Riesling cuttings, finally, made their way to California in 1857. "Mr. Charles Krug…a native of Mainz, on the Rhine…In 1858 he entered into the wine business in Sonoma…at present it [the vineyard] covers 68 acres and has 60,000 vines among which are Johannesburg Riesling" (Menefee 1879: 204). The Haraszthy wine company had a 360-acre Riesling Vineyard near Esperanza, Yolo County, California, in the late 1880s (Anon 1892: 245). Before Prohibition, Riesling was common in central California. The results were, and remain, generally disappointing.

> The Rhenish grapes were among the first to be tried in California, and they have been tested, perhaps, more widely than any other. Very little wine, however, has been made in California that at all resembles the true Rhine wines. Many of our Riesling wines bear a much closer resemblance to Sauternes than to Rhine wines. This is to some extent due… to the high percentage of sugar which the German grapes attain here at maturity… the slow cool fermentation which the wines of Johannesburg and Steinberg undergo… The best results are obtained with this variety in the coolest localities. (Hilgard 1896: 180)

Only small portions of California are suited to Riesling. It is too warm, and the grapes ripen too quickly. In the early 1980s, there were over 11,000 Riesling acres in California. By 2007, that acreage was reduced to 2,861 (USDA NASS 2008). The cool coastal areas of Monterey County are home to most of the remaining acreage. The Alexander and Anderson valleys on the North Coast alone appear capable of consistently producing late harvest Rieslings.

The first vines, Isabella cuttings, were brought to Oregon by an Iowan settler in 1847. Vinifera grapes, including Riesling, were planted in the Willamette Valley before 1870. Economic conditions, beverage preference, and Prohibition were leading factors in the demise of the early Oregon wine industry. In Washington, "The earliest record of grape growing … can be traced back to 1825" (Zraly 2008: 46). The first Riesling cuttings, planted near Walla Walla came to Washington State in 1871.

The first commercial winery in China, the Changyu Pioneer Wine Company, was established in 1892. Dragon Seal Winery was established in 1910 by a French friar. Subsequently, Lutheran and Catholic missionaries planted small vineyards, and Riesling was probably among the grapes grown. After the fall of the emperor and through the reign of Mao, the industry stagnated.

Riesling's Post-prohibition Spread

Prohibition led to the eradication of vineyards across the United States. Riesling was commercially eradicated in Oregon and Washington. With Prohibition's repeal, Californians began replanting, but the Depression and World War II slowed the effort. Riesling was not a popular variety amongst California growers and was sparingly planted there. The Monterey Peninsula alone has substantial Riesling vineyards. Over time, Riesling gained acceptance outside California.

The first successful adaptation of Riesling to the Finger Lakes of New York occurred in 1955. Dr. Konstantin Frank, a plant pathologist, determined that given very cold, hardy rootstocks, Riesling vines could succeed in the Finger Lakes region. He grafted the roots of vines native to Quebec to Riesling shoots. Important to his success were disease-free parent material, pesticides, and fungicides not previously available. From his successful vineyards along the southwestern shores of Lake Keuka, the modern eastern North American wine industry began. The Finger Lakes region is now home to over 100 wineries, most specializing in dry Rieslings. Small Riesling vineyards are currently located in Pennsylvania, Ohio, Maryland, and Virginia (Fig. 11.3).

Richard Sommer established Oregon's first modern Riesling vineyard in the Umpqua Valley in 1961. From this southern Oregon valley, Riesling production diffused contagiously northward through the Rogue, Willamette, and Columbia valleys in following years. Production of Oregon Riesling peaked in the 1980s. By 1995, there were 710 acres remaining. In 2005, statewide Riesling acreage dropped to 524. And again, interest in Riesling moved northward.

Washington's post-Prohibition Riesling production began in 1967 when the Chateau Ste. Michelle Winery opened. In 2008, Riesling surpassed Chardonnay as Washington's number one grape (Stegemoeller 2009). Chateau Ste. Michelle is the largest US Riesling producer with an annual output of 1.3 million cases (Gregutt 2007). Most vineyards are in eastern Washington which receives scant rainfall and where irrigation is required.

Across the border, British Columbia was home to 236 acres of Riesling or 4.3% of total BC acreage in 2003. Riesling production in BC since 1995 has been volatile. Peak production came in 1995 with 825 tons of grapes harvested. The minimum production level was in 2000 with only 433 tons. By 2003, tonnage had rebounded to 681 tons (BC Ministry of Agriculture 2004). The aridity of eastern British Columbia coupled with its winter cold makes it an ideal area for ice wine production. Climatic conditions make ice wine possible almost every year in the Okanagan Valley in eastern British Columbia.

Following Dr. Frank's lead, wineries specializing in Riesling began appearing on the Niagara Peninsula of

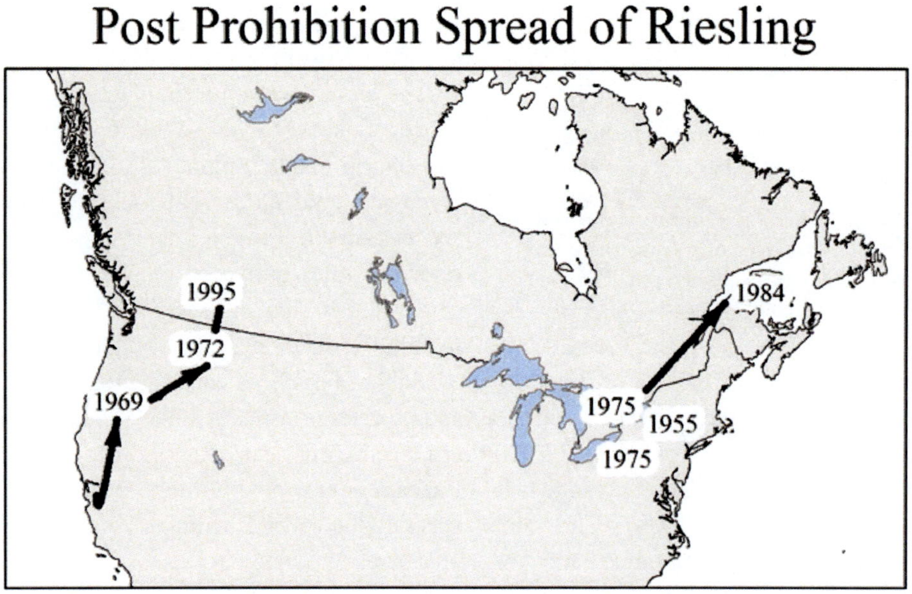

Fig. 11.3 Post-prohibition spread of Riesling in North America

Ontario in 1975. Inniskillin was the first. The many wineries now located there extend from the Niagara River – US–Canada border – 30 miles westward and inland from Lake Ontario for 3 miles. The eroded limestone slopes of the Niagara Peninsula vineyards face northward toward Lake Ontario. The base of the escarpment is covered with a deep layer of glacial till. The wineries of this region specialize in Vidal Blanc and Riesling wines, but the other noble varieties are being planted. Jost Vineyard, established 1984, is located adjacent to Prince Edward Island Sound and is the northernmost eastern Riesling producer in North America.

On New Zealand's South Island, the Montana Vineyards, established 1973, commenced New Zealand wine production. Over 100 years earlier, Riesling production from Australian cuttings failed on North Island. Riesling does not dominate any of the New Zealand regions but is present in nearly all. Between 1999 and 2009, Riesling acreage increased from 1,080 to 2,292 yet retained its relative position as the sixth most widely grown varietal. Of this acreage, nearly half is in the Marlborough district and another quarter in Waipara. The first Tasmanian Riesling vineyard at Pipers Brook was planted in 1974.

Several Riesling-producing wineries opened in China during the 1980s with international sponsorship. The first definitive record of Riesling production comes from the Huadong Winery, established in 1985. Hundreds of new ventures have been launched since then. Corporate vineyards in excess of 500 acres are typical. The total scale of the Chinese wine industry is immense, 178,000 ha with 3,000–4,000 additional hectares added annually, but the mix and scale of varietals is unavailable.

With nearly 900,000 ha of vineyards, Eastern European countries produce an unknown volume of Riesling. South America and South Africa have limited but growing Riesling acreage. Pioneer wineries in Bio Bio and Patagonia are beginning to appear on the international market. The same is true of South Africa.

Riesling Descendants

The Riesling grape has a large number of children and an increasing number of grandchildren. Instead of employing the grafting solution used in France, Rhinelanders sought to breed a Phylloxera-resistant variety. A number of Riesling-parented varieties emerged from this effort, none match Riesling's flavor. These include the Muller-Thurgau (Riesling × Madeleine Royale), Scheurebe (Riesling × Sylvaner), Kerner (Riesling × Trollinger), Ehrenfelser (Riesling × Sylvaner), Rieslaner (Riesling × Sylvaner), Rotberger (Riesling × Trollinger), and about a score more. The Muller-Thurgau proved the most famous, widespread – and detrimental. Muller-Thurgau (aka Riviner) was created in 1882 by Hermann Muller who lived in

Thurgau, Switzerland. Covering over 100,000 acres, Muller-Thurgau is Riesling's most successful descendant. It is a good producer of bulk wine. Public sentiment is strongly against this grape, and its presence is often hidden in proprietary-named wines. Muller-Thurgau is the principle variety used to make Blue Nun, Black Tower, and Liebfraumilch.

Today, genetically engineered Riesling grandchildren are under evaluation at university experiment stations. They are preselected and propagated with specific characteristics in mind. Breeders seek a variety that tastes as good (or better) than Riesling, is more disease resistant, and produces better. Disease resistance and increased production have proved easier than finding a better-tasting grape.

Selection and evaluation of a new vine can take decades. Seeds are assigned an identification number and planted. After a dozen or so years, the vine and its crop can be fully evaluated. If it has desirable characteristics, a larger planting block may be established. Most of the vines evaluated come to nothing, but every so often these experimental varieties bear fruit.

Globalization of Riesling

Globalization can take many forms. The success of Riesling as a marketable commodity grown around the globe is one form that began with the first sale of Riesling cuttings. The scale of international Riesling shipments speaks to its role in the global marketplace. Riesling, along with the other noble varieties, sets the rules for designing future varieties. Descendants capable of producing similar flavor profiles and of living in more extreme environmental conditions will allow a more widespread distribution of Riesling genetic material.

Riesling's diffusion represents the simplest form of globalization, an expansion into suitable locations. Europeans consciously brought vines to new places in the hope they would grow. After 500 years of effort, and with several successful colonizations, two-thirds of all Riesling still grows in Germany. Riesling descendants are rarely planted outside Germany.

As a global competitor for wine dollars, Riesling is a significant but minor player. In "2008 Riesling accounted for 2.2 percent of market sales and had a 1.5 percent share of total case volume" in the US market (Tinney 2008). Table 11.2 shows the acreage of Riesling vineyards by country. Nearly two-thirds of all Riesling vines grow in Germany (51,000 acres). Riesling occupies almost 11,000 acres in Australia, 8,300 in Alsace, 4,000 in Austria, and 2,800 in Washington State. Today, Riesling represents less than 3% of total Austrian acreage. Riesling plantings in Eastern Europe expanded during the 1990s with significant support from corporate wine investment. Numbers are unavailable.

Table 11.2 Riesling around the world (Schildknect 2008)

Country	Original year	Riesling acreage 2008	% world total
Germany	13??	54,250	60
France (Alsace)	14??	8,375	9
Austria	15??	4,375	5
Hungary	15??		
New Zealand	1970	1,925	2
Australia	1832	10,937	12
Chile	1851	725	>1
USA	1857/1955	7,500	8
Argentina	1870?	300	>1
South Africa	1870?	875	>1
China	1900?/1980	???	
Canada	1973	1,100	1
Total		90,362	

The International Riesling Foundation (IRF) was created in 2007 to improve the quality and visibility of Riesling and Riesling wines. It is a global effort including major wineries and regional wine organizations. No other grape has a varietal organization created to proselytize its virtues. In less than 2 years, the IRF succeeded in obtaining agreement on a graphic designed to inform consumers of wine sweetness. Sweetness, as defined, is determined by the sugar to acid ratio. Their four point scale allows consumers to easily identify the sweetness of a Riesling. The scale itself was recently co-opted to convey information for other varieties.

Summary

Riesling came from seed to international commodity in only 500 years (1350–1850). Today, it grows on over 90,000 ha (350 sq. mi.) of land on six continents. Mostly, that land is at the poleward fringe of grape viability. Riesling wine varies with *terroir*. Flavor profiles vary with soil type. Sweetness and intensity levels vary with ripening speed (climate). The most flavorful and sweetest Riesling generally comes from regions permitting the longest hang times. Fuller-bodied Riesling usually comes from fast-ripening regions.

The Riesling spread across the Rhine basin in the 1400s. Monks and nobles selected it for their own estates. The cuttings of Riesling were prized and became an item for sale. Our first definitive reference to Riesling's existence came from a sales receipt. Riesling plantings continued in Germany until Phylloxera and war brought about their devastation. Many of the German vineyards were replanted with Muller-Thurgau or other, lesser, Riesling descendants. About a third of German vine acreage is Riesling today. The worst viticultural decision was defining Riesling based on sweetness alone. Other more prolific but less desirable varieties like the Muller-Thurgau still cover vast portions of German vineyards.

The global diffusion of Riesling occurred in two waves. The first was initiated by three visionary winemakers, MacArthur, Haraszthy, and Echazarreta. Each brought a treasure trove of grape cuttings, including Riesling, from Europe to a new continent. The second wave began in 1955 with Dr. Frank's experiment on Lake Keuka. The second wave is characterized by the polar expansion of Riesling vineyards. Tasmania, Australia; South Island, New Zealand; Chile; Canada; New York; and Washington are second-wave successes.

Riesling is an endemic component of the wine's globalization. In the vineyard, it is at the poleward fringe of production. Riesling's ability to produce different flavor profiles from each region has led to its continual re-discovery among wine drinkers. Planted first in Washington, New York, Canada and New Zealand, its success has led to recent varietal diversification in those places. Riesling's descendents are designed to extend production even further poleward. In the marketplace, Riesling is praised by millions and despised by many reformed Riesling drinkers. The approachability of Riesling's flavor will keep it in the vineyards and shelves for generations to come.

References

Anon. (1892). *Master hands in the Affairs of the Pacific Coast.* San Francisco, CA: Western historical Publishing.

BC Ministry of Agriculture, F. a. (2004). *An Overview of the British Columbia Grape Industry 2003.* Retrieved Aug 25, 2009, from British Columbia Agriculture, Food and fisheries: www.agf.gov.bc.ca/grape/publications/documents/overview_grapes_dec2004.pdf

Beeston, John. 2001. A Concise History of Australian Wine. Allen and Unwin. Crows Nest NSW Australia.

California State Board of Agriculture. (1915). *Statistical Report of the California State Board of Agriculture.* Sacramento: State of California Printing Office.

Clarke, O. M. (2007). *Oz Clarke's Grape and Wines: The definitive Guide to the World's Great Grapes and the Wines they make.* Orlando, FL: Harcourt, Inc.

Coghlan, T. (1896). A Statistical Account of the Seven Colonies of Australia. William Applegate Gullick, Government Printer, Sydney.

Conigrave, J. F. (1886). *South Australia: a sketch of its history and resources* (2 ed.). London: Colonial and Indian Exhibition.

Deutches Weininstitut. (1997). *A short guide to German wine.* Mainz, Germany: Deutches Weininstitut.

Fischer, U., D. Roth, and M. Christmann (1999). The impact of geographic origin, vintage, and wine estate on sensory properties of Vitis Vinifera cv. Riesling Wines. *Food Quality and Preference* 10:281–8.

Gregutt, P. (2007). *Washington Wines and Wineries: The Essential Guide.* Berkeley: Univ California Press.

Hilgard, E. L. (1896). *Report of the Viticultural work during the seasons 1887–93 by the California Agricultural Experiment station.* Sacramento: State of California.

Johnson, H. (1989). *Vintage: the Story of Wine.* New York: Simon and Schuster.

Lamb, H. H. (1995). *Climate, History and the Modern World.* New York: Taylor & Francis.

Menefee, C. A. (1879). *Historical and descritive sketch book of Napa, Sonoma, Lake, and MendocinoMemefee, Campbell.* Napa, CA: Reporter Publishing House.

Pigott, S. (1991). *Riesling.* (H. Eyers, Ed.) London: Viking Books.

Price, F. (2004). *Riesling Renaissance.* London: Mitchell Beazley.

Regner, Ferdinand, H Steinkellner, J Glössl, S Kampfer. 1998. "Reconstruction of a grapevine pedigree by microsattelite analysis." Theoretical and Applied Genetics. Springer. Berlin 97:227–31.

Schildknect, D. (2008, Mar). *Riesling on a Roll.* Retrieved Sept 3, 2009, from wine-business-international.com: www.wine-business-international.com/mwbi_308_s12–13_ws_riesling.pdf

Schriener, 2001. *Ice wine: the Complete Story.* Warwick Publishing. Toronto.

Stegemoeller, I. (2009, Jan 25). *Washington wine grape harvest led by Riesling.* Retrieved sept 1, 2009, from winepress northwest: www.winepressnw.com/news/story2457.html

Tinney, M.-C. (2008, April 15). *Retail Sales Report: Riesling a Key Driver in Sales Growth.* Retrieved Aug 28, 2009, from Wine Business: www.winebusiness.com/wbm/?go=getArticle&dataId=55225

USDA NASS (2008) http://www.nass.usda.gov/Statistics_by_Subject/index.php. Last accessed August 14, 2008.

Wines, P. R. (2007). *Riesling Rules* (2 ed.). Pacific Rim Wines.

Zraly, K. (2008). *Windows on the World Complete Wine Course* (Vol. 2009). New York: Sterling Publishing Co.

Part V

Techniques

Remote Sensing for Viticultural Research and Production

12

Lee F. Johnson, Rama Nemani, John Hornbuckle, Wim Bastiaanssen, Bryan Thoreson, Bruno Tisseyre, and Lars Pierce

Abstract

Geospatial technologies continue to advance mapping methods across societal sectors. Remote sensing, or the collection of Earth-viewing digital images by satellite or aircraft, is increasingly used as a viticultural production tool. The images may be used in isolation, or analyzed in combination with other supporting spatial data layers within a computerized geographic information system. Physical geography and corresponding cultural management can affect grapevine productivity, fruit characteristics, and wine quality. The influence of these factors may be expressed as alterations in the biophysical or biochemical properties of the grapevine canopy in ways that are often amenable to detection by remote sensing systems. This chapter introduces remote sensing technology and surveys the field of applied viticultural research to include methods of development for on-farm management, so-called precision viticulture, and regional land cover mapping. Finally, an overview is provided of prototype remote sensing advisory systems that have been developed for operational production support in wine-growing regions worldwide.

Introduction

Geographic methods have been revolutionized by the advancement of geospatial technologies over the past few decades. Three technologies in particular have changed the way in which we measure, analyze, and visualize the Earth's surface: remote sensing, the global positioning system (GPS), and geographic information systems (GIS). Remote sensing refers to

L.F. Johnson (✉)
Division of Science and Environmental Policy, California State University, Monterey Bay, Seaside, CA, USA

NASA Ames Research Center, Moffett Field, CA, USA
e-mail: Lee.F.Johnson@nasa.gov

R. Nemani
NASA Ames Research Center, Moffett Field, CA, USA
e-mail: rama.nemani@nasa.gov

J. Hornbuckle
CSIRO Land and Water, Griffith, NSW, Australia
e-mail: John.Hornbuckle@csiro.au

W. Bastiaanssen
WaterWatch, Wageningen, The Netherlands
e-mail: w.bastiaanssen@waterwatch.nl

B. Thoreson
SEBAL North America, Davis, CA, USA
e-mail: bryant@de-water.com

B. Tisseyre
UMR ITAP Montpellier SupAgro/Cemagref, Montpellier, France
e-mail: tisseyre@supagro.inra.fr

L. Pierce
Division of Science and Environmental Policy, California State University, Monterey Bay, Seaside, CA, USA
e-mail: lpierce@csumb.edu

P.H. Dougherty (ed.), *The Geography of Wine: Regions, Terroir and Techniques*,
DOI 10.1007/978-94-007-0464-0_12,

capture of digital imagery of the Earth's surface by aircraft and satellite. The GPS is a satellite constellation that allows ready determination of the geo-coordinates of ground-based receivers, hence facilitating navigation as well as collection of spatially referenced field data. GIS enables the convenient capture, storage, manipulation, and management of spatially aligned, geographic data layers. This trio of geospatial technologies is complemented by continual advances in such areas as Internet-based mapping and visualization, image processing algorithms, hand-held computing, wireless sensor webs, as well as dramatic improvements of the price-performance curve for computers of all types. This chapter will feature the use of remote sensing in viticulture, while implicitly drawing on the allied technologies mentioned above.

Remote sensing, as used in this chapter, refers to the passive measurement of solar energy reflected and emitted from the Earth's surface. Multispectral sensors measures and records energy in discrete bands, or portions of the electromagnetic spectrum (Hall et al. 2002). Sensors discussed here generally record data in the blue, green, and red spectral regions, and at longer wavelengths in the reflected, or "near" infrared, and emitted or thermal infrared. The resulting digital images may be regarded by a human interpreter as an aerial photograph, or automatically processed by computer based on spectral or textural information content. Remotely sensed data are objective, systematic, and can be readily added to a GIS for comparison to other spatially referenced data layers. The ability to efficiently collect data over large areas is perhaps the greatest advantage of remote sensing. Yet, some amount of interpretation, fieldwork, or modeling is generally required to convert the spectral data into relevant information for the end-user.

Earth-resource research with remote sensing began in earnest with the launch of Landsat-1 in the early 1970s. Agriculture was one of the early focus areas, spawning significant joint NASA-USDA efforts in the LACIE and AgRISTARS programs for grain inventory and forecasting (MacDonald and Hall 1980). These and other efforts examined the physical principles relating spectral measurements to crop properties and pioneered methods for mapping crop type and condition from satellite. Today, a wide variety of spaceborne and airborne systems are available for multispectral sensing. Agronomic applications continue to evolve for such aspects as crop growth, nutrient status, disease, infestation, and evapotranspiration (Pinter et al. 2003).

Some 75,000 km^2 of the Earth's surface are engaged in grape production, the majority of which is used for wine. Geographic factors such as climate, soils, and topography affect grapevine viability and productivity (Dickenson and Salt 1982; Vaudour 2002; Tonietto 2008). Influence of these factors can be expressed as alterations in the biophysical or biochemical properties of the grapevine canopy in ways that may be amenable to spectral detection. The following two sections review some research aspects of viticultural remote sensing, starting with development of techniques for on-farm management and followed by regional land cover mapping. The final section provides an overview of prototype advisory systems that have been developed for production support in various parts of the world.

Site-Specific Viticulture

Site-specific viticulture, also termed precision viticulture (PV), refers to the collection and use of data relating to crop performance attributes of individual production units such as the vineyard block (Bramley and Hamilton 2004). Benefits can include increased productivity, enhanced environmental protection and higher profitability. In many regions, large capital investment now requires that growers strive to realize the maximum potential of the vineyard to include paying attention to marginal zones. Yet, as vine growth conditions can vary substantially within-block, inadequate characterization can lead to under-performance of production quality and volume. Site-specific practices use tools that provide information about the variable cropping system, and also afford greater ability to manage and perhaps modify variations more precisely.

A number of ground-based and remote sensing technologies are developing to support PV. Broadly speaking, these fall into the following categories: yield monitoring, in-vineyard fruit quality monitoring, GPS geo-referencing, soil monitoring, vine water status monitoring and variable-rate technology, and canopy vigor monitoring (Tisseyre et al. 2007). Research is being conducted in wine production areas worldwide to advance site-specific management. Some efforts have been made to develop sensing systems and associated measurement protocols to enable finer-resolution mapping than were previously available in viticulture. Other projects are exploring methods to quantify within-block variability and assist managers with data analysis and decision making, e.g., Johnson et al. 2001; Acevedo-Opazo et al. 2008. The goal is to allow viticulturists to improve and ultimately optimize wine grape cultivation systems.

Vigor

Perhaps the most widely studied aspect of PV remote sensing is with respect to vine vigor. Vineyard leaf area, for instance, is related to fruit ripening rate (Winkler 1958), infestation and disease (Wildman et al. 1983; English et al. 1989), water status (Smart and Coombe 1983), fruit characteristics, and wine quality (Smart 1985; Jackson and Lombard 1993). Canopy management methods for leaf area regulation include pruning, shoot thinning, leaf removal, cover crop cultivation, irrigation scheduling, and application of soil and crop amendments. Despite its importance, growers have no efficient way of regularly mapping and monitoring leaf area during the growing season. Direct measurement of leaf area by leaf removal is accurate but time consuming and destructive to the vine. Indirect methods include measurement of canopy-intercepted solar radiation, regressions based on shoot length and shoot number, and post-season collection of pruning weights (prior season woody production). While reasonably quick and accurate, these point-based techniques are not spatially exhaustive and hence require interpolation for mapping applications. Obviously, with the latter measurement, any resulting map will be strictly retrospective.

Remote sensing offers a means of vastly increasing sampling density with respect to that achievable by ground-based methods alone. Multispectral remote sensing is highly sensitive to quantitative expressions of plant canopy vigor, such as leaf area index (LAI; the ratio of leaf surface area to ground area), fractional cover, biomass, and shaded area. These quantitative variables can provide a basis for modeling vegetation canopy water and carbon flux, e.g., Heilman et al. (1982), Nemani and Running (1989), Allen et al. (1998). Various spectral indices can be formed from response in visible-region bands, where chlorophyll absorption dominates, and the near infrared (NIR) where vegetation is highly reflective. Perhaps most commonly used is the normalized difference vegetation index, or NDVI, which is formulated as (NIR-red)/(Knarred) (Tucker 1979).

The relationship between vegetation indices and leaf area has been explored for many annual crops (Wiegand et al. 1991). Somewhat less attention has been paid to this topic in vineyards and other perennial agricultural systems. Vineyard canopies can present some remote sensing challenges. The canopies are highly discontinuous with foliage clumped in individual vines or along rows with relatively low overall ground cover fraction. Soils may contain understory foliage as cover crop or weeds depending on cultivation practice and time of year. In addition, canopy architecture can vary among fields due to use of differing trellis (shoot positioning) systems.

Despite factors such as these, several studies over the past decade have demonstrated the use of multispectral remote sensing as a tool for vineyard vigor mapping. In Spain's Castilla-La Mancha region, for instance, strong linear relationships were reported between Landsat Thematic Mapper (30-m resolution) NDVI and vineyard LAI, fractional cover, and aboveground biomass production (Montero et al. 1999). The imagery was used to track regional phenology and provide insight to the effect of water availability on vine development.

In California's North Coast, Dobrowski et al. (2002) used a combination of field spectroscopy and 0.5 m resolution airborne data to examine three vegetation indices in relation to canopy density expressed as leaf area per meter of row. The indices were the NDVI, ratio vegetation index (RVI; NIR/red) and the perpendicular vegetation index (PVI) – a more complex formulation involving the difference in red and NIR response from both soil and vegetation (Richardson and Wiegand 1977). All indices were significantly correlated with canopy density. The NDVI and SR performed similarly, but the RVI maintained a linear relationship over a broader range of leaf area values. The PVI had weaker sensitivity and was more difficult to parameterize.

Additional work in California compared high-spatial resolution (4 m) satellite imagery compared with ground-based LAI measurements (Johnson et al. 2003). The ground data were taken at 23 sites in the Napa Valley and the adjacent, more maritime Carneros Region. Seven different red and white cultivars were included, some vertically shoot positioned and some untrained. As is typical of the region, LAI did not exceed 3, and ranged as low as 0.5 in the selected sites. A significant linear relationship ($R^2=0.73$) was

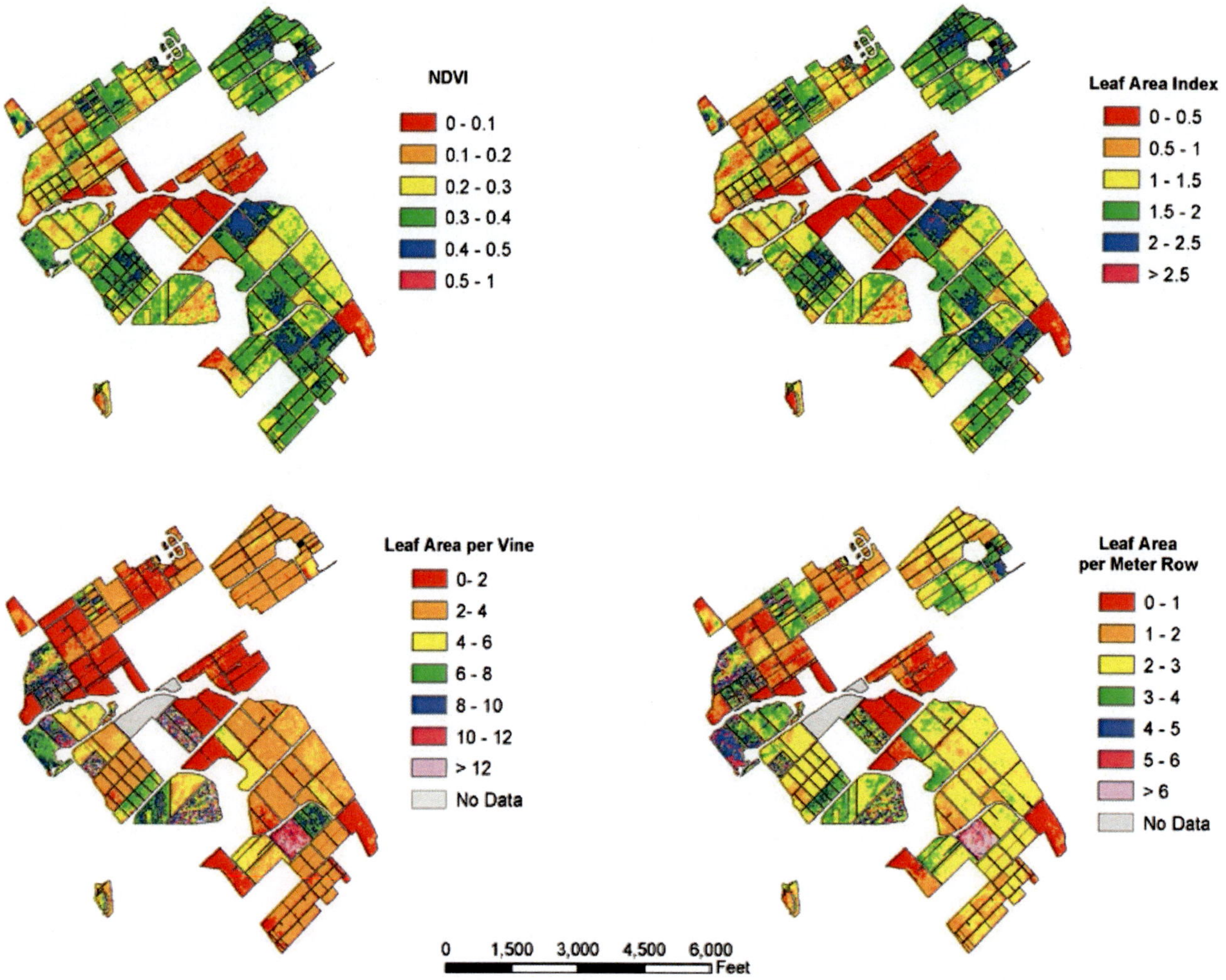

Fig. 12.1 Various expressions of vine vigor mapped in Napa Valley with 4 m resolution IKONOS satellite imagery, 2001

observed between the two variables and used to map LAI throughout two vineyard properties. Information on row- and vine-spacing per block was then incorporated to map leaf area on a per-vine and per-meter-of-row basis in order to facilitate grower interpretation (Fig. 12.1). Complementary satellite-based research on a single Napa Valley property found that the NDVI-LAI relationship was stable throughout the growing season, even in the absence of any sort of atmospheric correction procedure being applied to the imagery (Johnson 2003).

In Australia, airborne NDVI imagery was used for ultra-high resolution mapping (0.25 m) in a 1 ha commercial block of Cabernet Sauvignon (Hall et al. 2003). The imagery was processed with an algorithm called Vinecrawler to automatically separate pure vine pixels from mixed and soil pixels, then to extract metrics on canopy size, foliage density, and shape. Ultimately, a vigor map was produced at the per-vine level by combining the imagery with the block's row/vine coordinate system. An inverse relationship between vine size and anthocyanin content – an important fruit quality indicator – was further noted as was a positive relationship with plant nitrogen. Thus, supporting remote sensing technology is available for growers who have the desire and ability to manage blocks on a per-vine basis.

Trout et al. (2008) studied the relationship between NDVI and crop cover in California's San Joaquin Valley. Cover of grapes and several other horticultural crops was measured on the ground with a downward-looking camera suspended above the canopy and post-processed with software designed to automatically segment the photographs into vegetation and soil components. Landsat Thematic Mapper imagery was obtained on days corresponding crop cover measurement during mid-season 2005 and 2006 and converted to NDVI for each study field. The NDVI was strongly related to crop cover (R^2=0.95) for this diverse dataset, which included a range of different types, canopy structures, and growth stages. The NDVI began to lose sensitivity above 80% cover, a point generally regarded as effective full cover for water management applications (Neale et al. 2005).

Field research on a large-weighing lysimeter at the University of California's Kearney Agricultural Center showed that grapevine water use is linked to shaded area (radiation interception) cast by the canopy near solar noon (Williams and Ayars 2005). Lysimeters, which are considered the most precise way to measure crop evapotranspiration (ETc), are essentially plant containers resting on an underground scale. Short-term weight loss from the container is mainly due to leaf transpiration and soil evaporation; water is replaced by automated irrigation to maintain weight. The Kearney results showed a strong linear relationship (r^2=0.95) over two seasons between noontime shaded area and crop coefficient, which expresses ETc as a proportion of ET from a grass reference surface. In related airborne research, Johnson and Scholasch (2005) reported that 2 m resolution NDVI imagery was highly sensitive to shaded area in several vineyard blocks ranging from 15% to 50% shade, and thus offered potential for ET mapping. Whereas the foregoing study exploited correlation between vine vigor and shaded area, higher resolution imagery (0.25 m) can potentially support direct observation and mapping of shaded soil in the inter-row gaps (Wassenaar et al. 2001).

Viticulturists sometimes use dormant-season pruning weights – the removal of prior-season woody production – to characterize vine vigor. These measurements can provide a quantitative estimate of leaf area or biomass production for the prior season and can suggest management changes that might be made in the following season. Also, these data can be combined with fruit yields to evaluate vegetative and reproductive balance, a possible indicator of fruit quality (Smart and Robinson 1991). Dobrowski et al. (2003) explored the use of remote sensing for mapping pruning weights over a California North Coast vineyard containing five different within-row vine spacing ranging from 0.91 to 3.35 m. The RVI was significantly correlated with Cabernet Sauvignon pruning weights, and the relationship was consistent over two consecutive seasons. Stamatiadis et al. (2006) reported a consistent quadratic relationship between NDVI and pruning weights over two seasons using a tractor-mounted sensor over Merlot blocks in Greece.

Physiology

Chlorophyll is a key physiological indicator linked to plant productivity, developmental rate, and stress level. Remote sensing of chlorophyll in crops and other plant canopies has been an active research focus (Johnson et al. 1994; Daughtry et al. 2000; Haboudane et al. 2002; Gitelson et al. 2005), though viticultural application has been modest in comparison to efforts addressing vine vigor as summarized above. Airborne hyperspectral imagers were used to assess vineyard condition in northern Spain (Zarco-Tejada et al. 2005; Martin et al. 2007). As the name implies, hyperspectral sensors are specialized instruments that collect imagery in a greater number of contiguous spectral bands than multispectral sensors and with typically narrower bands. Several spectral vegetation indices were evaluated, including the NDVI. The best-performing single index was the transformed chlorophyll absorption reflective index (TCARI) using a relatively complex formulation including bands in the green (550 nm), red (670 nm) and red-NIR transition (700 nm) spectral regions. The best overall performer was the combined TCARI/OSAVI – the optimized soil-adjusted vegetation index – which additionally considered an NIR band at 800 nm. Success of these indices was linked to their suppression of soil background and LAI spectral influence. Indices designed for biomass monitoring, such as NDVI and RVI, had relatively low power for of leaf chlorophyll retrieval in the presence of these confusion factors.

Anthocyanin (Anth) is an accessory pigment that can be produced in response to a variety of stress factors including ultraviolet radiation, drought, herbivory, fungal pathogens, nutrient deficiency, salts, disease, and temperature (Chalker-Scott 1999; Close and Beadle 2003; Eryilmaz 2006). Spectral methods have recently been recommended for ground-based Anth estimation in individual grape leaves (Steele et al. 2009). The proposed indices use green reflectance corresponding to the Anth absorption peak, combined with reference measurements in the red-NIR transition (700 nm) and the NIR (780 nm), both of which have little sensitivity to Anth. These indices have yet to be tested in a remote sensing context. Others, however, have found negative correlation between canopy biomass (through NDVI) and Anth in fruit, a key red-grape quality indicator (Lamb et al. 2004; Stamatiadis et al. 2006; Hall et al. 2011). Grapes produce Anth and other flavonoids in reaction to direct sunlight exposure, and bigger canopies might be expected to provide a shadier micro-environment. Thus, NDVI or other vigor-related indices may provide insight to fruit composition and hence to potential wine quality.

Remote sensing has found some applicability with regard to pest infestation in vineyards. Wildman et al. (1983) used aerial color infrared photographs taken annually from 1978 to 1981 to monitor the spread of grape phylloxera in Napa Valley. Manual interpretation of the photographs was performed under magnification on a light table. In addition, the photographs were digitized to 256 shades of gray and submitted for computer analysis. Both methods were used to estimate annual counts of infested vines, and produced similar average annual increase factors over the study period. Computer processing was applied to digital multispectral imagery during a subsequent, more severe phylloxera outbreak (Johnson et al. 1996). NDVI imagery over two successive years, 1993–1994, was related to pruning weights taken from several study plots located within a single 5 ha block of Cabernet Sauvignon. To show year-to-year changes in canopy vigor, the 1994 scene was digitally superimposed upon and subtracted from the 1993 scene. The resulting image quantified the rate of decline in infested plots and showed some year-to-year vigor increases in healthy portions of the block. Phylloxera-related stress detection was found to enable ranking of fruit yields from different study plots up to 16 months before harvest (Baldy et al. 1996).

Land Cover Mapping

Land cover is a main driver of ecosystem functioning and biogeochemical fluxes, and the scientific community acknowledges the importance of land cover monitoring with respect to global change studies (National Research Council 1988; Townshend et al. 1991). Within viticulture, anticipated changes in climate may cause dislocations in grape production, with warmer conditions leading to poleward expansion and changes in cultivar or possible abandonment in areas that are currently near their climatic limit (Tate 2001; Jones 2005). Maps can be integrated with GIS databases to help land managers evaluate and cope with pollution, erosion, and flood risks.

Remotely sensed imagery can provide an efficient and repeatable means for creating and updating land cover maps. Orbital systems are particularly well-suited for regional mapping due to their frequent update and large field-of-view. An early study explored the potential use of Landsat Multispectral Scanner imagery (80 m resolution) as a complement to conventional aerial photography for U.S. Department of Agriculture (USDA) crop inventory in California's San Joaquin Valley (Hay 1974). Human photo-interpreters used a single hardcopy scene acquired in mid-summer to distinguish vineyards and orchards from other categories including field crops, pasture, and rangeland based on field size and color.

Various methods for automated image classification have been developed over the years with the objective of automatically categorizing all image pixels into land cover classes or themes. The most common category of algorithms involves spectral pattern recognition which assigns cover types based on their inherent reflectance properties within the spectral range of the sensor. Spectral recognition may be enhanced by multi-temporal imaging which is the use of sequential images to observe phenological progression as over a growing season. Typically the classes of interest are predefined, and representative samples of spectral patterns per category are provided to train the classification routine. A sampling effort is usually made to check map accuracy by comparing the output with ground reference data, for classes of interest.

A common approach is the maximum likelihood classifier. The probability of an input pixel belonging to each output class is calculated with respect to the training set statistics, and the assigned output class is simply that with the highest membership probability. Maximum likelihood classification was applied to early- and late-season Landsat Thematic Mapper data to map vineyards in New York (Trolier et al. 1989). Omission error, or the probability of omitting a vineyard pixel from the vineyard category, was just over 10% for the June scene and was judged as acceptable. Commission error, or the probability of assigning a non-vineyard pixel to the vineyard class, was deemed excessive (>20%), primarily due to confusion between minimally managed, weedy and unpruned vineyards, abandoned vineyards, and natural vegetation.

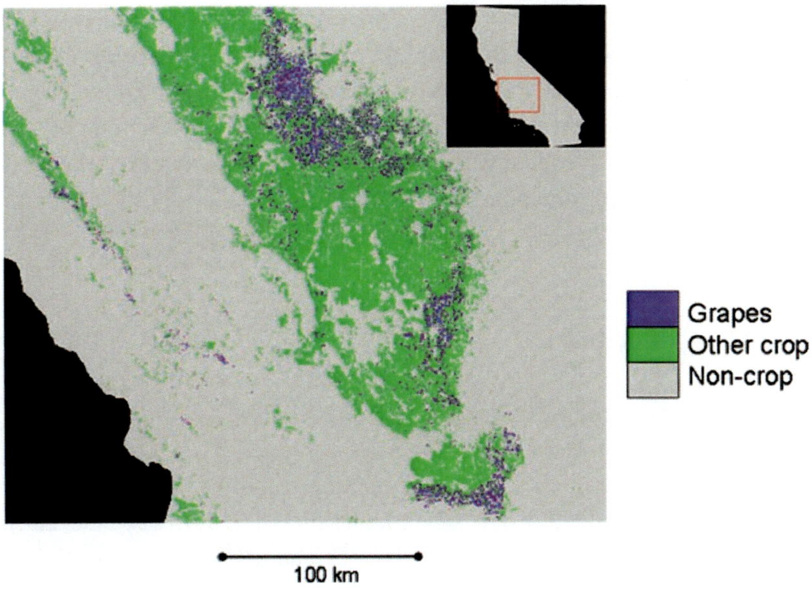

Fig. 12.2 California's San Joaquin Valley vineyards mapped by 2007 USDA/NASS Cropland Data Layer using Landsat Thematic Mapper imagery at 30 m resolution

Decision tree classification is a more recent algorithm based on advances in the field of machine learning, e.g., Friedl and Brodley 1997. The classifier recursively subdivides the input pixels by a series of tests devised by learning-algorithms applied to spectral patterns of the training data. Decision trees are fast, intuitive, nonparametric, somewhat noise-insensitive, and can provide higher classification accuracy than maximum likelihood. The USDA National Agricultural Statistics Service processes satellite imagery with decision trees to periodically generate state-wide crop maps under the Cropland Data Layer (CDL) program (Johnson 2008). The CDL exploits an extensive set of ground-based training and validation data compiled by the USDA June Agricultural Survey and USDA Farm Service Agency. A CDL map for California was compiled in 2007 with multi-temporal Landsat Thematic Mapper imagery and supporting spatial data from NASA's Moderate Resolution Imaging Spectroradiometer and the U.S. Geological Survey's National Land Cover Dataset (Fig. 12.2). About 60 crops were mapped statewide, including vineyards, with reported omission and commission errors for grape near 3% and 5%, respectively.

The ability to readily compare and monitor land cover changes over time is a strength of remote sensing. Change detection procedures were used to map dry-farmed vineyard conversions in an area of central Spain undergoing desertification (Lanjeri et al. 2004). While total vineyard loss in the region is generally quantified, spatially explicit information as to where changes occur is lacking, as is the specific nature of the conversions. Multitemporal Landsat Thematic Mapper scenes were used to establish typical seasonal NDVI profiles for active vineyards, abandoned vineyard, cereal, and bare soil. Early May was found to be the best time to differentiate vineyard from both cereals and abandoned vineyard, and mid-season, or July, was best for separating vineyard from bare soil. Comparison of NDVI-based classifications between 1991 and 1996 showed that 28% of the region's active vineyard area was lost during the 5-year period with about 17,000 ha converted to cereals or abandoned vineyard and another 7,000 ha converted to bare soil.

Another common classification approach is based on automated recognition and analysis of image spatial or textural patterns rather than on spectral content. These approaches exploit the distinctive periodic spatial patterns associated with vineyards and orchards. Textural methods require much finer spatial resolution (<1 m) than the spectral methods in order to resolve individual vines or vine rows. On the other hand, they use only a single band rather than multiple spectral channels; and, hence, may be of particular utility in instances where only black-and-white (panchromatic) photographs are available. Wassenaar et al. (2002) applied frequency analysis to digitized aerial photographs (0.25 m resolution) to automatically identify French vineyards and further, to extract textural descriptors related to row spacing, orientation, and trellis system. Warner and Steinmaus (2005) analyzed spatial autocorrelation in panchromatic IKONOS satellite imagery (1 m) to detect row structure in Washington

State vineyards, finding that classification accuracy of the textural approach greatly exceeded that achieved by maximum likelihood spectral classification. Research continues to produce textural algorithms that are robust with respect to variations in row spacing, grass or weed cover, missing vines, and image distortions, e.g., Da Costa et al. 2007; Delenne et al. 2008.

Production Support

Viticulturists and enologists worldwide are making increased use of remote sensing technology for practical management. Main uses include planning harvest at sub-block level to yield more wine lots for blending with improved lot consistency, establishing and adjusting micro-irrigation to improve block uniformity, timing the replacement of blocks afflicted by disease or infestation, delineating zones that need to be consistently disked to evaporate excess water or mulched to retain water, and early recognition of disease or other problem areas in the vineyard. These varied efforts are accomplished largely by qualitative inspection of color-coded NDVI images combined with on-ground investigation.

This section outlines some prototype advisory systems that are being developed for vineyard management. These services allow for exploitation of quantitative remote sensing aspects, as well as integration with ancillary spatial datasets and application models for generation of higher-level products. The first three systems address irrigation management by water-balance and energy-balance methods, while a fourth system characterizes vineyard variability in a non-irrigated region.

Australia

An advantage of growing grapes primarily under irrigated conditions is the ability to control timing and amount of water application in order to maintain root zone soil moisture in a preferred range. Irrigation scheduling, which involves determining when to irrigate and how much water to apply, can provide critical support to grape yield and quality objectives. With growing concern that water resources are insufficient in semi-arid regions to satisfy multiple users and interests, water supply must be well planned and efficiently consumed. Information on the amount of actual ET is essential for evaluating and monitoring crop water requirements, plant water stress, fruit production, irrigation supply, and water diversions. Winegrowers frequently practice deficit irrigation to enhance grape quality, and the resulting plant stress (stomatal regulation) causes ET to fall below what would be expected of a fully watered canopy. Of course, too great a stress level can adversely affect fruit maturation and vine health. The practical implication for vineyard stress management is simply that actual ET needs to be monitored, and ideally, mapped in a timely fashion.

Objective management tools are needed for irrigators to improve water use productivity and efficiency. Current indications are that only about 20% of vineyard irrigators in Australia's Murrumbidgee Irrigation Area use some form of quantitative irrigation scheduling method such as soil moisture monitoring. Adoption limitations include system cost and complexity of data or information provided, thus reducing the usability and practical application for irrigators.

The use of weather station information in irrigation scheduling has long been used in the scientific domain for predicting crop water requirements and scheduling irrigations (Allen et al. 1998). The general, widely accepted approach is based on application of reference station ET measured over a reference surface such as grass (ETo). This ETo is used to represent prevailing weather conditions and is modified by a crop coefficient (Kc) to estimate crop evapotranspiration (ETc). Plant canopy size, or fractional ground coverage, is a primary determinant of Kc for a particular crop type. Fractional coverage varies by time of year, canopy vigor, plant density, and, especially in the case of vineyards, method of vine training. Collecting sufficient plant canopy data to calculate Kc and thus ETc over a landscape can pose a challenge for ground-based methods.

An approach using satellite data and mobile phone delivery of information is being developed for Australian vineyards. The IrriSatSMS system aims to simplify input data collection requirements and reduce both costs and complexity of information output (Hornbuckle et al. 2009). At the center of the system is a server that acts as a data collection portal for various data feeds and a processing engine to convert these data into usable irrigation management information. The system relies mainly on a Short Messaging Service (SMS) interface to communicate with irrigators directly on their mobile phones. Satellite images

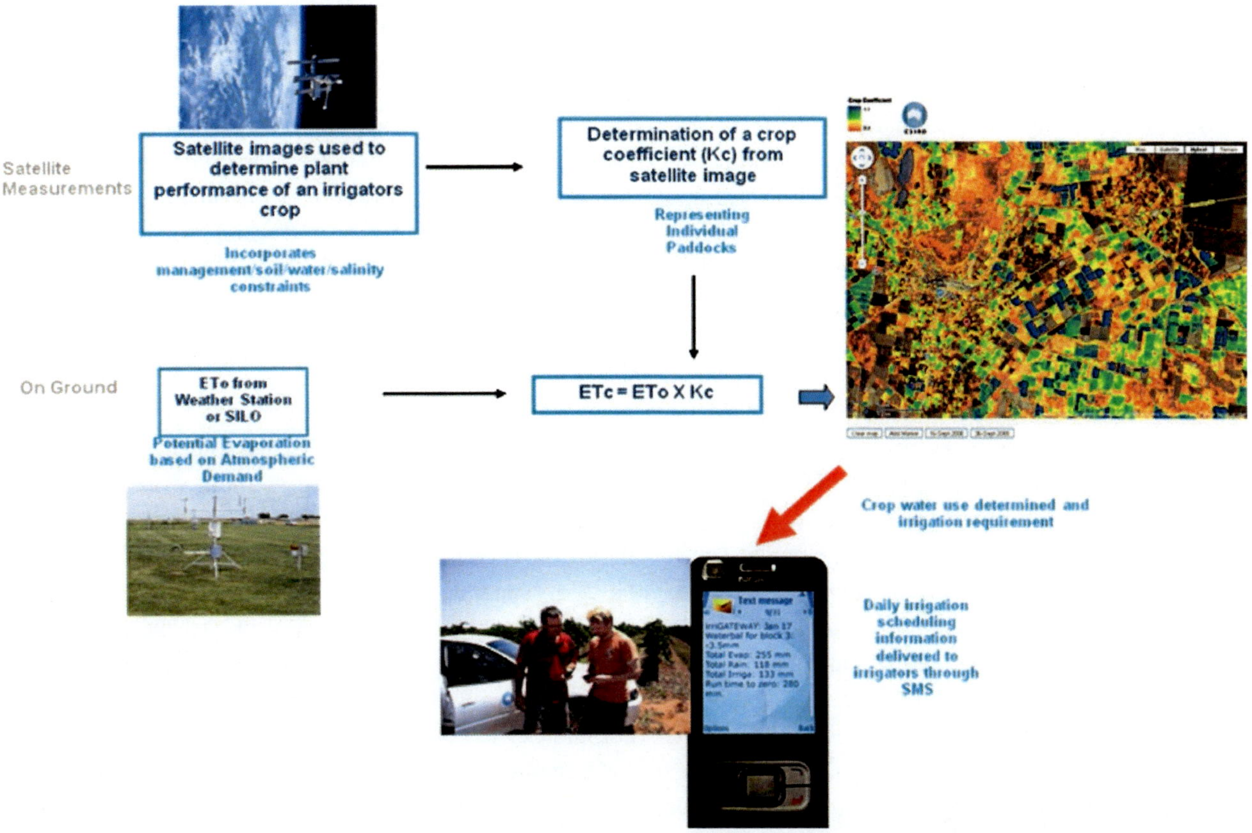

Fig. 12.3 Overview of the IrriSatSMS approach for irrigated water management being applied in Australian vineyards

collected across Australia every 14–20 days are retrieved from Landsat 5 Thematic Mapper within approximately 6 h of observation. Plant canopy size is estimated by NDVI and used to derive Kc, assuming non-stressed conditions, for individual irrigation management units. This information is then combined with real-time weather station data to determine maximum daily crop water use (Fig. 12.3).

Water balance (mm) for a particular irrigation block is calculated as $WB = R + I - ETc$, where R is rainfall depth, I is irrigation volume, and ETc represents daily water use absent stress effects. IrriSatSMS thus provides an indication of how much water the crop has potentially used since last irrigation. Past irrigation run times and rainfall data, required to monitor WB, are sent from irrigator to server by SMS or web interface. The system uses a known application rate of the specific irrigation system, as provided by the grower, to determine how long irrigation pumps/drippers should be operated in order to fully replace ETc. Brief, outgoing messages to growers' mobile phones or computers convey detail on water balance components – precipitation, applied irrigation, evaporation– along with associated pump run time needed to restore water balance. An additional web-based feature allows growers to anonymously compare their water use with that of nearby irrigators. Growers decide when and specifically how much to irrigate after consideration of all relevant agronomic, economic and social/cultural factors. Irrigators who wish to induce, maintain or modify vine water stress may intentionally opt for less than full water replacement at certain times of year as needed.

IrriSatSMS was developed and piloted in the Murrumbidgee Irrigation Area with vineyard irrigators. During the initial 2008–2009 season, about 20,000 mL of irrigation water was managed by the system, and the approach is now being extended to other irrigated regions and crops across the Murray Darling Basin in southeastern Australia. The IrriSatSMS approach is scalable as the server database and SMS systems can easily handle from hundreds to thousands of individual users or irrigation units. Geographic extent is non-limiting due to comprehensive availability of satellite data. There are several feedback, control, and input mechanisms provided during the process allowing targeted interactions between IrriSatSMS and the end-user. While the underlying processes are somewhat complex, the system is designed for ease of use and flexibility to the end-user in respect to how information is retrieved and communicated back to the system. The prototype system aims to provide growers with an easy to use and understandable daily irrigation water management service, and to provide a benchmarking and auditing mechanism to be used by growers and water providers.

California

The Vineyard-Soil Irrigation Model (VSIM) is another example of a water balance method under development in Napa Valley. The daily water balance model simulates soil moisture, vine water stress, and vine water use (ETc) based on spatial and temporal variations in vine density, soils, and weather. Input data include daily weather (ETo, precipitation, air temperature), soil properties (texture, rooting depth), and peak vine LAI, which may be estimated by remote sensing. The model computes daily vine canopy LAI, from bud break through canopy leaf fall using a growing-degree-day model that relates daily temperature summations to a peak temperature summation and peak LAI. Kc is estimated for both vine and any cover crop, e.g., grass and weeds. For vines, VSIM assumes a Beer's Law relationship between canopy light attenuation and LAI. Notably for deficit-irrigated vineyards, Kc is adjusted downward when estimated vine water stress increases in order to simulate the effect of stomatal regulation. Cover crop Kc is simply estimated as percent green cover which also may be estimated by remote sensing prior to vine bud-break. Soil properties influence storage and release of water from the soil moisture reserve. VSIM also accepts any management information that affects vine water balance, such as cover crop removal, applied irrigation, and vine pruning. The model has been validated with field measurements of soil moisture and stem water potential (vine stress level).

VSIM has been used to build comparative irrigation guidelines for major Oregon viticultural areas based on yearly and statewide variations in weather and climate. More recently, the model forecasted seasonal irrigation needs for a 300 ha Napa Valley vineyard that is subdivided into 75 different blocks based on vine spacing, trellis type, and variety. Typical peak LAI per block was estimated from airborne imagery by relating NDVI to ground-based measurements of LAI (after Johnson et al. 2003). Soil properties were interpolated from sample pit data and weather data were downloaded from the California Irrigation Management Information System. The model was used to estimate dates of first irrigation and total seasonal irrigation for each block based on measured weather through May 2009 and assuming that weather for the remainder of the season would be similar to that of the prior year.

VSIM was integrated with NASA's Terrestrial Observation and Prediction System (TOPS) (Nemani et al. 2009), which is designed to seamlessly combine data from remote and ground-based sensors with application models. The TOPS infrastructure supports the development of automated weekly irrigation forecasts based on U.S. National Weather Service 7-day forecasts. In specific, cloud cover forecasts are converted to ETo by calculating and applying an atmospheric transmittance factor to an estimate of potential radiation at the ground surface. On a weekly basis, TOPS automates the download and preprocessing of remote sensing and weather data, reads in the vineyard block data, initiates a VSIM model run for each block, and summarizes and archives the results. Water balance variables are provided through the current day and forecast for the following week. These estimates are made available over the Internet in map, graphical, and tabular formats for assistance in irrigation management (Fig. 12.4). Future versions of TOPS-VSIM will link directly to grower databases to incorporate management information, develop estimates of water balance variables for the entire season based on 3–6-month climate predictions and automate comparison of simulated and measured water balance variables.

The model can be used to develop vineyard irrigation requirements that are sensitive to variations in soils, vine density, and target levels of vine stress. The VSIM model is designed to help the vineyard manager consider all of the various factors that control vine water balance. Models such as VSIM can be used to help maintain strict irrigation controls based solely upon vine water needs, allowing growers to develop deficit irrigation schedules without compromising crop yields.

South Africa

Another commonly accepted method is derivation of ETc from surface energy balance. The latent heat flux

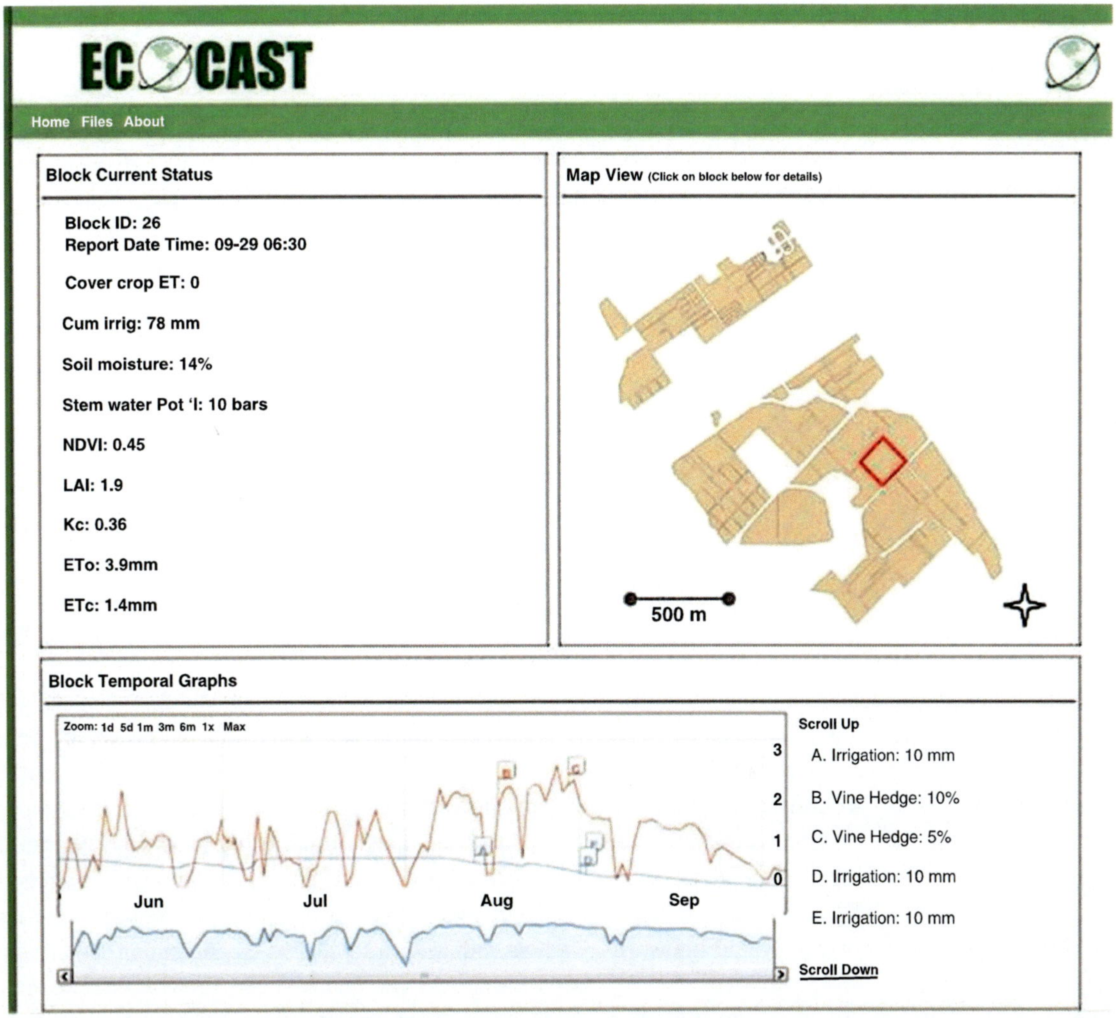

Fig. 12.4 Example user interface for VSIM water balance model in a Napa Valley vineyard

of the surface energy balance describes the amount of energy involved in transferring water from liquid to vapor phase. Energy flux (W/m^2) can be converted to water flux (mm/day) according to physical constants. Net solar radiation reaching the Earth's surface (Rn), balances with soil (G), sensible (H), and latent heat (ET) fluxes to conserve energy. Latent heat flux can be estimated as a closure term from estimates of the remaining fluxes: $ET = Rn - G - H$.

Remote sensing measurements of energy balance parameters can be used to estimate Rn, G, and H, and hence, ET fluxes for every pixel in an image. Thermal infrared measurements provide an estimate of crop canopy temperature which is an excellent indicator of evaporative cooling (Fuchs and Tanner 1966), and, thus, indirectly describes evaporative flux and root zone soil moisture. Well watered crops with high ET flux have a low canopy temperature (reduced H flux). Crops with water stress respond with low ET flux and greater H flux.

Researchers have developed numerous energy balance models that use canopy temperature and other remote sensing measurements to estimate ET at various spatial and temporal scales, e.g., reviews of Courault et al. (2005) and Kalma et al. (2008). Of these, the most widely applied and validated is the Surface Energy Balance Algorithm for Land (SEBAL). Multispectral imagery that includes a thermal band is used to parameterize SEBAL for surface albedo (reflectivity), NDVI, and surface temperature (Bastiaanssen et al. 2002; Teixeira et al. 2008). The model then estimates ETc from energy balance by applying radiative, aerodynamic, and energy balance physics in 25 computational steps. The approach inherently accounts

Table 12.1 Average water productivity in the various wine valleys of South Africa's Western Cape. Water productivity (kg/m³) expresses grape yield (kg/ha) per unit of ET (m³/ha)

	Table grapes				Wine grapes			
Name	2004–2005	2005–2006	2006–2007	Avg	2004–2005	2005–2006	2006–2007	Avg
Paarl South	6.19	5.54	2.71	4.82				
Paarl North	4.43	4.28	2.60	3.77				
Worcester (Hex River)	2.37	1.40	1.72	1.83	2.99	3.85	3.56	3.46
Worcester South	3.81	2.54	3.57	3.31	4.10	5.07	4.08	4.42
Worcester (Nuy)	2.10	1.34	2.71	2.05	3.10	3.87	3.68	3.55
Hex River Valley (Drieriviere)	3.18	2.45	3.64	3.09				
Hex River Valley (downstream)	4.01	3.40	4.88	4.10				
Hex River Valley (upstream)	4.08	5.18	5.39	4.88				
Hex River Valley (Groothoek)	2.99	3.25	3.94	3.39				
Average	3.69	3.26	3.46	3.47	3.40	4.26	3.77	3.81

for stomatal regulation in water-stressed canopies. Validations of SEBAL have shown that seasonal ETc results for a variety of land use classes and crop types generally fall within 5% of ground-based measurements (Bastiaanssen et al. 2005). Results for grapes in Spain, Turkey and Brazil show that SEBAL daily evaporative fraction, or ratio of latent heat flux to net available energy, agrees closely with ground measures (Bastiaanssen et al. 2008).

In South Africa as elsewhere, agriculture is seriously limited by water availability. The grape industry is vitally important to the economy of Western Cape Province, and a challenge is to maintain economically sustainable production while optimizing use of water resources. An overall objective is, thus, to improve water use efficiency (WUE) or fruit production per unit of water consumption (ETc). Since WUE can vary widely from year to year depending on site specific – soil, grape variety– and climatic – rainfall and temperature – conditions, as well as total irrigation water available from upstream reservoirs, regional averages are not particularly meaningful.

SEBAL was used to assess spatial and temporal variation of WUE for several vineyard regions including Hex Valley, Paarl, Worcester, Somerset, and Stellenbosch (Klaasse et al. 2008). A total of 35 Landsat Thematic Mapper scenes were used over three growing seasons (2004–2006). Water consumption was determined from surface energy balance as mentioned above by using thermal and optical spectral bands. The resulting seasonal ET estimates for each year were within 11% of ground-based water balance estimates derived at several experimental blocks. The agreement was deemed satisfactory due to some inherent limitations of the water balance data sampling process. Total biomass production (wood, leaves, fruit, roots) was estimated by converting NDVI to fraction of photosynthetically active radiation intercepted by the canopy, then applying a light use efficiency scalar (after Bastiaanssen and Ali 2003). Biomass was combined with stress factors and fruit production data from wine associations to parameterize a yield model for wine (Colombar) and table grapes (Dauphine, Sunred, Crimson). Finally, ET and yield estimations were combined to derive WUE.

The WUE maps were aggregated by region and year (Table 12.1) in order to examine overall trends. Perhaps most striking is the wide range in table grape WUE across regions (average 1.83–4.88 kg/m³). Due to prevalent use of irrigation, the range across years was relatively narrow, suggesting that geographic conditions play a bigger role in WUE than does annual weather. Table grape WUE Hex Valley upstream and Paarl South are the most efficient regions, approaching on average 5 kg of grapes produced per cubic meter of applied water. Wine grape WUE was more variable over time due to limited use of irrigation or dry-farming. Worcester South had the highest water productivity for wine grapes (4.42 kg/m³).

Spatially explicit, disaggregated information is required to promote local improvements and institute action programs to address WUE, e.g., Fig. 12.5. The

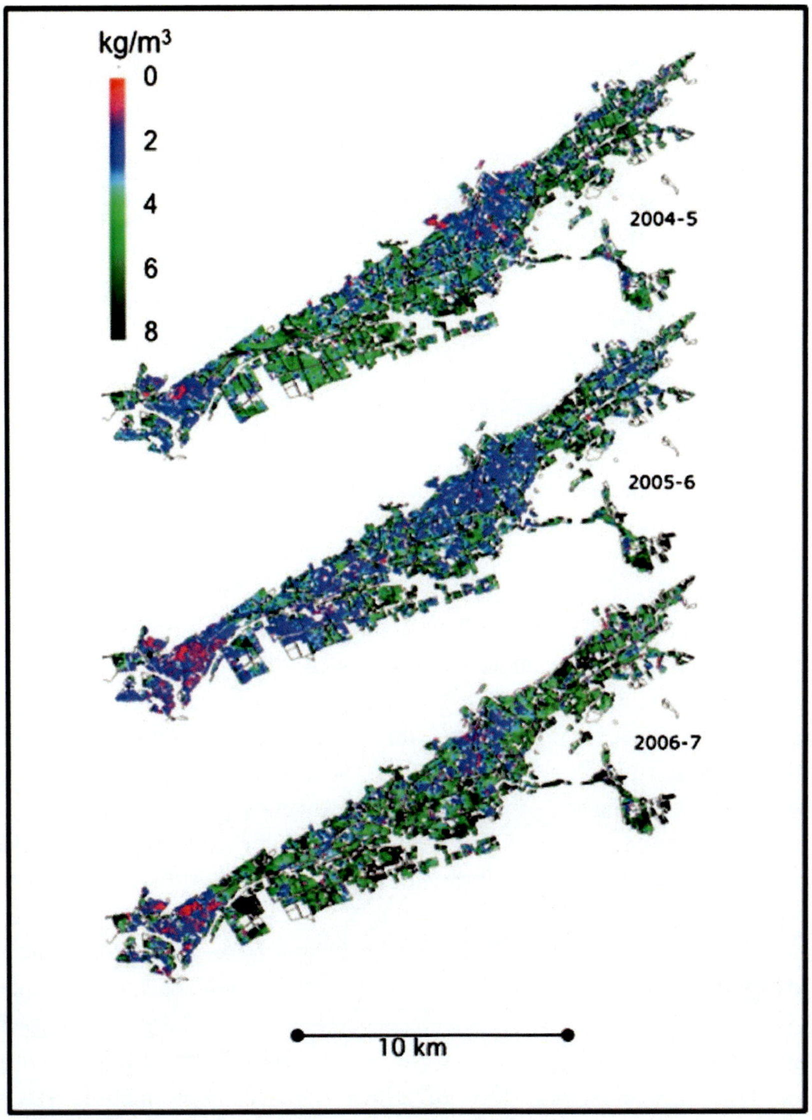

Fig. 12.5 Spatial variation of water use efficiency (fruit yield per cubic meter of applied water) evaluated by the SEBAL energy-balance model and Landsat for vineyards of Hex Valley, South Africa

relative constancy of spatial pattern during the 3-year period suggests that management – cultivar selection, farmer skills, irrigation, disease protection– and physical conditions – soil type, terrain slope, depth to water table– have great impact. The wide range of 1–7 kg/m^3 suggests that there is considerable room for improvement in specific properties and subregions. Hence, organized outreach and extension efforts would seem to be a worthy investment to narrow the gap between excellent and poor irrigation practices. The spatial variation within individual vineyards is of interest as well. Figure 12.6, for example, shows the detailed spatial patterns of crop water use (mm/season) at the 30-m resolution of the satellite image. It shows that irrigation management, especially, duration of irrigation, can have great impact on total water use, and thus indirectly on WUE.

To advance water-friendly grape cultivation practices, growers need up-to-date information on water consumption and biomass production during the season. Real-time information will enable farmers to adapt their water management to achieve more uniform results. The Department of Agriculture of the Western Cape is instituting an ET and biomass production monitoring and advisory system to provide this information during upcoming seasons. Growers can indicate the boundaries of vineyards for which they want to receive the information, and they will subsequently get weekly updates including maps of agronomic variables to include ETc, ET deficit, biomass production, WUE, leaf area index, crop nitrogen, soil moisture in the root zone, vineyard uniformity, and recommended irrigation schedule. The system will be similar to the FieldLook

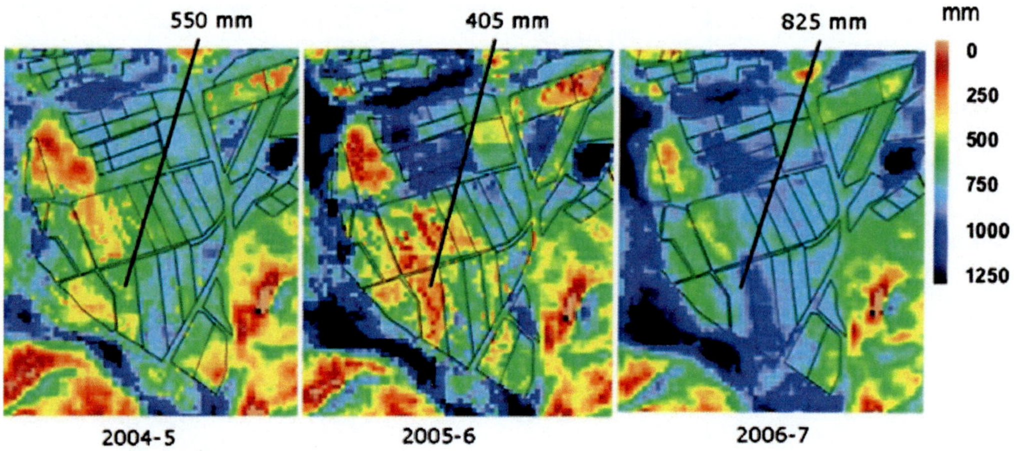

Fig. 12.6 Detailed pattern of seasonal on-farm water use in a South Africa vineyard modeled by SEBAL and Landsat

web-based information tool in The Netherlands and Belgium, an operational remote sensing service assisting farmers and advisors with irrigation and drainage management, fertilizer application, and pest/disease detection. More than 1,400 mainly wheat, potato, and sugar beet farmers have a subscription, and indications are that they are satisfied with this product. These concepts can be transferred to cultivation of grapes.

France and Italy

Although European vineyards constitute about 45% of wine areas globally and 60% of production, the European Union faces competition from several countries including the U.S., Australia, Chile, Argentina, and Canada. The European wine-making industry is highly fragmented, mainly based on family vineyards and other small landholdings. Grapes are collected at cooperatives and the collection area varies widely among regions. In contrast, innovative new-world producers are more aggregated and tend to be dominated by few, high production-volume leaders.

Due to the nature of its organization, the European production system faces several challenges. High *social variability* results from the great variety of winegrowers with which the cooperatives must deal. Growers are spread over large areas and are of varying age, and they have different equipment, computing ability, education, average area, and farm history. *Vine variability* can be high, and in some regions, it is possible to find up to 18 different varieties with different training systems. *Macro-scale variability* results from wide geographic dispersion and associated differences in climate, soil, elevation, slope, and other key environmental factors. *Micro-scale variability* can be encountered in vigor, ripening, and yield within a single vineyard block.

In order to maintain competitiveness, reorganization of Europe's vineyard industry is currently underway. These changes include merging of cooperatives, reduction of wine production in some marginal areas, and encouragement of vine planting in potentially favorable areas. Remote sensing and other geospatial technologies provide tools for managing production system complexity and improving production efficiency in viticultural regions. At European, national or regional, scales, these technologies may provide policy makers and other stewards with updated inventories of vineyard spatial distribution and condition, and facilitate land management decision making. At the cooperative/wine maker level, geospatial tools may enable managers to optimize production with respect macro-scale and social variability. At the individual vineyard level, these tools can provide information to more efficiently manage micro-scale variability for enhanced quality, yield, and revenue.

Two significant projects dedicated to viticultural remote sensing were recently launched. The Bacchus project was funded by the European Commission beginning in 2003 with a goal of demonstrating how geospatial technologies could provide relevant information to public administration or wine quality regulating bodies. Participants from leading wine producers – Italy,

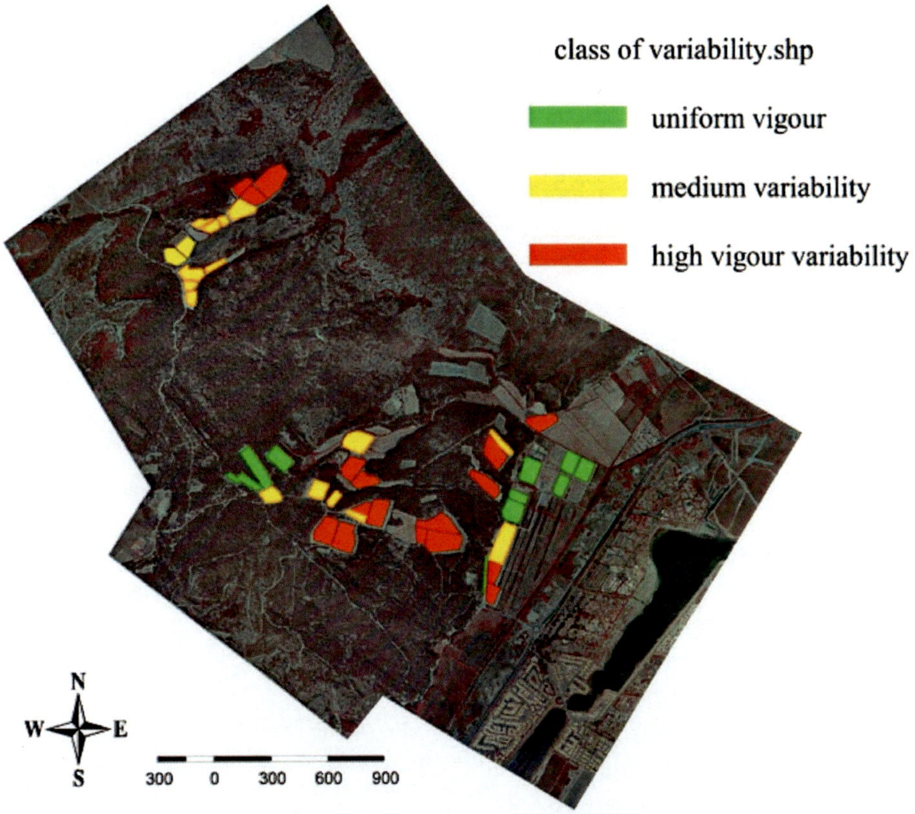

Fig. 12.7 Example Oenoview map of within-field vigor variability, derived from NDVI, for a 40-ha vineyard. Results for each vineyard block are superimposed on the original multispectral airborne imagery

France, Portugal, and Spain were included as vineyard managers in the private and public sectors teamed with scientists to define and implement a common approach to vineyard inventory and management. Two product types were developed: (1) an updated database of wine producing areas linked with information wine production such as grower name, location of estate, average production level, and (2) land management criteria according to vineyard spatial distribution and agroclimatic characteristics to provide managers with support for decisions concerning vineyard expansion or abandonment. A main remote sensing focus was development of a semiautomatic textural classification method for vineyard detection and mapping using high resolution satellites (IKONOS, QuickBird) and airborne imagery (Rabatel et al. 2008).

The Oenoview project is funded by France's Languedoc-Roussillon region, home to nearly 300,000 ha of vineyard, in collaboration with INRA (l'Institut National de la Recherche Agronomique). It involves research teams from Montpellier SupAgro and Cemagref, as well as companies like Institut Coopératif du Vin, Lattes (ICV) France, Infoterra (Toulouse, France), and l'Avion Jaune (Montpellier, France). Oenoview uses 8-m multispectral images from satellite (Formosat) or 0.5-m airborne imagery, depending on study area size and average block size. Images are converted to a vegetation index to derive green leaf cover fraction. Block boundaries, defined by GIS shape files provided by the end-user are used to extract block-by-block information from imagery. An index that accounts for both the magnitude and spatial pattern – autocorrelation of variability (Pringle et al. 2003; Tisseyre and McBratney 2008) is used to classify blocks into three categories: uniform, moderately variable, or highly variable (Fig. 12.7).

Oenoview aims to provide block variability information as early as possible, generally before véraison – the transition from fruit growth to fruit ripening. This information is intended to help managers of cooperatives to organize the harvest. Cooperatives deal with multiple growers, typically more than 100, and manage wine making from as many as 1,000 blocks. Block variability quantification will help to drive ripening controls via ground-based sampling in order to improve assessment of both yield and quality prior to harvest and can also be used to establish block selection for better quality management. Uniform blocks generally follow a standard scheme of selection based on classical ripening controls, and then grapes undergo an optimal vinification process according to predicted quality. Highly variable blocks follow a particular scheme of selection including complementary controls and block observations by viticulturists. Harvest from these blocks typically follows a basic vinification process dedicated to lower cost wines. Depending on demand and vintage, moderate blocks constitute a buffer class. This scheme of selection with Oenoview was tested for 2 years on several varieties at the cooperative level at Vignerons de MonTauch, Tuchan, France, where it significantly improved the quality of premium wines. Block classification is also relevant at the individual vineyard level, affording growers the opportunity to apply precision management techniques in regard to yield and quality targets. Oenoview was released in 2008, and the adoption rate is not yet established. The prototype system is still evolving, and improvements are under research and development.

References

Acevedo-Opazo, C., B. Tissseyre, S. Guillaume, and H. Ojeda. 2008. The potential of high spatial resolution information to define within-vineyard zones related to vine water status. *Precision Agriculture* 9:285–302.

Allen, R., L. Pereira, D. Raes, and M. Smith. 1998. *Crop evapotranspiration - Guidelines for computing crop water requirements*. Irrigation and Drainage Paper 56. Rome: United Nations Food and Agriculture Organization.

Baldy, R., J. DeBenedictis, L. Johnson, E. Weber, M. Baldy, and J. Burleigh. 1996. Relating chlorophyll and vine size to yields in a phylloxera-infested vineyard. *Vitis* 35 (4):201–205.

Bastiaanssen, W., M. Ahmad, and Y. Chemin. 2002. Satellite surveillance of evaporative depletion across the Indus Basin. *Water Resources Research* 38 (12):1273–82.

Bastiaanssen, W. and S. Ali. 2003. A new crop yield forecasting model based on satellite measurements applied across the Indus Basin, Pakistan. *Agriculture, Ecology and Environment* 94 (3):321–340.

Bastiaanssen, W., E. Noordman, H. Pelgrum, G. Davids, Thoreson, B., and R. Allen. 2005. SEBAL Model with Remotely Sensed Data to Improve Water Resources Management under Actual Field Conditions, *Journal of Irrigation and Drainage Engineering* 131 (1):85–93

Bastiaanssen, W., R. Allen, H. Pelgrum, A. Texeira, R. Soppe, and B. Thoreson. 2008. Thermal-infrared technology for local and regional scale irrigation analyses in horticultural systems. *Acta Horticulturae* 792 (V Int'l Symposium on Irrigation of Horticultural Crops, eds. I. Goodwin and M. O'Connel), ISHS.

Bramley, R. and R. Hamilton. 2004. Understanding variability in winegrape production systems. *Australian Journal of Grape and Wine Research* 10:32–45.

Chalker-Scott, L. 1999. Environmental significance of anthocyanins in plant stress responses. *Photochemistry and Photobiology* 79 (1):1–9.

Close, D. and C. Beadle. 2003. The ecophysiology of foliar anthocyanin. *The Botanical Review* 69 (2):149–161.

Courault, D., B. Seguin, and A. Olioso. 2005. Review on estimation of evapotranspiration from remote sensing data: from empirical to numerical modeling approaches. *Irrigation and Drainage Systems* 19:223–49.

Da Costa, J., F. Michelet, C. Germain, O. Lavialle, and G. Grenier. 2007. Delineation of vine parcels by segmentation of high resolution remote sensed images. *Precision Agriculture* 8:95–110.

Daughtry, C., C. Walthall, M. Kim, E. Brown de Colstoun, and J. McMurtrey. 2000. Estimating corn leaf chlorophyll concentration from leaf and canopy reflectance. *Remote Sensing of Environment* 74:229–39.

Delenne, C., S. Durrieu, G. Rabatel, M. Deshayes, J. Bailly, C. LeLong, and P. Couteron. 2008. Textural approaches for vineyard detection and characterization using very high spatial resolution remote sensing data. *International Journal of Remote Sensing* 29 (4):1153–67.

Dickenson, J. and J. Salt. 1982. In vino veritas: an introduction to the geography of wine. *Progress in Human Geography* 16 (2):159–89.

Dobrowski, S., S. Ustin, and J. Wolpert. 2002. Remote estimation of vine canopy density in vertically shoot positioned vineyards: Determining optimal vegetation indices. *Australian Journal of Grape and Wine Research* 8:117–125.

Dobrowski, S., S. Ustin, and J. Wolpert. 2003. Grapevine dormant pruning weight prediction using remotely sensed data. *Australian Journal of Grape and Wine Research* 9:177–82.

English, J., C. Thomas, J. Marois, and W. Gubler. 1989. Microclimates of grapevine canopies associated with leaf removal and control of Botrytis bunch rot. *Phytopathology* 79:395–401.

Eryilmaz, F. 2006. The relationships between salt stress and anthocyanin content in higher plants. *Biotechnology and Biotechnological Equipment* 20 (1):47–52.

Friedl, M. and C. Brodley. 1997. Decision tree classification of land cover from remotely sensed data. *Remote Sensing of Environment* 61:399–409.

Fuchs, M. and C. Tanner. 1966. Infrared thermography of vegetation. *Agronomy Journal* 58:597–601

Gitelson, A., A. Vina, V. Ciganda, D. Rundquist, and T. Arkebauer. 2005. Remote estimation of canopy chlorophyll content in grapes. *Geophysical Research Letters* 32:L08403, doi:10.1029/2005GL022688.

Haboudane, D., J. Miller, N. Tremblay, P. Zarco-Tejada, and L. Dextraze. 2002. Integrated narrow-band vegetation indices for prediction of crop chlorophyll content for application to precision agriculture. *Remote Sensing of Environment* 81:416–26.

Hall, A., D. Lamb, B. Holzapfel, and J. Louis. 2002. Optical remote sensing applications in viticulture – a review. *Australian Journal of Grape and Wine Research* 8:36–47.

Hall, A., J. Louis, and D. Lamb. 2003. Characterising and mapping vineyard canopy using high-spatial-resolution aerial multispectral images. *Computers and Geosciences* 29:813–22.

Hall, A., D. Lamb, B. Holzapfel, and J. Louis. 2011. Within-season temporal variation in correlations between vineyard canopy and winegrape composition and yield. *Precision Agriculture* 12:103–117.

Hay, C. 1974. Agricultural inventory techniques with orbital and high-altitude imagery. *Photogrammetric Engineering* 40:1283–93.

Heilman, J., W. Heilman, and D. Moore. 1982. Evaluating the crop coefficient using spectral reflectance. *Agronomy Journal* 74:967–71.

Hornbuckle, J., N. Car, E. Christen, T. Stein, and B. Williamson. 2009. *IrriSatSMS irrigation water management by satellite and SMS – A utilisation framework.* CRC for Irrigation Futures Technical Report No. 01/09, CSIRO Land and Water Science Report No. 04/09.

Jackson, D. and P. Lombard. 1993. Environmental and management practices affecting grape composition and wine quality: A review. *American Journal of Enology and Viticulture* 44:409–430.

Johnson, D. 2008. A comparison of coincident Landsat-5 TM and Resourcesat-1 AWIFS imagery for classifying croplands. *Photogrammetric Engineering and Remote* Sensing 74 (11):1313–23.

Johnson, L., C. Hlavka, and D. Peterson, 1994. Multivariate Analysis of AVIRIS data for canopy biochemical estimation along the Oregon Transect. *Remote Sensing of Environment* 47:216–230.

Johnson, L., B. Lobitz, R. Armstrong, R. Baldy, E. Weber, J. DeBenedictis, and D. Bosch. 1996. Airborne imaging aids vineyard canopy evaluation. *California Agriculture* 50 (4):14–18.

Johnson, L., D. Bosch, D. Williams, and B. Lobitz. 2001. Remote sensing of vineyard management zones: implications for wine quality. *Applied Engineering in Agriculture* 17 (4):557–60.

Johnson, L., D. Roczen, S. Youkhana, R. Nemani, and D. Bosch. 2003. Mapping vineyard leaf area with multispectral satellite imagery. *Computers and Electronics in Agriculture* 38 (1):37–48.

Johnson, L. 2003. Temporal stability of the NDVI-LAI relationship in a Napa Valley vineyard. *Australian Journal of Grape and Wine Research* 9:96–101.

Johnson, L, and T. Scholasch. 2005. Remote sensing of shaded area in vineyards. *HortTechnology* 15:859–863.

Jones, G. 2005. Climate change and global wine quality. *Climatic Change* 73 (3):319–43.

Kalma J., T. McVicar, and M. McCabe. 2008. Estimating land surface evaporation: A review of methods using remotely sensed surface temperature data. *Surveys in Geophysics* 29(4–5):421–69.

Klaasse, A., W. Bastiaanssen, C. Jarmain, and A. Roux. 2008. *Water use efficiency of table and wine grapes in Western Cape, South Africa.* Wageningen: WaterWatch.

Lanjeri, L., D. Degarra, and J. Melia. 2004. Interannual vineyard crop variability in the Castilla-La Mancha region during the period 1991–1996 with Landsat Thematic Mapper images. *International Journal of Remote Sensing* 25 (12):2441–57.

Lamb, D., M. Weedon, and R. Bramley. 2004. Using remote sensing to predict grape phenolics and colour at harvest in a Cabernet Sauvignon vineyard. *Australian Journal of Grape and Wine Research* 10:46–54.

MacDonald, R. and F. Hall. 1980. Global crop forecasting. *Science* 208 (Issue 4445):670–79.

Martin, P., P. Zarco-Tejada, M. Gonzalez, and A. Berjon. 2007. Using hyperspectral remote sensing to map grape quality in Tempranillo vineyards affected by iron deficiency chlorosis. *Vitis* 46 (1):7–14.

Montero, F., J. Melia, A. Brasa, D. Segarra, A. Cuesta, and S. Lanjeri. 1999. Assessment of vine development according to available water resources by using remote sensing in La Mancha, Spain. *Agricultural Water Management* 40:363–375.

National Research Council. 1988. *Toward an understanding of global change: Initial priorities for U.S. contributions to the International Geosphere Biosphere Program.* Washington, DC: National Academy Press.

Neale, C., H. Jayanthi, and J. Wright. 2005. Irrigation water management using high resolution airborne remote sensing. *Irrigation and Drainage Systems* 19:321–26.

Nemani, R. and S. Running. 1989. Testing a theoretical climate-soil-leaf area hydrologic equilibrium of forests using satellite data and ecosystem simulation. *Agricultural and Forest Meteorology* 44:245–60.

Nemani, R., H. Hashimoto, P. Votava, F. Melton, W. Wang, A. Michaelis, L. Mutch, C. Milesi, S. Hiatt, and M. White. 2009. Monitoring and forecasting ecosystem dynamics using the Terrestrial Observation and Prediction System (TOPS). *Remote Sensing of Environment* 113 (7):1497–1509

Pinter, P., J. Hatfield, J. Schepers, E. Barnes, M. Moran, C. Daughtry, and D. Upchurch. 2003. Remote sensing for crop management. *Photogrammetric Engineering and Remote Sensing* 69 (6):647–64.

Pringle, M., A. McBratney, B. Whelan, and J. Taylor. 2003. A preliminary approach to assessing the opportunity for site-specific crop management in a field, using yield monitor data. *Agricultural Systems* 76:273–292.

Rabatel, G., C. Delenne, and M. Deshayes. 2008. A non-supervised approach using Gabor filters for vine-plot detection in aerial images. *Computers and Electronics in Agriculture* 62:159–68.

Richardson, A. and C. Wiegand. 1977. Distinguishing vegetation from soil background information. *Photogrammetric Engineering and Remote Sensing* 43:1541–1552.

Smart, R. and B. Coombe. 1983. Water relations of grapevines. In *Water Deficits and Plant Growth*, vol. 2, ed. T. Kozlowski, 137–196. New York: Academic.

Smart, R. 1985. Principles of grapevine canopy microclimate manipulation with implications for yield and quality. *American Journal of Enology and Viticulture* 36:230–239.

Smart, R. and M. Robinson. 1991. *Sunlight into wine: A handbook for winegrape canopy management.* Adelaide: Winetitles.

Stamatiadis, S., D. Taskos, C. Tsadilas, C. Christofides, E. Tsadila, and J. Schepers. 2006. Relation of ground-sensor canopy reflectance to biomass production and grape color in two Merlot vineyards. *American Journal of Enology and Viticulture* 57 (4):415–22.

Steele, M., A. Gitelson, D. Rundquist, and M. Merzlyak. 2009. Non-destructive estimation of anthocyanin content in grapevine leaves. *American Journal of Enology and Viticulture* 60 (1):87–92.

Tate, A. 2001. Global warming's impact on wine. *Journal of Wine Research* 12 (2):95–109.

Teixeira, A., W. Bastiaanssen, M. Ahmad, and M. Bos. 2008. Reviewing SEBAL input parameters for assessing evapotranspiration and water productivity for the Lower-Middle Sao Francisco river basin, Brazil, Part A: Calibration and validation. *Agriculture and Forest Meteorology* 149 (3–4):462–476.

Tisseyre, B., H. Ojeda, and J. Taylor. 2007. New technologies and methodologies for site-specific viticulture. *International Journal of Wine and Vine Research* 41 (2):63–76.

Tisseyre B. and A. McBratney. 2008. A technical opportunity index based on mathematical morphology for site-specific management using yield monitor data: Application to viticulture. *Journal of Precision Agriculture* 9 (1–2):101–113.

Tonietto, J. 2008. Geographical indicators for grapes. *Acta Horticulturae* 785:467–76.

Townshend, J., C. Justice, W. Li, C. Gurney, and J. McManus. 1991. Global land cover classification by remote sensing: Present capabilities and future possibilities. *Remote Sensing of Environment* 35:243–255.

Trolier, L., W. Philipson, and W. Philpot. 1989. Landsat TM analysis of vineyards in New York. *International Journal of Remote Sensing* 10 (7):1277–1282.

Trout, T., L. Johnson, and J. Gartung. 2008. Remote sensing of canopy cover in horticultural crops. *HortScience* 43(2):333–337.

Tucker, C. 1979. Red and photographic infrared linear combinations for monitoring vegetation. *Remote Sensing of Environment* 8:127–50.

Vaudour, E. 2002. The quality of grapes and wine in relation to geography. *Journal of Wine Research* 13 (2):117–41.

Warner, T. and K. Steinmaus. 2005. Spatial classification of orchards and vineyards with high spatial resolution panchromatic imagery. *Photogrammetric Engineering and Remote* Sensing 71 (2):179–87.

Wassenaar, T., F. Baret, J. Robbez-Masson, and P. Andrieux. 2001. Sunlit soil surface extraction from remotely sensed imagery of perennial, discontinuous crop areas: The case of Mediterranean vineyards. *Agronomie* 21:235–45.

Wassenaar, T., J. Robbez-Masson, and P. Andrieux. 2002. Vineyard identification and description of spatial crop structure by per-field frequency analysis. *International Journal of Remote Sensing* 23 (17):3311–25.

Wiegand, C., A. Richardson, D. Escobar, and A. Gerbermann. 1991. Vegetation indices in crop assessments. *Remote Sensing of Environment* 35:105–19.

Wildman, W., R. Nagaoka, and L. Lider. 1983. Monitoring spread of grape phylloxera by color infrared aerial photography and ground investigation. *American Journal of Enology and Viticulture* 34 (2):83–94.

Williams, L. and J. Ayars. 2005. Grapevine water use and the crop coefficient are linear functions of the shaded area measured beneath the canopy. *Agricultural and Forest Meteorology* 132:201–11.

Winkler, A. 1958. The relation of leaf area and climate to vine performance and grape quality. *American Journal of Enology and Viticulture* 9:10–23.

Zarco-Tejada, P., A. Berjon, R. Lopez-Lozano, J. Miller, P. Martin, V. Cachorro, M. Gonzalez, and A. de Frutos. 2005. Assessing vineyard condition with hyperspectral indices: Leaf and canopy reflectance simulation in a row-structured discontinuous canopy. *Remote Sensing of Environment* 99:271–87.

Geospatial Tools and Techniques for Vineyard Management in the Twenty-First Century

13

David R. Green

Abstract

Tools and techniques to monitor and map the environment are now very well developed and widely used in practice. One area in which information technology has been increasingly and successfully applied over the past 20 years is in agricultural and horticultural applications. A specialised area of application of horticulture is Precision Viticulture (PV), a subject that has been increasingly documented in the USA and Australia. Precision Viticulture is now a very well developed approach to vineyard monitoring, mapping and management, and one that has been successfully demonstrated through many studies and practical applications leading to greatly improved efficiency and effectiveness in the day-to-day operation of the vineyard and, ultimately, improved fruit quantity, quality and wine production. This has been particularly true for the larger commercial vineyards with both the financial resources to utilise such technologies and operating over relatively large areas of grapevines. This chapter seeks to provide an up-to-date overview of the role of some of the geospatial and associated technologies in Precision Viticulture (PV).

Introduction

Tools and techniques to monitor and map the environment are now very well developed and widely used in practice. Remote sensing, in the form of aerial photography and a range of airborne sensors, and more recently satellite imagery, has provided an effective way to gather data and information about the environment at many different spatial, spectral, temporal and radiometric resolutions. Digital mapping and cartography, and latterly Geographical Information Systems (GIS), have also provided the means to map and visualise the environment at a variety of different scales, to create environmental databases and to analyse and integrate a wide range of spatial and temporal datasets. Global Positioning Systems (GPS), and more recently the associated mobile technologies such as the Personal Digital Assistant (PDA) and mobile GIS software, have provided the capability to gather, process and display information whilst working in a field environment. Small microprocessors have also yielded miniature environmental sensors that can be deployed in arrays and wireless networks to gather multiple sources of environmental data at a number of locations. Software developments, too, have provided the basis to model the environment, to generate

D.R. Green (✉)
Department of Geography and Environment, University of Aberdeen, St. Mary's, Elphinstone Road, Aberdeen AB24 3UF, Scotland, UK
e-mail: d.r.green@abdn.ac.uk

P.H. Dougherty (ed.), *The Geography of Wine: Regions, Terroir and Techniques*,
DOI 10.1007/978-94-007-0464-0_13,

visual scenarios and to visualise both data and information with the added benefit of being able to undertake exploratory and interactive studies of the environmental data collected. The Internet now offers rapid access to information in a variety of different forms, and integrated applications of these technologies have provided the means to create Decision Support and Planning Systems (DSS).

One area in which information technology has been increasingly and successfully widely applied over the past 20 years is in agricultural and horticultural applications at the farm level. Both are by their very nature spatial (Smith and Whigham 1999). Precision agriculture and precision horticulture utilise geospatial technologies, including remote sensing (Liaghat and Balasundram 2010) for data collection, positioning, analysis, display, visualisation and communication for the monitoring, mapping and management of a wide range of agricultural and horticultural crops from alfalfa and corn to fruit and nut orchards and tomato crops (http://www.vineview.com).

A specialised area in horticultural applications is Precision Viticulture (PV), a subject that has been increasingly well documented in the USA and Australia and most recently in a volume by Proffit et al. (2006). To date, however, Precision Viticulture (PV) has been somewhat slower to develop than its counterpart in agriculture as the nature of the grapevine and its management are very different from most other crops, being a row-spaced canopy, and implementation of management techniques is generally different and considered somewhat more difficult. Furthermore, viticultural tradition has probably delayed or restricted the implementation of such novel tools and techniques in most European vineyards as compared to the widespread use and development of PV in the New World countries of North America, Australia and New Zealand. Lamb et al. (2008) also argue that there is still a very large knowledge gap between developers and users of IT-based tools and techniques in this field which is partly responsible for the slow uptake of these technologies in practice.

However, Precision Viticulture (PV) is now a very well-documented approach to vineyard monitoring, mapping and management, and one that has been successfully demonstrated through many studies and practical applications leading to greatly improved efficiency and effectiveness in the day to day operation of the vineyard and ultimately the impact upon improved fruit quantity, quality and wine production (Ryerson et al. 2008; Fraigneau 2009). This has been particularly true for the larger commercial vineyards with both the financial resources to utilise such technologies and operating over relatively large areas of grapevines.

This chapter seeks to provide an up-to-date overview of the role of some of the geospatial and associated technologies in Precision Viticulture (PV). Although remote sensing typically forms a major part of Precision Viticulture (PV), as evidenced by the number of publications in this field, a separate chapter in this volume by Johnson et al. covers this aspect of vineyard monitoring and management in more detail than will be considered here. In this chapter, therefore, only minimal coverage will be given to remote sensing, aiming only to touch upon the more specialised remote sensing tools and techniques that will complement those discussed elsewhere in the book. This chapter also aims to highlight some of the more affordable geospatial technologies and techniques that are now currently available for use in the smaller scale vineyard as opposed to solely focusing on the larger commercial-scale vineyard.

Precision Viticulture

Precision Viticulture (PV) is generally considered to be the application of one or more of the geospatial technologies and other related devices, tools and techniques that use spatial location for the purposes of collecting, processing, analysing and visualising environmental data collected in a vineyard. It is considered to be a cyclical process by Bramley (2001) involving observation and data collection, e.g. using remote sensing and mapping, data interpretation and evaluation with the aid of a GIS, and finally, implementation of a management plan. Such information is used for management of the vines, the vineyard, and decision making and planning in the context of vineyard management.

As discussed by Proffit et al. (2006), there is now a wide range of computer-based monitoring and management tools that offer the potential to improve the availability of up-to-date data and information collection, processing, analysis and visualisation to assist in the management of a vineyard. In practice, precision viticulture (PV) provides researchers and winegrowers with a suite of tools to measure and manage the vineyard. However, it should be noted that not all of these tools and techniques can be directly utilised yet by the vineyard manager as their use is still constrained by costs, expertise and familiarity with the different IT technologies, and a general awareness and acceptance of the potential benefits. Indeed, even in the USA and Australia many of the tools and techniques available still often require the expertise and equipment of specialist technology companies to carry out the analysis and visualisation in the data into information pathway.

PV is also considered to be an approach to wine production based on recognition of the fact that the productivity of vineyards and individual blocks within a vineyard can be inherently variable over space and time because of differences in topography, soils, soil moisture, slope and aspect, plant health and vigour, microclimate and the management practices (Proffit et al. 2006). Variability in the environment ultimately leads to variability in grape yield (quantity) and quality and ultimately table grape and wine production. Vineyard management with the aid of PV is therefore considered to be a targeted approach rather than one that will be implemented uniformly over all areas within the vineyard. It therefore seeks to help a vineyard manager gain increased control over an inherently variable production system so that any given management decision has an increased chance of delivering the desired or expected outcomes (Bramley 2001).

The Geospatial Technologies

The geospatial technologies typically include Global Positioning Systems (GPS) and mobile Geographical Information Systems (GIS) for data collection and mapping; remote sensing for monitoring and mapping; desktop GIS for spatial data processing, analysis, modelling and mapping and environmental databases for data storage and retrieval. Today this suite of tools would also consider including modelling, simulation and geovisualisation software tools as well as the use of the Internet, e.g. website and web access, online GIS, mapping and geovisualisation.

As discussed in the chapter by Johnson et al. (this volume), remote sensing, ranging from aerial photography to numerous airborne and satellite sensors, has provided the means to gather up-to-date imagery, data and information about vineyards. Through visual photo-interpretation, and more often now the use of Digital Image Processing (DIP) software for electronic image enhancement and classification, a considerable amount of useful information about a vineyard environment, both canopy and surroundings such as the soil and cover crop, e.g. phenological status, plant vigour, soil moisture and health/disease, can be relatively easily and repeatedly extracted from the remotely sensed image data. Attempts to automate the process for some regularly required information have perfected the information extraction technology to the stage where the imagery can be uploaded and processed online to extract key information which can be used in decision making and planning, e.g. Normalised Difference Vegetation Index (NDVI), an approach similar in many ways to the convenience of taking digital image storage media from a digital camera to the supermarket or pharmacy to process, enhance and print the photographs. This has been particularly successful where regular daily or weekly overflights of imagery are used for real time monitoring, and rapid processing of the data yields much needed information for planning and decision making.

In recent years, the Internet and Google Earth (GE) have provided the capability for improved access to online information by a wider audience. For vineyards, data and information collected using field-based data acquisition systems and low-cost wireless networks can be displayed against background image and terrain information in GE. This opens up many new opportunities to display and share spatial information, to communicate spatial information to a wider audience and to form the basis for constructing local (geospatial-based) information systems for a vineyard without the need for a desktop GIS or knowledge on how to use it.

The following sections of this chapter provide a brief overview of some of the recent developments in the geospatial and related technologies and their applications to vineyard management. Space is also given to ground-based measurements that now benefit from the convenience of spatial or locational information provided by a GPS and a GIS.

Remote Sensing

As stated earlier in this chapter, the application of remote sensing to vineyard monitoring and management is more than adequately covered elsewhere in a separate chapter in this volume. This section will therefore seek only: (a) to cover some of the less well-known examples of platforms and sensors that have been used, (b) to consider some additional aspects of remote sensing for data collection and processing through the use of digital image processing software for information extraction and (c) to integrate remotely sensed data with other sources of spatial information in a GIS environment. Furthermore, these examples will focus specifically on their use in the smaller, non-commercial vineyard.

Airborne Imagery

Aircraft have been widely used for a number of years to capture airborne imagery of vineyards, e.g. Wassenaar et al. (2000), Lamb et al. (2002), Hall et al. (2002), Da Costa et al. (2007), Pitcher-Campbell et al. (2001), Chanussot et al. (2005), Delenne et al. (2010), Ryerson (1974), and Brown (2008). In the beginning, the most useful imagery was either true colour or colour infrared (CIR) aerial photography (Hall et al. 2002). A number of well-known studies by Wildman (1979, 1981, 1987), for example, promoted the use of CIR photography for monitoring outbreaks of phylloxera disease and the phenological status of vineyards. With the advent of digital aerial photography and imagery for monitoring and management purposes, processing data into information has become much faster with the aid of Digital Image Processing (DIP) software, many of which are now available for the vineyard manager, being cheaper and easier to use.

Other remote sensing studies have made use of airborne video (Grierson and Bolt 1995) and high resolution multi-spectral airborne imagery (Hall et al. 2002; Da Costa et al. 2007; Johnson and Lobitz No date) to provide detailed descriptions of vines in a vineyard, information about plant vigour, the identification of dead vines (Chanussot et al. 2005) and biomass, together with ground measurements to define plant water restriction zones within a vineyard, e.g. Acevedo-Opazo et al. (2008). Others have made use of SAR (Burini et al. 2008). Some researchers have made use of CASI hyperspectral imagery, e.g. Hall et al. (2002) and Zarco-Tejada et al. (2005), and Lidar imagery of vineyards to derive information about the vegetation surface and the terrain (Fig. 13.1). Thermal information has also been used on a number of occasions; for example, Jia et al. (2006) used thermal imagery from the Airborne Hyperspectral System (AHS) airborne platform (to complement ground imagery and ASTER satellite imagery) to estimate sensible and latent heat fluxes.

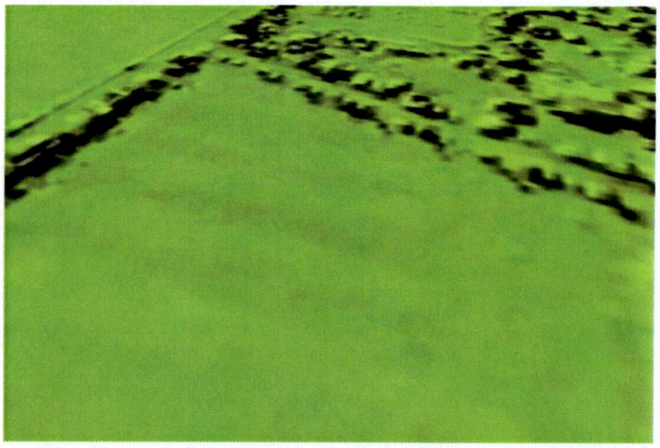

Fig. 13.1 Airborne imagery of a vineyard

Lidar, like all other sensors, provides unique information about the environment that cannot be obtained from any other source. The advantage of this particular source of imagery is the information that it contains about the local variations in vineyard terrain, e.g. hollows and ridges that can be useful when trying to identify potential cold spots and air drainage problems and barriers in a vineyard, but, moreover, valuable information about the vegetation canopy structure can also be obtained (Dubayah et al. 2000). The resolution of such data is also significantly greater than can be obtained from most published topographic contour maps. In addition, Digital Elevation Models (DEM) or Digital Terrain Models (DTM) can be used to explore the relationship between topography and other variables, e.g. soils, soil moisture and drainage. DEMs can also be overlaid with other remotely sensed imagery and digital map data from a GIS to provide a semi-realistic visualisation of the vineyard. In the UK, Lidar imagery is currently being used by some vineyard managers to aid in monitoring and management of the vineyard (Meadowgrove Vineyard, UK).

Remotely sensed imagery from many different sources is often used together and each source is complementary. Most GIS can also handle remotely sensed data as a layer within the GIS and, more often than not, many different sources of geospatial data are integrated as part of the vineyard management task.

But, despite the wealth of imagery and image data now available there are still some potential limitations to its wider use, particularly by the smaller vineyard with limited financial resources and technical expertise. Image acquisition can be expensive particularly if it is

Fig. 13.2 Large-scale model airborne platform

to be acquired, purchased and used on a regular basis. Furthermore, image processing and information extraction still requires expertise, and whilst fairly low-cost software can now be purchased, e.g. Adobe Photoshop and Idrisi from Clark Labs, the learning curve can be quite steep for the vineyard manager, and the expertise and facilities of a commercial company may therefore be required to generate the information needed. This, too, will incur a significant cost for the small vineyard with limited resources.

Small-Scale Airborne Platforms

Though not often taken seriously as a platform for capturing remotely sensed imagery for environmental applications in the past, large-scale model airborne platforms, e.g. fixed wing (aircraft) and rotary wing (helicopters) have, on a number of occasions, been considered as a low-cost alternative to capturing imagery flown by larger-scale platforms (*see* http://www.spaceref.com/news/viewpr.html?pid=12357). With recent technology developments such platforms, therefore, provide a powerful basis to gather data and information that can be tied into a vineyard GIS database and in-situ data collection on the ground.

There are now a wide range of small-scale platforms available to anyone considering the acquisition of low-cost aerial remotely sensed imagery. These include kites, model aircraft, balloons, helicopters, Unmanned Aerial Vehicles (UAVs) and microlights (Johnson et al. 2003a, b; Johnson and Herwitz 2004; Berni et al. 2009). Far from being 'toys,' many of these airborne platforms can now be equipped with traditional but lightweight SLR and digital cameras as well as video cameras for the collection of panchromatic, true colour and colour infrared photography and video footage. Additionally multi-spectral imagery can easily be captured through repeated overflights of the vineyard with the aid of camera filters, or more recently through the use of on-board hyperspectral systems. Some of the so called model platforms are also large enough to carry multiple payloads and now even make use of wireless transmission for real-time data capture (http://www.uav-applications.org/projects/vineyards_1.html). Johnson et al. (2003a, b), Johnson and Herwitz (2004) and Berni et al. (2009) have all used UAVs and lightweight sensors to collect multispectral and thermal imagery of vineyards

Whilst the photography acquired can then be manually interpreted using the techniques of aerial photo-interpretation, Digital Image Processing (DIP) software, much of it now low-cost, can also be used to geo-correct and mosaic the photographic prints or images together as the basis for onscreen interpretation and the mapping of thematic information for input to a GIS. DIP software has considerable potential for the small scale vineyard, allowing the vineyard manager to overfly a vineyard on a regular basis for monitoring and mapping (Fig. 13.2).

These smaller platforms offer a relatively low-cost and flexible means for the frequent capture of airborne imagery to both monitor and map a small vineyard whilst the larger ones can be used for larger area coverage. The lower cost of PCs and accompanying hardware, e.g. storage media, scanners, printers and software, also provides an opportunity for the small vineyard to be able to capture, store, process and map the data on a regular basis in-house.

Satellite Imagery

A number of authors have, over the years, considered and evaluated the role of different sources of satellite imagery for the monitoring of vineyards (Philipson et al. 1980; Minden 1982; Trolier et al. 1989; Chelli et al. 2005). Ducati and Da Silva (2006) used ASTER data for Precision Viticulture. Brancorado et al. (2006) used IKONOS data for vineyard studies in Italy. Early examples of satellite imagery and data often had relatively limited use for vineyard monitoring because of the relatively low spatial and limited spectral resolutions of the data. However, greater success has since been achieved with the newer and higher resolution satellite sensors that now also provide a much better combination of spectral, spatial, temporal and radiometric resolutions for monitoring vineyards and vine plants.

For the most part, however, many of these studies have often been, and still are largely experimental academic research, in terms of their being widely applicable to different locations on a repetitive basis. Some have been more successful and usable than others. Practical use of the image data has also been constrained by the success with which it is possible to extract information about the vine canopy on a reliable and regular basis. Establishing the relationship between the image and the ground information of interest has frequently been quite difficult because of the complexity of the cause-effect relationship between the remote sensing signal and the vegetation and/or ground surface which is also complicated by the intervening atmosphere.

One of the most successful outcomes of image analysis has been the generation of Normalised Difference Vegetation Index (NDVI) maps from imagery, and it has been possible to produce relative and normalised NDVI maps and change detection maps from image analysis on a regular basis (http://www.vineview.com). This provides the basis for providing the information that can be used to develop a management plan, fertiliser strategy, a cultivation and harvest plan to facilitate frost protection, pruning, irrigation and replanting decisions. This information also provides the basis for problem identification, practice improvement, precision application, sampling, harvest planning and profit improvement (http://www.grayhawk-imaging.com).

However, not all information extraction has been so straightforward. A great deal of effort and energy, for example, has been put into academic plant canopy modelling studies, e.g. Suits (1983), Verhoef and Bunnik (1976), Smith (1982), Jaquemoud and Baret (1990) and Jacquemoud et al. (2009), to model the spectral reflectance of plant canopies, and more specifically vine canopies (a row canopy). These studies have formed the basis for extending our knowledge and understanding of the reflectance and emittance characteristics of a vine canopy. Inversion models have also been developed to extract information about the canopy from image data. Recently statistical reconstructions of vine canopies have been explored to assist in more accurately characterising and simulating vine structure and canopy light interception and transmittance (Louarn et al. 2007).

Geographic Information Systems (GIS)

Geographic Information Systems (GIS) are frequently defined as computer-based software and hardware to collect, store, manipulate, analyse and display map and image-based data (Burrough 1986). Today there are many examples of GIS software available for various computer platforms including the desktop, laptop, and even Personal Digital Assistant (PDA) platforms. Ranging from simple field data collection-based software to mapping toolboxes, advanced spatial analysis systems and increasingly sophisticated display and geovisualisation software tools, GIS has become more widely available, easier to use and more affordable over time. Today, GIS software suitable for use in a vineyard environment can be purchased for around $600–$1,000, and there are also many freely available systems that are also perfectly adequate for most purposes.

Commonly available GIS software packages include free open source software such as MapWindow (http://www.mapwindow.org/), GeoDa (http://geodacenter.asu.edu/), TNT-Lite (http://www.microimages.com/

Fig. 13.3 Mobile GIS device

tntlite/) and GRASS (http://grass.itc.it/), to low-cost mapping software, e.g. MapViewer (http://www.goldensoftware.com/) and the more costly and sophisticated spatial analysis software, e.g. Global Mapper (http://www.globalmapper.com/), Manifold (http://www.manifold.net), ClarkLabs Idrisi (http://www.clarklabs.org) and ESRI's ArcExplorer, ArcView and ArcGIS (http://www.esri.com). Some of the software packages available also include powerful integrated digital image processing tools, e.g. Idrisi Taiga, thereby providing the capability for the vineyard manager to enhance, analyse and classify remotely sensed images on site.

Plugin Global Positioning System (GPS) units and various options also allow locational data to be added to the base maps in a GIS, whilst mobile GIS software such as ESRI's ArcPad (http://www.esri.com) and PocketGIS (http://www.pocket.co.uk/) (Fig. 13.3) provide the means to log positional information in the field together with point, line, or area (polygon) feature attributes in the form of, for example, digital photographs and field sketches.

GIS, coupled with Global Positioning Systems, a wide range of environmental sensors (see later in this chapter), increasingly powerful PDAs with field mobility and image processing capabilities, therefore, provide a powerful and relatively low cost suite of easy-to-use tools with which to collect spatial data and to process it into information which can then be used as the basis for decision making and planning.

Many vineyards, particularly those operating on a commercial scale, now make use of GIS in one form or another to map the attributes of the vineyard, from defining and plotting the vineyard boundary, row positions and blocks to the inclusion of the vineyard infrastructure, e.g. access routes and buildings (Merenlender 2000). Over a relatively short period of time vineyard managers can quickly and easily build a detailed spatial database containing a temporal record of the vineyard that can be further analysed using GIS-based spatial analysis techniques to help monitor and map spatial patterns over time, e.g. soil moisture, drainage, phenology, disease and biomass. GIS is also a powerful modelling tool that can be used to integrate data to better understand the relationships between grapevine growth and environmental variables (Castel et al. 2007).

Coupled with image analysis tools, GIS, therefore, provides a comprehensive and integrated spatial data and information toolbox to capture, store, analyse and visualise vineyard data. Dainelli et al. (2005) describe some of the different ways in which very high resolution remotely sensed imagery and GIS has been used to study site variability in Europe, e.g. the Bacchus Project. Reductions in the cost of hardware and software now means that such technology is even more widely available to the smaller vineyards to aid in utilising the tools of PV for day to day management. Furthermore, nearly all GIS and image processing software now comes with tutorials plus data and web-based support to assist in reducing the learning-curve needed to become familiar with the tools, techniques and systems. Better user-interfaces and more user-friendly hardware and software have also considerably aided in the use of off-the-shelf systems helping the non-specialist to make use of sophisticated technology and techniques without the need to be an expert in the field.

Global Positioning Systems (GPS)

A fundamental part of PV is to have an accurate spatial location of features at any one time within a vineyard. Various different types of Global Positioning Systems (GPS) are now widely available to give locational positions of varying accuracies for any individual, feature, or object on the Earth's surface. These systems use satellites to provide a 3D position: longitude, latitude and elevation.

Fig. 13.4 GPS unit

GPS technology has become increasingly sophisticated and accurate, as well as lower in cost. Whilst single, non-differential, GPS receivers probably have limited use in a vineyard for data collection with accuracies that are inadequate for mapping vineyards. Differential GPS (DGPS) provides a positioning accuracy of 2 m or better. Although more costly, Real Time Kinematic (RTK) Differential GPS, using a local base station with known reference coordinates, offers the sort of accuracy (2 cm) that is required for the accurate mapping of vineyard features, as well as for mounting on planting and harvesting vehicles to provide the basis for the navigational control of post and trellis erection, vine planting and other management systems such as fertiliser application and spraying and grape picking (Tisseyre et al. 2007), as well as the capture of photographic and other attribute data about the canopy.

GPS are clearly at the heart of the practice of vineyard spatial data collection and mapping (Fig. 13.4) and a growing number of devices incorporate GPS to provide a spatial location for each measurement required. The data collected can now be plugged directly into a GIS thereby making the process of data collection and mapping much more direct than was possible in the past.

Visualisation

Many GIS software packages now come with 3D visualisation capabilities. This allows for the display of terrain and surfaces in the form of Digital Terrain Models (DTM) or Digital Elevation Models (DEM), and Digital Surface Models (DSM). Whilst some topographic information is available to purchase for vineyard terrain, GIS toolboxes now allow one to generate DTMs from Lidar and contoured map data. More sophisticated (and expensive!) toolboxes that come with image processing software systems, e.g. Erdas Imagine (http://www.erdas.com/) and soft photogrammetry software, e.g. BAESocetSet (http://www.baesystems.com) allow one to derive detailed DTMs from stereoscopic aerial imagery. The task is not, however, trivial and it is more usual to purchase the data.

Aerial photography and satellite imagery, together with map layers from a GIS, can then easily be overlaid on a DTM (usually with some vertical exaggeration, e.g. 3X) to create a semi-realistic view of the vineyard site, and often providing a new perspective of the site for the vineyard manager, and one that can be examined from any view angle. Some GIS software also allows one to fly through the landscape using predefined flight lines and to capture the flight in the form of a video file, which can be used for the purposes of communication or for scientific visualisation. Draping other map layers over the DTM also provides the possibility to examine relationships between the terrain slope, aspect, soils, illumination and accessibility.

In Niagara, Canada, Duncan (2007) has also developed a new type of vineyard management system which involves the collection of GPS data linked to a database containing coordinates, production yields, soil analysis, a digital photograph and vine health data. Using this information, GIS maps showing slope, aspect, contours, wind patterns and cold air pooling are created to study the effects of vineyard topography. An integral part of this study involves the integration of 3D imaging, visualisation and GIS technologies to generate a realistic 3D map and fly-throughs.

Internet and Google Earth

A related development of the benefits of using geovisualisation in vineyard management is the use of Google Earth for terrain and information visualisation. The Internet or World Wide Web (WWW) offers the vineyard manager a number of possibilities in relation to vineyard monitoring and management. This includes access to vineyard information through websites, e.g. local information systems for a vineyard and winery and online mapping or GIS systems allowing thematic

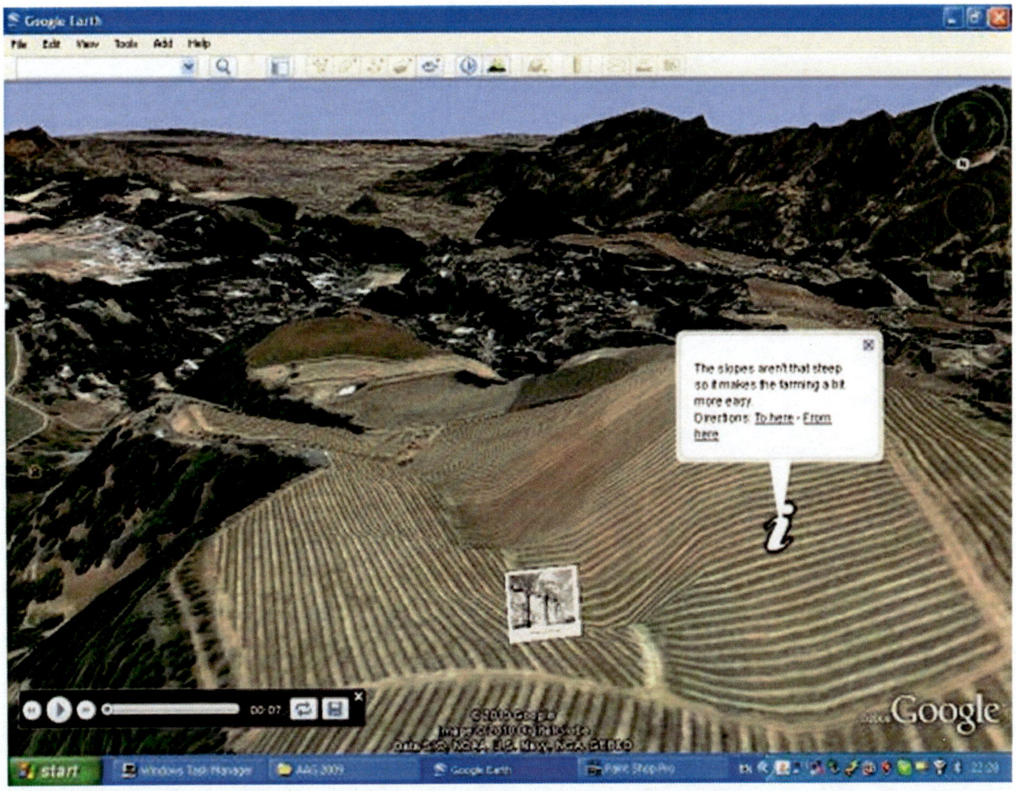

Fig. 13.5 Google Earth

layers to be accessed and displayed (much as they would be in a desktop GIS), as well as the capability to monitor vineyard information from wireless sensor networks placed around the vineyard. An example of the use of the web and GIS is the Catalan Viticultural Register (Lloret and Jurado 2004).

Smart networks allow data to be captured, recorded and displayed or visualised in a graphic format on the web in Google Earth and accessible in the vineyard office providing a wide range of local environmental information about the vineyard, e.g. temperature, soil moisture and humidity which can be logged and displayed (Tongrod et al. 2009; Baggio 2005).

Combined with data captured with the aid of a GPS, the Internet can now also be used to add spatial data and information to Google Earth (GE) (http://earth.google.co.uk/). With the advent of publicly available geographical software applications, such as Google Earth (GE), NASA's Worldwind (http://worldwind.arc.nasa.gov) and Microsoft's Virtual Earth (http://www.viawindowslive.com/VirtualEarth.aspx), there are now opportunities to visualise GIS and image data within this software. Google Earth in particular has become very popular and can be used as the basis to construct a simple Local Information System, virtual fieldtrips or tours and even fly-throughs using the 3D terrain viewing capability.

With the aid of a combination of utility software, e.g. MapWindow, Sketchup (http://sketchup.google.com/) and Shape2Earth (http://shape2earth.com/), it is possible to add GIS data to the GE view, thereby creating a Local Information System (LIS) and Virtual Reality field environment (Green 2009). This information can then be displayed and shared as well as included with more general vineyard information systems. There are already a number of examples where Local Information Systems have been created for a vineyard providing access to information about the vineyard sufficient for a virtual tour of the vineyard and including more detailed and specific layers about various aspects of the vineyard (Fig. 13.5). There are now also many low-cost or free software utilities that can be used to assist in the preparation of a visualisation for a viticultural landscape.

In the context of geovisualisation, Mike Bobbitt Associates (http://www.mikebobbitt.com) have used various software tools to illustrate vine row illumination and the sunlit and shadow areas of the canopy and

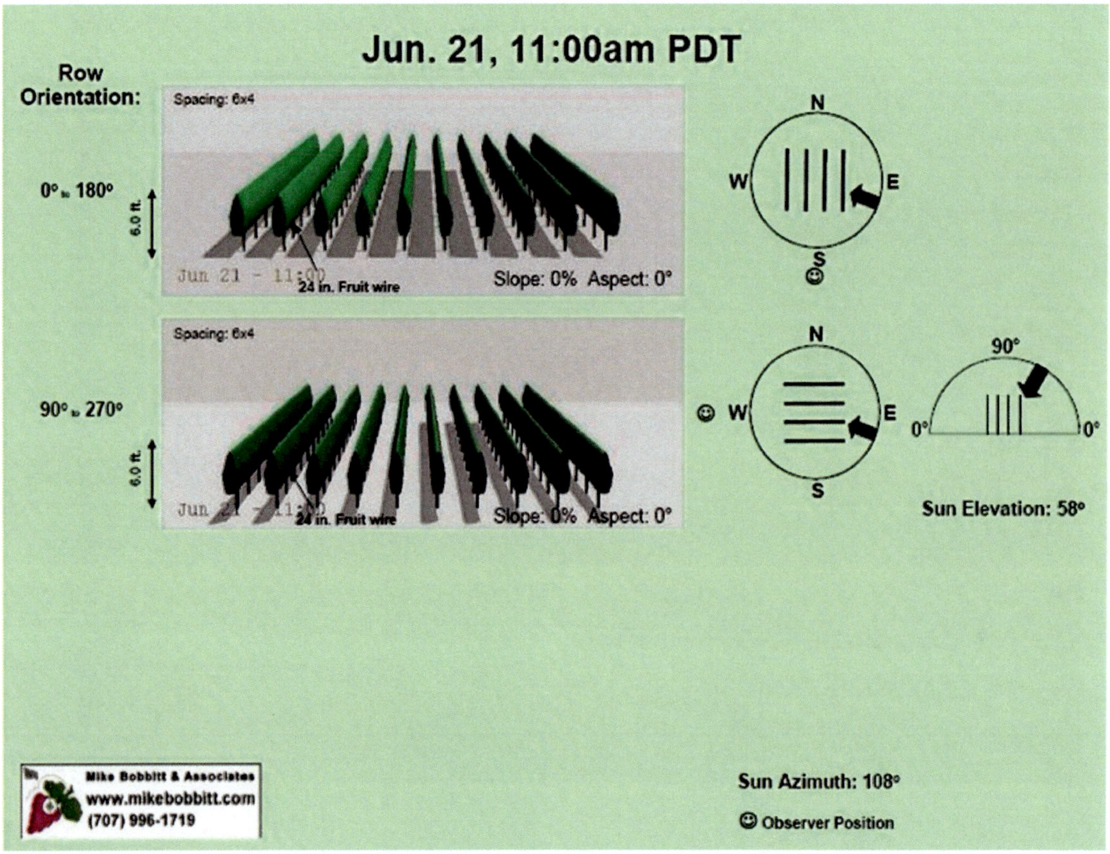

Fig. 13.6 Modelling vine row illumination and sunlit and shadow areas of a vine canopy (Mike Bobbit Associates)

the area in between the rows, and the effects of equipment side slope and row orientation (Fig. 13.6). Other examples displayed on their website include interactive maps that show the vineyard block map, a 3D view of the vineyard and a regional map showing the AVA (American Viticultural Area) boundaries. These combine GIS-based maps with GE views of the vineyard and have the potential to be very powerful communication tools

Developments in Ground-Based Monitoring and Measurement

Traditionally, ground-based measurements have always been part of vineyard management to provide the data and information needed to manage the crop. As already discussed, PV has enhanced our ability to do this more quickly and efficiently with the assistance of a range of new tools and techniques such as remotely sensed imagery, GPS and GIS. Aside from the ease with which it is now possible to collect data more quickly than in the past and more frequently, ground-based data collection is also essential to provide the information necessary for management as well as to assist in the calibration of any remotely acquired data and information.

Over time, the ground-based devices available for collecting such data have also evolved from the once relatively crude and home-crafted DIY tools devised for measuring leaf angles, leaf area and solar radiation to much more sophisticated and accurate devices available off-the-shelf and out of the box, e.g. a range of electronic and digital sensors, light metres and leaf area metres.

Today such field-based equipment is also far more finely tuned to the needs of vineyard management than was the case in the past, being smaller, more portable and catering to the information needs of the commercial vineyard, as well as being commercially available. Such equipment is also cheaper (though not always necessarily affordable if several devices are needed), is designed to collect data more quickly, and includes the capability to capture locational information, thereby allowing it to be plotted, along with many other variables, on a map and correlated with other spatial information. Advances in battery power have also helped to make the equipment more practical in the field, and improved interfaces mean that the operational learning curve is much shorter.

Essential to the practice of PV is reliance on a detailed knowledge of the growth stages of grapevines (phenology) in order to provide a guide for when certain measurements need to be carried out. For example, the vine growth stages described by Coombe and Dry (2004) and the photographs of Institut Francais de la Vigne et du Vin (www.vignevin-sudouest.com) are often cited. Knowledge of grapevine phenology is essential, for example, to help develop a model for the estimation of potential grape yield and quality. Differential GPS, for example, can be used to show the differences in the phenology of the vineyard by recording the location versus the observation that may be based on visual comparison with the reference information or a photographic record (Mariani et al. 2007).

This section of the chapter briefly describes some of the most commonly found examples of ground-based measurement tools often included in the realm of Precision Viticulture that can help to provide the means to generate a more complete picture of all the important variables that are needed to provide the vineyard manager with the basis for a more informed decision-making process.

Field-Based Wireless Sensors and Sensor Networks

Our ability to design and produce ever smaller and lower-cost micro-processor-based sensors and to overcome some of the battery and power management constraints of the past has now enabled vineyard managers to purchase and deploy comprehensive sensor networks in vineyards. Many of these utilise power saving and wireless radio technology; these are generally referred to as Wireless Sensor Networks (WSN) (Morais et al. 2008).

Sensor arrays provide the opportunity to monitor key information of interest to vineyard management. These sensors include the monitoring of soil properties such as moisture and temperature, humidity, wind speed and incoming and outgoing spectral radiation. They have the potential to improve the quantity and quality of products produced by a vineyard using sensors that monitor environmental, climatic and physiological parameters (Morais et al. 2008), and can help to understand vineyard variability to allow appropriate management practices to be implemented (Matese et al. 2009). The design of optimum wireless sensors for such purposes is discussed by Morais et al. (2008), Agrosense (2008).

Wireless networks also provide the means to gather data in real time, and to assist in the identification of areas of disease and the application of fungicides. They can measure temperature, wind speed, humidity, light, wind direction and leaf wetness, all in the form of a so-called weather chip (Fig. 13.7) (McCauley 2006). Practically, such information can be used throughout the growing season to help in the timing of irrigation and fungicide applications (Morais et al. 2008).

Providing the means to capture and display vineyard information in real-time can be a bonus for the vineyard manager who can then monitor the environmental variables from the office computer and, if appropriate and useful, to place such information onto a map of the vineyard. This sort of capability is probably even more important in the larger vineyard than the small one where it is not practically possible to visit all areas within the vineyard on a regular basis.

There are now many examples of such sensors available on the market, and the decreasing costs have significantly impacted upon the frequency and regularity of vineyard monitoring capability. Using sensors costing as little as $50.00 per unit (Hobo data loggers and wireless sensors: http://www.onsetcomp.com/) in a wireless network allows a vineyard to be monitored at any time interval desired, the data being fed back wirelessly to a desktop computer in the manager's office (Dering 2007), and transformed into information in the form of graphs and maps, plotted on a GIS map, or overlaid on Google Earth. Coupled with information from remotely sensed imagery, GPS and GIS, such sensors provide the modern vineyard manager with a wealth of information that can easily be integrated in a GIS to aid in management, planning and decision making.

Hemispherical Photography

Hemispherical photography (making use of a camera with a fish-eye lens) has been used for sometime to capture information about the distribution of leaves within a plant or tree canopy and radiation in numerous

Fig. 13.7 Wireless-based data collection

agricultural and horticultural crops. Photography can be taken from above, within and below the canopy to provide information about the canopy or the light regime within the canopy.

Whilst visual analysis and analysis and interpretation of the photography was once quite a time consuming task, a number of computer-based software products have since enabled faster and more accurate analysis of the images. One of the first was a public domain programme CANOPY developed by Rich (1989) that allowed capture of the images, separation of the canopy foliage from the canopy gaps, and calculation of measures of the canopy architecture and geometry and light penetration. An example focusing on a vine canopy in Vineland, Ontario, in Canada is shown in Fig. 13.8.

Fig. 13.8 Hemispherical photograph of a vine canopy

Today the traditional camera/film combination has been replaced by the HemiView System hardware and software produced by Delta-T Devices (www.delta-t.co.uk). HemiView is a software package designed to analyse hemispherical photographs, and can be used to predict radiation levels above and below a canopy, as well as the direct and diffuse components of the incoming radiation. A significant development has been the digital camera option HEMI-DIG that allows for the capture and transfer of digital hemispherical imagery for input to the HemiView software. Along with the camera there is also available a specially designed Nikon fisheye lens and a self-levelling camera mount. Outputs from HemiView include information on sky geometry, gap fraction, leaf area index, solar radiation, site factors, time-series and sunflecks.

Yield Monitoring and Yield Mapping

Winegrowers usually require a measurement of the grape crop yield. Several yield monitoring systems have been developed for this purpose. Tisseyre et al. (2007) mention a number of these. One such system comprises an apparatus which measures the quantity of grapes harvested with a DGPS recording the coordinates. An advantage of yield monitoring systems is that they are non-intrusive (Proffit et al. 2006). The sonic beam system uses ultrasonic sensors to measure the volume of grapes passing along the discharge conveyor. An alternative is the gravimetric yield-monitoring system that weighs the grapes passing across load cells below the discharge conveyor belt of the harvester. The yield monitors can record about 2,000 measurements per hectare with an average speed of 3 km/h (Tisseyre et al. 2007). From both the yield and coordinates data collected, it is then possible to generate yield maps for the vineyard, typically the product of GIS software, as described by Bramley and Williams (2001). The cost and operation of the specialised equipment, however, means that such technology is largely confined to large vineyards and wineries in Australia, Spain, or California. In France, by comparison, this kind of management is still not in widespread in use.

Selective Harvesting

Whilst grape yield maps have their uses, and are frequently a product of PV, they are not always considered particularly useful when it comes to undertaking vineyard management in an environment which is considered very variable. Tisseyre and McBratney (2008) therefore proposed the development of a Technical Opportunity Index (TOI) to see if it was possible to develop Site Specific Management (SSM) from a yield map and on this basis, for example, to determine selective treatments and harvesting within the vineyard. This method seeks to extend the capability of GIS to provide a Decision Support System (DSS) to aid the vineyard manager based on whether the observed variation within a field is manageable for some proposed field operation (p. 112).

Canopy Measurements

Characterising the vine canopy can be useful to estimate productivity, based on the canopy architecture and density (Schultz et al. 2001). Whilst some canopy measurements can be derived from aerial photography and satellite imagery, as mentioned earlier in this chapter, other ground-based systems using remote sensing monitoring systems, have been developed to assess and map canopy properties. Measurements of the canopy often include leaf area and Leaf Area Index (LAI) which are often made by destructive sampling of the leaves in the vine canopy and with the aid of statistically established correlations estimating the number of leaves and finally the leaf area. Leaf area measurements are usually achieved with the aid of a leaf area metre, and besides being a destructive sampling method, are costly and time consuming.

Today many of the canopy measurement systems are based on digital imaging systems which allow the measurement of several canopy parameters such as canopy height and canopy porosity (gap fraction). Such equipment can usually be mounted on typical vineyard machinery, e.g. a tractor otherwise used for trimming and spraying. Measurements can then be carried out several times during the season to assist in better management of the vineyard (Tisseyre et al. 2007; Polo et al. 2009).

Canopy density can also be measured with handheld apparatus such as a ceptometer, e.g. the AccuPAR Model LP-80 (http://www.decagon.com/canopy/accuparlp80/) that records Photosynthetically Active Radiation (PAR) and Leaf Area Index (LAI). Measurements made in real-time can be stored on the device. Vineyard leaf area is a key determinant of grape characteristics and wine quality established through relationships between NDVI and LAI (Johnson et al. 2001; 2003a, b).

Tuohy (2005) and Bramley et al. (2007) describe a new sensor developed in New Zealand called Grapesense (Grapesense Brochure, (no date)), a video-based image analysis system designed specifically to measure canopy density and porosity for monitoring and mapping of canopy size and growth. The video camera equipment is mounted on a tractor and uses a pink or magenta screen placed behind the vine row of interest to provide a measurement of canopy porosity.

Work by Goutouly et al. (2006) has also led to the development of alternatives to NDVI derived from airborne or spaceborne sensors using ground-based methods. An active sensor, known as Greenseeker, has been developed to measure NDVI which is sensitive to porosity variations of the canopy, and can also be used to establish the growth and vigour of the vine.

Pruning Weight

Strong relationships between leaf area and pruning weight have been identified by Johnson et al. (2003a, b). Others such as Dobrowski et al. (2003) have found good relationships between pruning weight and Ratio Vegetation Indices (RVI) derived from aerial photographs. The measurement of pruning weight is relatively simple and relies on weighing the vine canes cut after pruning to estimate vine vigour. This involves the use of a simple top-load scale combined with the GPS co-ordinates of the sample weighed. The relationship between pruning weights and RVI is also easier than collecting canopy density data. This can be useful early in the season when it would be possible to identify areas on an aerial photograph where pruning weights are expected to be low which would potentially mean insufficient vegetation to ripen the fruit. Action could then be taken to undertake crop thinning or apply Nitrogen to boost growth.

Trunk Diameter Measurements

The diameter of the trunk of a vine plant can easily be measured with either a vernier or a digital calliper. The former provides 1/10 mm accuracy for daily monitoring whilst the latter provides 1/100 mm accuracy. Trunk circumference measurements are a useful measure because they show a strong correlation with vine vigour.

An electronic device known as a dendrometer can also be used to measure trunk diameter. An accuracy of 1/100 mm can be achieved and the device can be used in wireless mode. Matching the measurements to certain growth stages can be helpful when establishing the correlation between trunk diameter and grapevine water status (Selles et al. 2007). There is also a good relationship between average Maximum Daily Trunk Growth (MDG) and average Berry Growth Rate (BGR) from fruit set to veraison, leading to a bigger berry size being obtained at harvest. However, the cost of such a device usually means that only a few will be deployed in the vineyard.

Stem Water Potential (SWP), usually measured at midday, can also be correlated with irrigation treatments. SWP is also correlated to BGR from fruit set to veraison. However, MDG is more related to soil water content than SWP. Therefore MDG and SWP can be used as a tool for irrigation scheduling in vineyards.

Light Within the Canopy

Aside from the hemispherical photography often used for canopy radiation measurement, there are many other sensors that have been developed over the years for measuring incoming and outgoing radiation and hence reflectance or albedo. In the past, monitoring light above, within and below the canopy was undertaken using a range of directional and hemispherical sensors including solar trackers and directional and hemispherical radiometers. The use of spectral filters allowed for the collection of Photosynthetically Active Region PAR and NIR reflectance.

Today these sensors have been modernised and replaced by more portable and easy to use sensors. Delta-T Devices (www.delta-t.co.uk), for example, provides a range of PAR (Photosynthetically Active Region) PAR sensors, sunshine sensors which can be attached to data loggers, including a device known as a SunScan Probe Canopy Analysis System which can be used to acquire PAR interception and Leaf Area Index (LAI). The latter can also be hooked up to a portable data collection terminal. Such sensors can also be attached to a probe so that they can be inserted in the canopy at different heights to establish the extinction of radiation within the vine canopy.

Soil Measurements

Soil type and soil properties, e.g. moisture, are often quite variable across a vineyard site. The variability of soil can be established through conductivity measurements (or its reverse, resistivity). Conductivity correlated with several soil parameters, and these can therefore be mapped. Two types of equipment are used: an ElectroMagnetic Induction (EMI) sensor, and an electrical resistivity sensor. Different models are available to measure conductivity to different soil depths, e.g. from 0 to 75 cm, 25–120 cm and up to 3 m (Tisseyre et al. 2007). Care needs to be used when operating the device as metal objects such as trellis structures in the vineyard can influence the readings, and frequent calibration is often needed. Ideally the measurements have to be made at the following vine growth stages: budburst, shoots (10 cm), flowering or post-harvest.

The electrical sensitivity sensor is used in PV for a one dimensional survey known as Vertical Electrical Sounding (VES). The measurement points are the same as with the EMI. In some cases, the sensors can be pulled behind a quad bike to speed up the time taken to complete a survey (Bramley 2001; Samouelian et al. 2005).

The result of such surveys is the production of soil conductivity or resistivity maps that can be correlated with other soil properties such as solid constituents, water content, cation exchange capacity (CEC), pore fluid composition or soil temperature (Samouelian et al. 2005). Such maps reveal the spatial patterns of soil conductivity and therefore soil properties within the vineyard and can be used within a GIS for correlation with other thematic layers. This information can provide a useful tool to help zone the vineyard into high and low vigour zones.

Soil temperatures measured at different depths and during different growth stages can be used to monitor the influence of temperature on grape maturity. Soil temperature sensors can be placed at different depths, e.g. 10, 50 and 80 cm (Burgos et al. 2007). High soil temperatures at 10 cm depth appear to be correlated with the maturity of grapes. Moreover, the number of days between bud and flowering, and between flowering and veraison decreases when the soil temperature is warmer, and the maturity of the grapes is accelerated. Measurement of soil temperatures and their correlations can help wine growers when they establish their vineyard revealing that the soil environment plays a major role in the development of the vine. However, the relationship between soil moisture and temperature still needs to be better understood so as to be able to predict how the wine *terroirs* will respond to future predictions about global warming and climate change.

Other Sensors

Other sources of information are also important to the vineyard manager. At present there are still no harvester-mounted or hand-held sensors specifically designed to measure quality parameters such as Baume and Brix, pH, titratable acidity (TA) and anthocyanins or phenolics (Tisseyre et al. 2007). However, near-infrared (NIR) spectroscopy can be used to find out the composition of the grapes. A NIR spectrometer can collect very accurate measurements compared to reference methods (Tuohy 2005), and the correlations between NIR measurements and reference methods are very good for Brix, pH, and TA.

Mechanical Tools

Variable Rate Technology (VRT) has not been used widely in viticulture to date because Decision Support Systems (DSS), which allow one to decide how much input to apply at a location, are not well developed. Nevertheless, where the technology has been applied, albeit in a simple way, it has been found to have a direct financial benefit. One of the main applications of VRT mentioned by Tisseyre et al. (2007) is for weeding operations. VRT weeding systems have been commercially available since 2003 and are used to selectively apply herbicides to the vineyard. Detection of the weeds by an optical sensor is the basis for the application of the spray. It has been found that this technology can save up to 75% in the use of herbicides (Tisseyre and Taylor 2005). A similar system, Weedseeker, can also be used on discontinuous vertical canopies to apply fungicides. This also reduces the use of chemical products as the spray is only applied to the canopy and not the gaps (Fig. 13.9).

Current Research and Future Developments

There are a number of sensor developments in the pipeline for vineyards. These include yield, soil property and sprayer sensors. A number of yield sensors are now under development in various parts of the world for placement on grape harvesters. Soil property sensors such as the apparent Electrical Conductivity (EC) are being developed. Tisseyre et al. (2007) mention the Time Domain Reflectometry (TDR) sensor for mobile apparatus. This sensor uses radar. The dielectric reading from the sensor is converted to water saturation providing reliable and accurate soil moisture monitoring capability. It can be used either vertically or horizontally. A new mobile 'on-the-go' soil sensor

Fig. 13.9 Weedseeker

is also mentioned by Tisseyre et al. (2007). Others sensors being developed include soil ion sensors (mainly nitrogen and potassium sensors) and near infrared and mid-infrared sensing systems. Finally, ground penetrating radar and gamma radiometrics are being trialled in vineyards to determine their potential role. Spraying sensors are also being developed that will measure the flow or pesticides from vineyard sprayers (Gil et al. 2007; Llorens et al. 2010).

Summary and Conclusions

Whilst vineyard management has always relied upon the collection of data about the vine canopy, the terrain and the environment, the tools and techniques developed in the context of PV have facilitated greater knowledge and understanding of the vineyard environment and its management for wine production. The tools and techniques now used in PV are quite wide ranging and can provide very accurate information about all aspects of the vineyard. For the most part, yield sensors, vigour maps using indices such as NDVI or PCD or soil sensing surveys are most widely used. Other measurements such as pruning weight and trunk diameter can also be made in relation to spatial location.

However, whilst it is clear that there are many tools and techniques now available, most are still only used in the larger commercial vineyards that have the financial resources and the expertise to hand to justify the use of expensive and high-tech devices and systems. Although some of the cheaper kits are now in use in smaller vineyards, many vineyards are in any case physically just too small to practically make use of the technology, let alone afford it. As noted by Lamb et al. (2008), there are also other factors that limit the widespread use of the technologies ranging from a reluctance to adopt the technology, justification of the costs versus benefits and the expertise needed to make use of the many different tools and techniques, some of which still need to be simplified to encourage their operational use, and accompanied by opportunities for training.

But, as can be seen from the quantity of information now available in the form of journal papers, web pages and factsheets, together with books on precision agriculture and Precision Viticulture, it should not be long before people are more aware of the role and value of PV in vineyard monitoring and management.

References

Acevedo-Opazo, C., B. Tisseyre, S. Guillaume, and H. Ojeda. 2008. The potential of high spatial resolution information to define within-vineyard zones related to vine water status. *Precision Agriculture* 9:285–302.

Agrosense. 2008. Wireless sensor networks and remote sensing foundation of a modern agricultural infrastructure in the region. *Presentation*. Serbia and Slovenia: ENR, EU.

Baggio, A. 2005. Wireless networks in precision agriculture. Poster presentation. *Proceedings of the First Realwsn 2005 Workshop on Real-World Wireless Sensor Networks*. Stockholm, Sweden: SICS Technical Report T2005:09.

Berni, J. A.J., P. J. Zarco-Tejada, L. Suárez, V. González-Dugo, and E. Fereres. 2009. Remote sensing of vegetation from UAV platforms using lightweight multispectral and thermal imaging sensors. *Proceedings ISPRS Workshop*. Hannover 2009. High-Resolution Earth Imaging for Geospatial Information Hannover, Germany, June 2–5, 2009

Bramley, R.G.V. 2001. Progress in the development of precision viticulture: Variation in yield, quality and soil properties in contrasting Australian vineyards. In: Currie, L.D. and P. Loganathan. (Eds.). *Precision tools for improving land management. Occasional report No. 14*. Fertilizer and Lime Research Centre, Massey University, Palmerson North, New Zealand.

Bramley and Williams. 2001. A protocol for the construction of yield maps from data collected using commercially available grape yield monitors. *CSIRO*. 22p.

Bramley, R., D. Gobbett, and J-P. Praat. 2007. A proximal canopy sensor: A tool for managing vineyard variability and adding value to viticultural research. *Final Report on Project No. CSL06/01* to the Grape and Wine Research and Development Corporation. CSIRO/GWRDC, Adelaide.

Brancorado, L., O. Failla, P. Dosso, and F. Serina. 2006. Use of satellite in precision viticulture: The Franciacorta experience. VIe Congrès International desTerroirs Viticoles 2006 - *VIth International Terroir Congress* 6:276–279.

Brown, R. 2008. Airborne remote sensing for precision viticulture in Niagara. *Presentation*. School of Engineering, University of Guelph, Canada. 36 slides,

Burini, A., G. Schiavon and D. Solimini. 2008. Fusion of high resolution polarimetric SAR and multi-spectral optical data for precision viticulture. *Proceedings 2008 International Geoscience and Remote Sensing Symposium. IGARSS'08*. Boston, MA, USA.

Burrough, P.A. 1986. *Principles of Geographical Information Systems for land resources assessment*. Oxford: Clarendon Press.

Burgos, S., N. Dakhel, and M. Docourt. 2007. Le comportement thermique des sols: Caractérisation et influence sur la vigne. Global warming, which potential impacts on the vineyards? *Proceedings of the colloquium "Global warming, with potential impacts on the vineyards?"* Dijon, France, March 28–30, 2007, 1–3.

Castel, T., J. N. Dodane, J. M. Brayer, P. Curmi, F. Andreux, and D. Peyron, 2007. Spatial analysis of within-vineyard grapevine sensitivity to solar irradiance patterns: A case study using Pinot Noir in a complex terrain. *Global Warming, Which Potential Impacts On The Vineyards?* 28–30 Mars 2007/March 28–30, 2007. 9p

Chanussot, J., P. Bas, and L. Bombrun. 2005. Airborne remote sensing of vineyards for the detection of dead vine tress. *Proceedings of IGARSS'05*, 3090–3093.

Chelli, S., L. Fosco, and S. Sandrelli. 2005. Precision vineyard management from space: From Bachhus to Eneide. *ESA Bulletin* 123:34–39.

Coombee, B. G., and P. R. Dry. 2004. Eds. *Viticulture 1: Resources*. 2nd ed. Adelaide, Australia: Winetitles.

Da Costa, J.P., F. Michelet, C. Germain, O. Lavialle, and G. Grenier. 2007. Delineation of vine parcels by segmentation of high resolution remote sensed images. *Precision Agriculture* 8:95–110.

Dainelli, N., C. Conese, B. Rapi, and M. Romani. 2005. Integration of GIS and RS techniques for canopy variability evaluation in vineyards. COST 718 – New technologies for agrometeorological model applications. CNR Research Area – Polo Scientifico di Sesto Fiorentino. 14th February 2005. *PPT presentation*. 32p.

Delenne, C., S. Durrieu, G. Rabatel, and M. Deshayes. 2010. From pixel to vine parcel: A complete methodology for vineyard delineation and characterization using remote-sensing data. *Computers and Electronics in Agriculture* 70:78–83.

Dering, G.M. 2007. Geomorphological controls on air temperature in the Walla Walla Valley American Viticultural Area. *Unpublished Undergraduate Thesis* submitted in partial fulfilment of the requirements for graduation with Honors in Geology. Whitman College, Walla Walla, U.S.A... 33p.

Dobrowski, S., S. Ustin, and J. Wolpert. 2003. Grapevine dormant pruning weight prediction using remotely sensed data. *Australian Journal of Grape and Wine Research* 9:177–182.

Dubayah, R., R. Knox, M. Hofton, J. B. Blair, and J. B. Drake. 2000... Land surface characterization using LIDAR remote sensing. In, *Spatial Information for Land Use Management*, ed. M. Hill, & R. Aspinall, Singapore: International Publishers Direct. 25–38.

Ducati, J.R., and P. Rodirigues Da Silva. 2006. Remote sensing and radiometric techniques applied to vineyards in two regions of Rio Grande do Sul, Brazil *VIe Congrès International des Terroirs Viticoles 2006 - VIth International Terroir Congress 2006*. 226–231.

Duncan, A. 2007. Vineyard management system using GIS modelling. MARCH, 2007. Canada: Surveylab *Factsheet*. http://www.ikegps.com/Downloads/vineyard.pdf

Fraigneau, C. 2009. Precision viticulture: Tools to measure and manage vineyard variability. *Unpublished Report* submitted to David R. Green (University of Aberdeen) as part of an Internship (SupAgro Montpelier, France and Department of Geography and Environment, University of Aberdeen, Scotland, UK). August 2009. 48p.

Gil, E., A. Escola, J. R. Rosell, S. Planas, and L. Val. 2007. Variable rate application of plant protection products in vineyard using ultrasonic sensors. *Crop Protection* 26: 1287–1297

Goutouly, J.P., R. Drissi, D. Forget, and J. P. Gaudillère. 2006. Characterization of vine vigour by ground based NDVI measurements. *VIth International Terroir Congress 2006* 6: 237–241

Grapesense Brochure. No Date. Breakthrough technology for measuring canopy density in real time. New Zealand Lincoln Technology. 2p.

Green, D. R. 2009. Creating a Google Earth-based virtual field-trip visit to a vineyard. *Paper presented to AAG 2009*, Las Vegas Nevada. April 2009.

Grierson, I. and S. Bolt. 1995. Aerial video: A new effective method for vineyard planning. *The Australian Grapegrower and Winemaker* 8–9.

Hall, A., D. W. Lamb, B. Holzapfel, and J. Louis. 2002. Optical remote sensing applications in viticulture: A review. *Australian Journal of Grape and Wine Research* 8:36–47.

Jacquemoud, S. and F. Baret. 1990. Prospect: A model of leaf optical PROPERTIES SPECTRA. *Remote Sensing of Environment* 34: 75–91.

Jacquemoud, S., Verhoef, W., Baret, F., Bacour, C., Zarco-Tejada, P.J., Asner, G.P., Francois, C., and S.L. Ustin. 2009. Prospect + Sail models: A review of use for vegetation characterization, *Remote Sensing of Environment* 113: 56–66.

Jia, L., Z. Su, J. Elbers, A. Gieske, W. Timmermans, A. Olioso, F. Nerry, and J. Sobrino. 2006. Study on thermal heterogeneity of vineyard canopy in complex terrain: Measurements and modelling. *Geophysical Research Abstracts 8*: 10084.

Johnson, L.F., Roczen, D., and S. Youkhana. 2001. Vineyard Canopy Density Mapping with IKONOS Satellite Imagery. *Paper Presented at the Third International Conference on Geospatial Information in Agriculture and Forestry*, Denver, Colorado, 5–7 November 2001. 10p.

Johnson, L.F., D. E. Roczen, S. K. Youkhana, R. R. Nemani, and D. F. Bosch. 2003a. Mapping vineyard leaf area with multispectral satellite imagery. *Computers and Electronics in Agriculture* 38:33–44.

Johnson, L. F., S. R. Herwitz, S. E. Dunagan, B. M. Lobitz, D. V. Sullivan, and R. E. Slye. 2003b. Collection of ultra high spatial and spectral resolution image data over California vineyards with a small UAV. *Proceedings of the 30th International Symposium on Remote Sensing of Environment*, Honolulu, HI

Johnson, L.F., and S. R. Herwitz. 2004. UAVs for collection of farm imagery. *Online Document.* 3p, http://www.uav-applications.org/gallery/doc/FieldTalk.pdf

Johnson, L.F., and B. Lobitz. No Date. CRUSH Project. Earth Science and Enterprise. Ecosystem Science and Technology.

Lamb, D.W., P. Frazier, and P. Adams. 2008. Improving pathways to adoption: Putting the right ps in precision agriculture. *Computers and Electronics in Agriculture* 61:4–9.

Lamb, D.W., A. Hall, and J. Louis. 2002. Airborne Remote Sensing of Vines for Canopy Variability and Productivity. *Australian Grapegrower and Winemaker* 449a, 89–92.

Liaghat, S., and S. K. Balasundram. 2010. A review: the role of remote sensing in precision agriculture. *American Journal of Agricultural and Biological Sciences* 5 (1): 50–55.

Llorens, J., E. Gil, J. Llop, and A. Escola. 2010. Variable rate dosing in precision viticulture: Use of electronic devices to improve application efficiency. *Crop Protection* 29: 239–248.

Lloret, J. R., and D. P. Jurado. 2004. An Innovative GIS in a Web Environment for Vineyards: The Viticultural Inventory for Catalunya. *Mapping: Map and SIG Consulting 94*: 74–77.

Louarn, G.T., J. Lecoeur, and E. Lebon. 2007. A three-dimensional statistical reconstruction model of grapevine (Vitis vinifera): Simulating canopy structure variability within and between cultivar/training system pairs. *Annals of Botany* 101(8)1167–1184.

Mariani L., O. Failla, G. Dal Monte, and D. Facchinetti. 2007. IPHEN: A model for real time production of grapevine phonological maps. In: *Congress on Climate and Viticulture*, Zaragoza, 2007: 272–278.

Matese, A., S. F. Gennaro, A. Zaldei, L. Genesio, and F. P. Vaccari. 2009. A Wireless sensor network for precision viticulture: The nav system. *Computers and Electronics in Agriculture* 69: 51–58.

McCauley, I. 2006. Pigs, penguins, poultry and peaches: four courses of wireless sensing. *MS Powerpoint Presentation*. Sensor Network Forum, National ICT Australia. Department of Primary Industries, University of Melbourne, Australia.27th November 2006

Merenlender, A.M. 2000. Mapping vineyard expansion provides information on agriculture and the environment. *California Agriculture* 54(3): 7–12.

Minden, K. 1982. The application of remote sensing to vineyard management. *Unpublished Masters Thesis*. Cornell University: Dept of Engineering.

Morais, R., M. A. Fernandes, S. G. Matos, C. Serodio, P. J.S.G. Ferreira, and M. J.C.S. Reis. 2008. A zigbee multi-powered wireless acquisition device for remote sensing applications in precision viticulture. *Computers and Electronics in Agriculture* 68:94–106.

Philipson, W. R.; T. L. Erb; D. Fernandez; J. N. Mcleester.1980. Remote Sensing for Vineyard Management. In: *American Society of Photogrammetry, Annual Meeting*, 46th, St. Louis, MO, March 9–14, 1980, ASP Technical Papers. (A82–10027 01–43) Falls Church, VA, American Society of Photogrammetry, 1980, 371–378.

Pitcher-Campbell, S., M. Tuohy and I. Yule. 2001. The application of remote sensing and GIS for improving vineyard management. In: *Proceedings 1st National Conference on Geospatial Information and Agriculture*. Sydney, Australia I: 573–585. http://www.regional.org.au/au/gia/18/573pitchercampbell.htm.

Polo, J.R.R., R. Sanz, J. Llorens, J. Arno, E. Escola, M. Ribes-Dasi, J. Masip, F. Camp, F. Gracia, F. Solanelles, T. Palleja, L. Val, S. Planas, E. Gil, and J. Palacın. 2009. A tractor-mounted scanning LIDAR for the non-destructive measurement of vegetative volume and surface area of tree-row plantations: A comparison with conventional destructive measurements. *Biosystems Engineering* 102:128–134.

Proffit, T., R. Bramley, D. W. Lamb, and E. Winter. 2006. *Precision viticulture: a new era in vineyard management and wine production*. Adelaide: Winetitles.

Rich, P.M. 1989. A Manual for Analysis of Hemispherical Canopy Photography. Los Alamos, NM: Los Alamos National Laboratory Report LA-11733-M.

Ryerson, R.A. 1974. Remote sensing to detect frost prone areas in the Niagara Region. *Report for Vineland Research Station, Ontario Ministry of Agriculture and Food. Ottawa, Canada*: Canada Centre for Remote Sensing, Energy Mines and Resources.

Ryerson, R.A., S. Schwebs, R. B. Brown, S. Boles, M. and Duncan. 2008. The commercialization of remote sensing and GIS for vineyard management: A simple but powerful application of change detection. In: *Proceedings ASPRS 2008 Annual Conference*. Portland, Oregon. April 28th – May 2nd 2008. 10p.

Samouelian, A., I. Cousin, A. Tabbagh, A. Bruand, A., and G. Richard. 2005. Electrical resistivity survey in soil science: A Review. *Soil and Tillage Research* 83: 173–193.

Schultz, H.R., P. Pieri, S. Poni, and E. Lebon. 2001. Modelling water use and carbon assimilation of vineyards with different canopy structures and varietal strategies during water deficit. In: *Proceedings. 12th International Congress on Photosynthesis*. Paper Number S34–002.

Selles, G., R. Ferreyra, I. Munoz, and H. Silva. 2007. Use of trunk growth rate as criteria irrigation scheduling on table grapes CV. Crimson Seedless irrigated by drip. Paper presented at *5th International Table Grape Symposium South African Society for Enology and Viticulture*. 13–21 November 2007, Somerset West, South Africa. 4p.

Smith, J.A. 1982. Role of scene radiation models in remote sensing. In: *Proceedings of 8th Machine Processings of Remotely Sensed Data Symposium* 1982:546–549.

Smith, L., and P. Whigham. 1999. Spatial aspects of vineyard management and wine grape production. *Paper Presented at SIRC 99 – The 11th Annual Colloquium of the Spatial Information Research Centre*. Dunedin, New Zealand: University of Otago.

Suits, G.H. 1983. Extension of a uniform canopy reflectance model to include row effects. *Remote Sensing of Environment* 13:113–129.

Tisseyre, B. and J. Taylor. 2005. An overview of methodologies and technologies for implementing precision agriculture in viticulture. In: Proceedings of XII Congresso Brasileiro de Viticultura e Enologia – Anals. 45–54.

Tisseyre, B., H. Ojeda, and J. Taylor. 2007. New technologies and methodologies for site-specific viticulture. *International Journal of Vine and Wine Sciences* 41 (2): 63–76.

Tisseyre, B., and A. McBratney. 2008. A technical opportunity index based on mathematical morphology for site specific management: An application to viticulture. *Precision Agriculture* 9 (1–2):101–113.

Tongrod, N., A. Tuantranont, and T. Kerdcharoen. 2009. Adoption of precision agriculture in vineyard..In: *Sixth International Conference on Electrical Engineering/Electronics, Computer, Telecommunications and Information Technology*, ECTI-CON 6th -9th May 2009. Pattaya, Thailand.

Trolier, L. J., W. R. Philipson, and W. D. Philpot. 1989. Landsat TM analysis of vineyards in New York. *International Journal of Remote Sensing* 10 (7): 1277–1281.

Tuohy, M. 2005. Remote sensing in the vineyard management. www.nzwine.com/assets/Mike_Tuohy.pdf.

Verhoef, W. and N. J. J. Bunnik. 1976. *The spectral directional reflectance of row crops*. The Netherlands: NIWARS Publication, No. 35,.

Wassenaar, T., J., M. Robbez-Masson, P. Andrieux, and F. Baret. 2000. Vineyard spatial structure analysis by per-field aerial photograph processing. *International Archives of Photogrammetry and Remote Sensing* XXXIII (B7): 1692–1699.

Wildman, W.E. 1979. Color infrared: A valuable tool in vineyard management. In: *Workshop Proceedings Color Aerial Photography in the Plant Sciences and Related Fields*. 7: 229–238.

Wildman, W.E., K. W. Bowers, and R. A. Naja. 1981. Aerial photography in vineyard pest, soil and water management. In: *Grape Pest Management*, ed. D. L. Flaherty. University of California, Division of Agricultural Science Publication No. 4105. Berkeley, California: University of California.

Wildman, W.E., R. T. Nagaoka, and L. A. Lider. 1987. Monitoring the spread of grape phylloxera by color infrared aerial photography and ground investigation. *American Journal of Enology and Viticulture* 34(2): 83–94.

Zarco-Tejada, P.J., A. Berjon, R. Lopez-Lozano, J. R. Miller, P. Martın, V. Cachorro, M. R. González, and A. de Frutos. 2005. Assessing vineyard condition with hyperspectral indices: leaf and canopy reflectance simulation in a row-structured discontinuous canopy. *Remote Sensing of Environment* 99: 271–287.

Internet

Agricultural Electronics Corporation (Automatic Dendrometer): http://www.phytogram.com/cost.htm

Avidor High Tech: http://www.avidorhightech.com/html/weedseeker.html

Campbell Scientific (Datasheets on Temperature Sensors): http://www.campbellsci.com/soiltemperature

Delta-T Devices: http://www.delta-t.co.uk

Hobo Data Loggers: http://www.onsetcomp.com/

ICT International (Datasheet on AccuPAR LP-80 Ceptometer): http://www.ictinternational.com.au/lp-80.htm

Institut Cooperattif du vin (ICV): http://www.icv.fr

Kite Aerial Photography: http://www.arch.ced.berkeley.edu/kap/kaptoc.html

Meadowgrove Vineyard (Andy McLeod, personal communication): http://www.meadowgrove.com/

Precision Viticulture Tools: http://www.cse.csiro.au/client_serv/paresources.htm#resources

VineView: http://www.vineview.com

Visualisation: http://www.mikebobbitt.com/3D.htm

GrayHawk Imaging: http://www.grayhawk-imaging.com

VitSmarter: http://www.vitsmarter.com

Index

P.H. Dougherty (ed.), *The Geography of Wine: Regions, Terroir and Techniques*, DOI 10.1007/978-94-007-0464-0, © Springer Science+Business Media B.V. 2012